Neuroergonomics

T0276110

Neuroergonomics
The Brain at Work and in Everyday Life

Edited by

Hasan Ayaz
Drexel University, Philadelphia, PA, United States

Frédéric Dehais
ISAE-SUPAERO, Université de Toulouse, Toulouse, France

ACADEMIC PRESS
An imprint of Elsevier

Academic Press is an imprint of Elsevier
125 London Wall, London EC2Y 5AS, United Kingdom
525 B Street, Suite 1650, San Diego, CA 92101, United States
50 Hampshire Street, 5th Floor, Cambridge, MA 02139, United States
The Boulevard, Langford Lane, Kidlington, Oxford OX5 1GB, United Kingdom

Copyright © 2019 Elsevier Inc. All rights reserved.

No part of this publication may be reproduced or transmitted in any form or by any means, electronic or mechanical, including photocopying, recording, or any information storage and retrieval system, without permission in writing from the publisher. Details on how to seek permission, further information about the Publisher's permissions policies and our arrangements with organizations such as the Copyright Clearance Center and the Copyright Licensing Agency, can be found at our website: www.elsevier.com/permissions.

This book and the individual contributions contained in it are protected under copyright by the Publisher (other than as may be noted herein).

Notices
Knowledge and best practice in this field are constantly changing. As new research and experience broaden our understanding, changes in research methods, professional practices, or medical treatment may become necessary.

Practitioners and researchers must always rely on their own experience and knowledge in evaluating and using any information, methods, compounds, or experiments described herein. In using such information or methods they should be mindful of their own safety and the safety of others, including parties for whom they have a professional responsibility.

To the fullest extent of the law, neither the Publisher nor the authors, contributors, or editors, assume any liability for any injury and/or damage to persons or property as a matter of products liability, negligence or otherwise, or from any use or operation of any methods, products, instructions, or ideas contained in the material herein.

Library of Congress Cataloging-in-Publication Data
A catalog record for this book is available from the Library of Congress

British Library Cataloguing-in-Publication Data
A catalogue record for this book is available from the British Library

ISBN: 978-0-12-811926-6

For information on all Academic Press publications visit our website at
https://www.elsevier.com/books-and-journals

Working together
to grow libraries in
developing countries

www.elsevier.com • www.bookaid.org

Publisher: Nikki Levy
Acquisition Editor: Natalie Farra
Editorial Project Manager: Kathy Padilla
Production Project Manager: Poulouse Joseph
Designer: Mark Rogers

Typeset by TNQ Technologies

Dedication

This book is dedicated to Professor Raja Parasuraman who unexpectedly passed on March 22, 2015. Raja Parasuraman's pioneering work led to the emergence of Neuroergonomics as a new scientific field. He made significant contributions to a number of disciplines from human factors to cognitive neuroscience. His early work included important contributions to topics such as vigilance and human interaction with automated systems. He later consolidated his interests in human factors and cognitive neuroscience to develop a new discipline called Neuroergonomics, which he defined as the study of the brain and behavior at work.

His advice to young researchers was to be passionate in order to develop theory and knowledge that can guide the design of technologies and environments for people. His legacy, the field of Neuroergonomics, will live on in countless faculties and students whom he advised and inspired with unmatched humility throughout the span of his distinguished career. Raja Parasuraman was an impressive human being, a very kind person, and an absolutely inspiring individual who will be remembered by everyone who had the chance to meet him.

Contents

Section IV
Neurostimulation Applications

31. Hybrid Collaborative Brain–Computer Interfaces to Augment Group Decision-Making

Davide Valeriani, Caterina Cinel and Riccardo Poli

32. How to Recognize Emotions Without Signal Processing: An Application of Convolutional Neural Network to Physiological Signals

Nicolas Martin, Jean-Marc Diverrez, Sonia Em, Nico Pallamin and Martin Ragot

Section VI
Entries From the Inaugural International Neuroergonomics Conference

33. Technical Manifestations of the Everted Brain: The Impact and Legacy of Raja Parasuraman

Peter A. Hancock

34. Can We Trust Autonomous Systems?

Peter A. Hancock, Kimberly L. Stowers and Theresa T. Kessler

35. Learning and Modulating Spatial Probabilities in Virtual Environments

Amy L. Holloway, Peter Chapman and Alastair D. Smith

36. Physiological Markers for UAV Operator Monitoring

Raphaëlle N. Roy, Thibault Gateau, Angela Bovo, Frédéric Dehais and Caroline P.C. Chanel

37. Estimating Cognitive Workload Levels While Driving Using Functional Near-Infrared Spectroscopy (fNIRS)

Anirudh Unni, Klas Ihme, Meike Jipp and Jochem W. Rieger

List of Contributors

Nounagnon F. Agbangla Université de Poitiers, Poitiers, France

Atahan Agrali Drexel University, Philadelphia, PA, United States

Cédric T. Albinet Université de Poitiers, Poitiers, France; Université de Toulouse, INU Champollion, Albi, France

Awad Aljuaid Mechanical Engineering Department, Taif University, Saudi Arabia

Guillaume Andéol Institut de Recherche Biomédicale des Armées, Brétigny sur Orge, France

Jean M. André ENSC-Bordeaux INP, Bordeaux, France

Pietro Aricò BrainSigns srl, Rome, Italy; IRCCS Fondazione Santa Lucia, Rome, Italy; University of Rome "Sapienza", Rome, Italy

Branthomme Arnaud Dassault Aviation, Saint-Cloud, France

Romain Artico Université Paris Sud, Université Paris-Saclay, CIAMS, Orsay, France; CIAMS, Université d'Orléans, Orléans, France

Michel Audiffren Université de Poitiers, Poitiers, France

Hasan Ayaz Drexel University, Philadelphia, PA, United States; University of Pennsylvania, Philadelphia, PA, United States; Children's Hospital of Philadelphia, Philadelphia, PA, United States

Fabio Babiloni BrainSigns srl, Rome, Italy; University of Rome "Sapienza", Rome, Italy; Hangzhou Dianzi University, Hangzhou, China

Wendy Baccus George Mason University, Fairfax, VA, United States

Carryl L. Baldwin George Mason University Fairfax, VA, United States

Hubert Banville Université du Québec, Montreal, Canada

Klaus Bengler Technical University of Munich, Garching, Germany

Bruno Berberian Office National d'Etudes et de Recherche Aérospatiales, Salon-de-Provence, France

Jérémy Bergeron-Boucher Université Laval, Québec, QC, Canada

Ali Berkol Baskent University, Ankara, Turkey

Pierre Besson EuroMov, University of Montpellier, Montpellier, France

Siddharth Bhatt Drexel University, Philadelphia, PA, United States

Arianna Bichicchi University of Bologna, Bologna, Italy

Martijn Bijlsma TNO Netherlands Organization for Applied Scientific Research, Soesterberg, The Netherlands

Nikolai W.F. Bode University of Bristol, Bristol, United Kingdom

Vincent Bonnemains ONERA/DTIS, Université de Toulouse, France

Gianluca Borghini BrainSigns srl, Rome, Italy; IRCCS Fondazione Santa Lucia, Rome, Italy; University of Rome "Sapienza", Rome, Italy

Guillermo Borragán Université Libre de Bruxelles (ULB), Brussels, Belgium

Marc-André Bouchard Université Laval, Québec, QC, Canada

Angela Bovo ISAE-SUPAERO, Université de Toulouse, Toulouse, France

Eric Brangier Université Libre de Bruxelles, Bruxelles, Belgique; Université de Lorraine, Metz, France

Anne-Marie Brouwer TNO Netherlands Organisation for Applied Scientific Research, Soesterberg, The Netherlands

Heinrich H. Bülthoff Max-Planck Institute for Biological Cybernetics, Tübingen, Germany

Christopher Burns Liverpool John Moores University, Liverpool, United Kingdom

Vincent Cabibel EuroMov, University of Montpellier, Montpellier, France

Tuna E. Çakar MEF University, İstanbul, Turkey

Daniel Callan Center for Information and Neural Networks (CiNet), National Institute of Information and Communications Technology (NICT), Osaka University, Osaka, Japan; ISAE-SUPAERO, Université de Toulouse, Toulouse, France

Aurélie Campagne Université Grenoble Alpes, CNRS, LPNC UMR 5105, F-38000, Grenoble, France

Travis Carlson Office of Naval Research, Arlington, VA, United States

William D. Casebeer Lockheed Martin Advanced Technology Lab, Arlington, VA, United States

Deniz Zengin Çelik Galatasaray University, İstanbul, Turkey

Cindy Chamberland Université Laval, Québec, QC, Canada

Caroline P.C. Chanel ISAE-SUPAERO, Université de Toulouse, Toulouse, France

Peter Chapman University of Nottingham, Nottingham, United Kingdom

Luc Chatty Houston Methodist Hospital, Houston, TX, United States; University of Houston, Houston, United States

Laurent Chaudron ONERA Provence Research Center, Salon-de-Provence, France

Philippe Chevrel CNRS, Central Nantes & IMT-Atlantique, Nantes, France

Lewis L. Chuang Max-Planck Institute for Biological Cybernetics, Tübingen, Germany

Caterina Cinel University of Essex, Colchester, United Kingdom

Bernard Claverie ENSC-Bordeaux INP, Bordeaux, France

Antonia S. Conti Technical University of Munich, Garching, Germany

Yves Corson University of Nantes, LPPL, Nantes, France

Johnathan Crépeau Université Laval, Québec City, QC, Canada

Adrian Curtin Drexel University, Philadelphia, PA, United States; Shanghai Jiao Tong University, Shanghai, China

Frédéric Dehais ISAE-SUPAERO, Université de Toulouse, Toulouse, France

Arnaud Delafontaine University Paris-Sud, Orsay, France

Gaétane Deliens Université Libre de Bruxelles (ULB), Brussels, Belgium

Arnaud Delorme Office National d'Etudes et de Recherche Aérospatiales, Salon-de-Provence, France

Stefano I. Di Domenico Institute for Positive Psychology and Education, Australian Catholic University, North Sydney, NSW, Australia; University of Toronto Scarborough, Toronto, ON, Canada

Gianluca Di Flumeri BrainSigns srl, Rome, Italy; IRCCS Fondazione Santa Lucia, Rome, Italy; University of Rome "Sapienza", Rome, Italy

Jean-Marc Diverrez Usage and Acceptability Lab, Cesson-Sévigné, France

Manh-Cuong Do University Paris-Sud, Orsay, France

Mengxi Dong University of Toronto Scarborough, Toronto, ON, Canada

Andrew T. Duchowski School of Computing, Clemson University, Clemson, SC, United States

Anirban Dutta EuroMov, University of Montpellier, Montpellier, France

Lydia Dyer University of Nottingham, Nottingham, United Kingdom

Sonia Em Usage and Acceptability Lab, Cesson-Sévigné, France

Kate Ewing BAE Systems, Wharton, United Kingdom

Stephen Fairclough Liverpool John Moores University, Liverpool, United Kingdom

Brian Falcone George Mason University, Fairfax, VA, United States

Tiago H. Falk Université du Québec, Montreal, Canada

Sara Feldman Drexel University, Philadelphia, PA, United States

Ying Xing Feng Universiti Teknologi Petronas, Perak, Malaysia

Victor S. Finomore West Virginia University, Morgantown, WV, United States

Nina Flad Max-Planck Institute for Biological Cybernetics, Tübingen, Germany; IMPRS for Cognitive and Systems Neuroscience, Tübingen, Germany

Alice Formwalt George Mason University, Fairfax, VA, United States

Alexandra Fort Université de Lyon, IFSTTAR, TS2, LESCOT, Lyon, France

Paul Fourcade Université Paris Sud, Université Paris-Saclay, CIAMS, Orsay, France; CIAMS, Université d'Orléans, Orléans, France

Marc A. Fournier University of Toronto Scarborough, Toronto, ON, Canada

Jérémy Frey University of Bordeaux, Bordeaux, France; INRIA Bordeaux Sud-Ouest/LaBRI, Talence, France

C. Gabaude Université de Lyon, IFSTTAR, TS2, LESCOT, Lyon, France

Olivier Gagey University Paris-Sud, Orsay, France; C.H.U Kremlin Bicêtre, Kremlin Bicêtre, France

Marc Garbey Houston Methodist Hospital, Houston, TX, United States; LaSIE UMR 7356 CNRS, University of La Rochelle, La Rochelle, France

Liliana Garcia ENSC-Bordeaux INP, Bordeaux, France

Thibault Gateau ISAE-SUPAERO, Université de Toulouse, Toulouse, France

Lukas Gehrke Biological Psychology and Neuroergonomics, Technische Universitaet Berlin, Berlin, Germany

Nancy Getchell University of Delaware, Newark, DE, United States

Evanthia Giagloglou University of Kragujevac, Kragujevac, Serbia

Christiane Glatz Max-Planck Institute for Biological Cybernetics, Tübingen, Germany; International Max Planck Research School, Tübingen, Germany

Kimberly Goodyear Brown University, Providence, RI, United States; National Institutes of Health, Bethesda, MD, United States

Robert J. Gougelet Cognitive Science Department, University of California, San Diego, San Diego, CA, United States

Jonas Gouraud Office National d'Etudes et de Recherche Aérospatiales, Salon-de-Provence, France

Klaus Gramann Biological Psychology and Neuroergonomic, Technische Universitaet Berlin, Berlin, Germany; Center for Advanced Neurological Engineering, University of California San Diego, San Diego, CA, United States

Dhruv Grewal Babson College, Boston, MA, United States

Carlos Guerrero-Mosquera Universitat Pompeu Fabra, Barcelona, Spain

Céline Guillaume Université Libre de Bruxelles (ULB), Brussels, Belgium

Martin Hachet INRIA Bordeaux Sud-Ouest/LaBRI, Talence, France

Alain Hamaoui Institut National Universitaire Champollion, Albi, France

Gabriella M. Hancock California State University, Long Beach, Long Beach, CA, United States

Peter A. Hancock University of Central Florida, Orlando, FL, United States

Ahmad Fadzil M. Hani Universiti Teknologi Petronas, Perak, Malaysia

Amanda E. Harwood George Mason University, Fairfax, VA, United States

Mitsuhiro Hayashibe EuroMov, University of Montpellier, Montpellier, France

Terry Heiman-Patterson Drexel University, Philadelphia, PA, United States

Girod Hervé Dassault Aviation, Saint-Cloud, France

Maarten A.J. Hogervorst TNO Netherlands Organisation for Applied Scientific Research, Soesterberg, The Netherlands

Amy L. Holloway University of Nottingham, Nottingham, United Kingdom

Jean-Louis Honeine University Paris-Sud, Orsay, France; University of Pavia, Pavia, Italy

Keum-Shik Hong Pusan National University, Busan, Republic of Korea

Klas Ihme DLR, Institute of Transportation Systems, Braunschweig, Germany

Kurtulus Izzetoglu Drexel University, Philadelphia, PA, United States

Meltem Izzetoglu Villanova University, Villanova, PA, United States; Yeshiva University, Bronx, NY, United States

Philip L. Jackson Université Laval, Québec, QC, Canada

Christophe Jallais Université de Lyon, IFSTTAR, TS2, LESCOT, Lyon, France

Christian P. Janssen Utrecht University, Utrecht, The Netherlands

Branislav Jeremic University of Kragujevac, Kragujevac, Serbia

Meike Jipp DLR, Institute of Transportation Systems, Braunschweig, Germany

Evelyn Jungnickel Biological Psychology and Neuroergonomics, Technische Universitaet Berlin, Berlin, Germany

Hélio Kadogami ONERA Provence Research Center, Salon-de-Provence, France

Gozde Kara Steinbeis Advanced Risk Technologies GmbH, Stuttgart, Germany

Waldemar Karwowski Department of Industrial Engineering & Management Systems, University of Central Florida, United States

Quinn Kennedy Naval Postgraduate School, Monterey, CA, United States

Theresa T. Kessler University of Central Florida, Orlando, FL, United States

Muhammad J. Khan Pusan National University, Busan, Republic of Korea

Rayyan A. Khan Air University, Islamabad, Pakistan

Marius Klug Biological Psychology and Neuroergonomics, Technische Universitaet Berlin, Berlin, Germany

Amanda E. Kraft Lockheed Martin Advanced Technology Lab, Cherry Hill, NJ, United States

Michael Krein Lockheed Martin Advanced Technology Lab, Cherry Hill, NJ, United States

Ute Kreplin Massey University, Auckland, New Zealand

Bartlomiej Kroczek Jagiellonian University, Krakow, Poland

Lauens R. Krol Technische Universitaet Berlin, Berlin, Germany

Frank Krueger George Mason University, Fairfax, VA, United States

Ombeline Labaune Université Paris Sud, Université Paris-Saclay, CIAMS, Orsay, France; CIAMS, Université d'Orléans, Orléans, France

Daniel Lafond Thales Research & Technology, Québec, QC, Canada

Claudio Lantieri University of Bologna, Bologna, Italy

Paola Lanzi DeepBlue srl, Rome, Italy

Amine Laouar ISAE-SUPAERO, Université de Toulouse, Toulouse, France

Dargent Lauren ISAE-SUPAERO, Toulouse, France

Rachel Leproult Université Libre de Bruxelles (ULB), Brussels, Belgium

Véronique Lespinet-Najib ENSC-Bordeaux INP, Bordeaux, France

Ling-Yin Liang University of Evansville, Evansville, IN, United States

Fabien Lotte INRIA Bordeaux Sud-Ouest/LaBRI, Talence, France

Ivan Macuzic University of Kragujevac, Kragujevac, Serbia

Nicolas Maille ONERA Provence Research Center, Salon-de-Provence, France

Horia A Maior University of Nottingham, Nottingham, United Kingdom

S. Malin Université de Lyon, IFSTTAR, TS2, LESCOT, Lyon, France

Alexandre Marois Université Laval, Québec City, QC, Canada

Franck Mars CNRS, Central Nantes & IMT-Atlantique, Nantes, France

Nicolas Martin Usage and Acceptability Lab, Cesson-Sévigné, France

Nadine Matton Ecole Nationale de l'Aviation Civile (ENAC), Université de Toulouse, France

Magdalena Matyjek Jagiellonian University, Krakow, Poland

Kevin McCarthy Drexel University, Philadelphia, PA, United States

Ryan McKendrick Northrop Grumman, Redondo Beach, CA, United States; George Mason University, Fairfax, VA, United States

Tom McWilliams Massachusetts Institute of Technology, AgeLab, Cambridge, MA, United States

Bruce Mehler Massachusetts Institute of Technology, AgeLab, Cambridge, MA, United States

Ranjana Mehta George Mason University, Fairfax, VA, United States

Ranjana K. Mehta Texas A&M University, College Station, TX, United States

Mathilde Menoret ENSC-Bordeaux INP, Bordeaux, France

Yoshihiro Miyake Tokyo Institute of Technology, Yokohama, Japan

Alexandre Moly ISAE-SUPAERO, Université de Toulouse, Toulouse, France

Rabia Murtza George Mason University, Fairfax, VA, United States

Makii Muthalib Silverline Research, Brisbane, Australia

Mark Muthalib EuroMov, University of Montpellier, Montpellier, France; Deakin University, Melbourne, Australia

Noman Naseer Air University, Islamabad, Pakistan

Jordan Navarro Université de Lyon, Bron, France

Roger Newport University of Nottingham, Nottingham, United Kingdom

Anton Nijholt University of Twente, Enschede, The Netherlands

Michal Ociepka Jagiellonian University, Krakow, Poland

Morellec Olivier Dassault Aviation, Saint-Cloud, France

Ahmet Omurtag Nottingham Trent University, Nottingham, United Kingdom

Banu Onaral Drexel University, Philadelphia, PA, United States

Hiroki Ora Tokyo Institute of Technology, Yokohama, Japan

Bob Oudejans TNO Netherlands Organisation for Applied Scientific Research, Soesterberg, The Netherlands

Özgürol Öztürk Galatasaray University, İstanbul, Turkey

Martin Paczynski George Mason University, Fairfax, VA, United States

Nico Pallamin Usage and Acceptability Lab, Cesson-Sévigné, France

Raja Parasuraman George Mason University, Fairfax, VA, United States

Mark Parent Université Laval, Quebec City, Canada

René Patesson Université Libre de Bruxelles, Bruxelles, Belgique

Kou Paul Dassault Aviation, Saint-Cloud, France

Philippe Peigneux Université Libre de Bruxelles (ULB), Brussels, Belgium

Matthias Peissner University of Stuttgart, Stuttgart, Germany; Fraunhofer Institute for Industrial Engineering IAO, Stuttgart, Germany

G. Pepin Université de Lyon, IFSTTAR, TS2, LESCOT, Lyon, France

Stephane Perrey EuroMov, University of Montpellier, Montpellier, France

Vsevolod Peysakhovich ISAE-Supaéro, Université de Toulouse, Toulouse, France

Markus Plank Biological Psychology and Neuroergonomic, Technische Universitaet Berlin, Berlin, Germany

Riccardo Poli University of Essex, Colchester, United Kingdom

Kathrin Pollmann University of Stuttgart, Stuttgart, Germany; Fraunhofer Institute for Industrial Engineering IAO, Stuttgart, Germany

Simone Pozzi DeepBlue srl, Rome, Italy

Nancy M. Puccinelli University of Bath, Claverton Down, Bath, United Kingdom

Jean Pylouster Université de Poitiers, Poitiers, France

Kerem Rızvanoğlu Galatasaray University, İstanbul, Turkey

Martin Ragot Usage and Acceptability Lab, Cesson-Sévigné, France

Bryan Reimer Massachusetts Institute of Technology, AgeLab, Cambridge, MA, United States

Emanuelle Reynaud Université de Lyon, Bron, France

Joohyun Rhee Texas A&M University, College Station, TX, USA

Jochem W. Rieger University of Oldenburg, Oldenburg, Germany

Anthony J. Ries Army Research Laboratory, Aberdeen, MD, United States

Benoit Roberge-Vallières Université Laval, Québec, QC, Canada

Achala H. Rodrigo University of Toronto Scarborough, Toronto, ON, Canada

Anne L. Roggeveen Babson College, Boston, MA, United States

Ricardo Ron-Angevin University of Málaga, Málaga, Spain

Guillaume Roumy French Air Force, France

Raphaëlle N. Roy ISAE-SUPAREO, Université de Toulouse, Toulouse, France

Anthony C. Ruocco University of Toronto Scarborough, Toronto, ON, Canada

Bartlett A. Russell Lockheed Martin Advanced Technology Lab, Arlington, VA, United States

Jon Russo Lockheed Martin Advanced Technology Lab, Cherry Hill, NJ, United States

Richard M. Ryan University of Rochester

Amanda Sargent Drexel University, Philadelphia, PA, United States

Kelly Satterfield Oak Ridge Institute for Science and Education, Wright-Patterson Air Force Base, OH, United States

Ben D. Sawyer Massachusetts Institute of Technology, AgeLab, Cambridge, MA, United States

Sébastien Scannella ISAE-SUPAERO, Université de Toulouse, Toulouse, France

Menja Scheer Max-Planck Institute for Biological Cybernetics, Tübingen, Germany

Melissa Scheldrup George Mason University, Fairfax, VA, United States

Alex Schilder University of Delaware, Newark, DE, United States

Nicolina Sciaraffa BrainSigns srl, Rome, Italy; IRCCS Fondazione Santa Lucia, Rome, Italy; University of Rome "Sapienza", Rome, Italy

Lee Sciarini Naval Postgraduate School, Monterey, CA, United States

Magdalena Senderecka Jagiellonian University, Krakow, Poland

Sarah Sharples University of Nottingham, Nottingham, United Kingdom

Tyler H. Shaw George Mason University, Fairfax, VA, United States

Patricia A. Shewokis Drexel University, Philadelphia, PA, United States

Andrea Simone University of Bologna, Bologna, Italy

Hichem Slama Université Libre de Bruxelles (ULB), Brussels, Belgium; Erasme Hospital, Brussels, Belgium

Alastair D. Smith University of Plymouth, Plymouth, United Kingdom

Bertille Somon Office National d'Etudes et de Recherche Aérospatiales, Salon-de-Provence, France; Université Grenoble Alpes, CNRS, LPNC UMR 5105, F-38000, Grenoble, France

Hiba Souissi Université de Nice Sophia-Antipolis, Nice, France

Moritz Späth Technical University of Munich, Garching, Germany

Kimberly L. Stowers University of Central Florida, Orlando, FL, United States

Clara Suied Institut de Recherche Biomédicale des Armées, Brétigny sur Orge, France

Junfeng Sun Shanghai Jiao Tong University, Shanghai, China

Rajnesh Suri Drexel University, Philadelphia, PA, United States

Tong Boon Tang Universiti Teknologi Petronas, Perak, Malaysia

Yingying Tang Shanghai Jiao Tong University School of Medicine, Shanghai, China

Emre O. Tartan Baskent University, Ankara, Turkey

Nadège Tebbache Institut National Universitaire Champollion, Albi, France

Franck Techer Université de Lyon, IFSTTAR, TS2, LESCOT, Lyon, France; University of Nantes, LPPL, Nantes, France

Cengiz Terzibas Multisensory Cognition and Computation Laboratory, Universal Communication Research Institute, National Institute of Information and Communications Technology, Kyoto, Japan

Catherine Tessier ONERA/DTIS, Université de Toulouse, France

Claudine Teyssedre Université Paris Sud, Université Paris-Saclay, CIAMS, Orsay, France; CIAMS, Université d'Orléans, Orléans, France

Hayley Thair University of Nottingham, Nottingham, United Kingdom

Jean-Denis Thériault Université Laval, Québec, QC, Canada

Alexander Toet TNO Netherlands Organization for Applied Scientific Research, Soesterberg, The Netherlands

Shanbao Tong Shanghai Jiao Tong University, Shanghai, China

Jonathan Touryan Army Research Laboratory, Aberdeen, MD, United States

Amy Trask University of Delaware, Newark, DE, United States

Sébastien Tremblay Université Laval, Quebec City, Canada

Anirudh Unni University of Oldenburg, Oldenburg, Germany

François Vachon Université Laval, Québec, QC, Canada

Davide Valeriani University of Essex, Colchester, United Kingdom

Benoît Valéry ISAE-SUPAERO, Université de Toulouse, Toulouse, France; Ecole Nationale de l'Aviation Civile (ENAC), Université de Toulouse, France

Helma van den Berg TNO Netherlands Organization for Applied Scientific Research, Soesterberg, The Netherlands

Valeria Vignali University of Bologna, Bologna, Italy

Mathias Vukelić University of Stuttgart, Stuttgart, Germany; Fraunhofer Institute for Industrial Engineering IAO, Stuttgart, Germany

Jijun Wang Shanghai Jiao Tong University School of Medicine, Shanghai, China

Max L. Wilson University of Nottingham, Nottingham, United Kingdom

Emily Wusch University of Delaware, Newark, DE, United States

Petros Xanthopoulos Decision and Information Sciences, Stetson University, United States

Eric Yiou Université Paris Sud, Université Paris-Saclay, CIAMS, Orsay, France; CIAMS, Université d'Orléans, Orléans, France

Amad Zafar Pusan National University, Busan, Republic of Korea

Thorsten O. Zander Technische Universitaet Berlin, Berlin, Germany

Matthias D. Ziegler Lockheed Martin Advanced Technology Lab, Arlington, VA, United States

Ivana Živanovic-Macuzic University of Kragujevac, Kragujevac, Serbia

Section I

Introduction

Chapter 1

Progress and Direction in Neuroergonomics

Frédéric Dehais[1], Hasan Ayaz[2,3,4]

[1]*ISAE-SUPAERO, Université de Toulouse, Toulouse, France;* [2]*Drexel University, Philadelphia, PA, United States;* [3]*University of Pennsylvania, Philadelphia, PA, United States;* [4]*Children's Hospital of Philadelphia, Philadelphia, PA, United States*

INTRODUCTION

Recent advances in cognitive neuroscience and progress in neuroimaging have radically changed our understanding of the neural mechanisms underlying human perceptual, cognitive, and motor functioning. These findings are of great importance for applied scientific disciplines concerned with the evaluation of human performance. Since the early 2000s, Neuroergonomics, the intersection of Neuroscience, Cognitive Engineering, and Human Factors, proposes to examine the brain mechanisms and underlying human–technology interaction in increasingly naturalistic settings representative of work and everyday-life situations. The objective of merging these disciplines into a single field of research is to encourage cross-fertilization and provide new tools at the epistemological, methodological, and technical levels (see Fig. 1.1)[1a]. This approach, known as Neuroergonomics, was initially proposed by Prof. R. Parasuraman (1998), progressively conceptualized[1–3] and then formalized by Profs. Parasuraman and Rizzo in their book "Neuroergonomics: The Brain at Work."[4] This discipline is defined by its founder, Prof. Parasuraman, as the "scientific study of the brain mechanisms and psychological and physical functions of humans in relation to technology, work and environments." The postulate is that the understanding of the underlying neurocognitive processes that occur during complex real-life activities such as human–technology interaction could be used to improve safety and efficiency of the overall human–machine teaming. Thus, the objective of Neuroergonomics, consistently with Human Factors and Ergonomics, is to enhance the integration of the human by fitting machine with human and fitting the human to machine. This innovative approach has found several applications ranging from the operation of complex systems (e.g., flying aircraft, supervising nuclear power plants, driving autonomous vehicles, surgeons in the operating room) to the improvement of the performance of disabled patients or elderly people in their daily interaction with their environment.[5]

These goals are achieved by improving the design of the complex system to human cognition, adapting the interface dynamically during use for augmentation of human performance and its transfer to improved functioning at work or in everyday-life situations.

UNDERSTANDING THE BRAIN IN EVERYDAY ACTIVITIES

Neuroergonomics promotes the use of various brain-imaging techniques and psychophysiological techniques. A challenge of great importance for Neuroergonomics is to succeed in reproducing ecological conditions in a well-controlled laboratory. Thus, Neuroergonomics proposes to conduct a "gradient" of experiments starting with well-controlled protocols with high spatial resolution devices that are constrained by the use of low-fidelity simulators, progressing to more ecological experiments in dynamic microworlds using devices that are portable but with lower accuracy, to eventually conducting less-controlled experiments in simulators and real ecological conditions (see Fig. 1.2).

Indeed, functional Magnetic Resonance Imaging (fMRI) or Magnetoencephalography (MEG) provides precious insight into the neural mechanisms underpinning cognitive processes. However, these techniques have several drawbacks that prevent from designing ecological experiments to examine the brain "at work." Despite these apparent limitations, the use of such techniques with advanced signal processing allowed investigation of drivers'[6] or pilots' neural activation[7]; Durantin et al. (2017)[7a], while performing simulated tasks. An alternative approach to overcome the aforementioned limitations is to consider the use of field-deployable portable modalities such as Electroencephalography (EEG) or functional Near Infra-Red Spectroscopy (fNIRS), which allows the noninvasive examination of brain function under realistic settings. Although EEG allows assessment of the electrical activity of the neurons, fNIRS is a noninvasive optical brain-monitoring technology

Neuroergonomics. https://doi.org/10.1016/B978-0-12-811926-6.00001-4
Copyright © 2019 Elsevier Inc. All rights reserved.

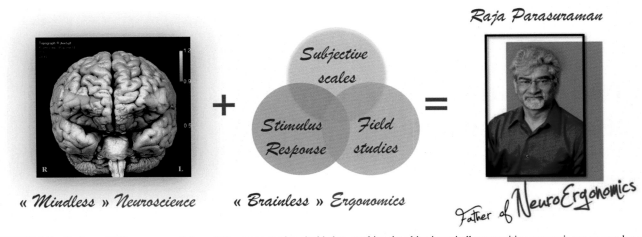

FIGURE 1.1   Professor R. Parasuraman, father of Neuroergonomics, decided to combine the objective mindless cognitive neuroscience approach and the subjective brainless Ergonomics approach.

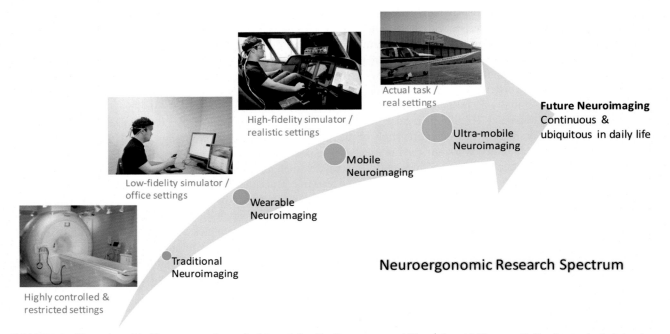

FIGURE 1.2   Illustration of the Neuroergonomics methodology defined by Parasuraman and Rizzo[4]: from highly controlled but less ecological situations to highly ecological but less-controlled situations. Cerebral and autonomous nervous system activations are compared across different situations to ensure the validity of the measurements. This methodology allows to tackle human cognition through an experimental continuum.

that measures the cerebral hemodynamics associated with neural activity.[8] fNIRS and EEG are complementary as they overcome each other's measurement weaknesses in terms of spatial and temporal information (Liu, Ayaz and Shewokis 2017)[8a]. Moreover, correlational and causal analyses between the two signals can provide a deep understanding of the neurovascular coupling and brain dynamics offering interesting prospects for Neuroergonomics.[9]

Adapting Interaction

The aforementioned neurophysiological measures can be collected and computed in an off-line manner to assess the user's experience and evaluate system design. Recent progress in signal-processing and machine-learning techniques has also opened promising solutions for human–machine interaction. Indeed, another facet of Neuroergonomics is the design of an "active" or a "passive" Brain–Computer Interface (BCI) based on the online processing of the neurophysiological signals.[10] "Active" BCI allows a user to control artifacts with his brain wave without requiring any physical actions on the user interface. Different paradigms have been implemented so far, such as P300 spellers,[11] and Steady State Visually Evoked Potential (SSVEP).[12] and allow the user to drive a car,[13] operate robots,[14] fly a helicopter[15] or use a wheelchair.[16] However, such BCIs need to be improved as they require extensive training and lead the users to focus on controlling their own brain waves, leaving few cognitive resources to monitor or interact with their systems. Alternatively, "Passive" BCIs are not meant to directly control a device (e.g., a mouse) via brain activity but to support "implicit interaction."[17,18] Research on "passive" BCIs provides interesting insight as they aim to infer the human operator's mental state (Gateau, Ayaz, and Dehais, 2018[18a]; Gateau, Durantin, Lancelot, Scannella, and Dehais, 2015)[18b] and may either provide "neurofeedback" to user (Grozea, Voinescu, and Fazli, 2011)[18c] or adapt the nature of the interactions to overcome cognitive bottlenecks. For instance, adaptive automation, task reallocation, or the triggering of cognitive countermeasures are potential solutions to assist humans and optimize human-system performance (Dehais, Causse, and Tremblay, 2011[18d]; Szalma and Hancock, 2008)[18e]. Thus, the design of the neuroadaptive user interface represents a growing field full of promise for Neuroergonomics.

Augmenting Cognition

The rediscovery, over two decades ago,[19] of transcranial brain stimulation has led to a proliferation of research on brain and cognitive augmentation, in both healthy adults and patients with neurological or psychiatric disease.[20,21] Noninvasive brain stimulation techniques, such as transcranial direct current stimulation (tDCS), transcranial alternating current stimulation (tACS), and transcranial magnetic stimulation (TMS) provide researchers with unique opportunities to alter brain activity in both clinical and healthy groups and study causal mechanism of brain activity and behavior as well as clinical outcome measures. Moreover, utilizing simultaneous neuroimaging together with such neurostimulation has shown to be possible, for example fNIRS and tDCS,[22] EEG and tDCS,[23] fNIRS and TMS,[24] and fMRI and tDCS[25] are some examples of multimodal studies.

Noninvasive brain stimulation has also been shown to accelerate learning and enhance human performance in healthy individuals with complex natural tasks.[26,27] Such approaches present unique opportunities for research as well as field deployment, because most of the hardware is portable and miniaturized. Further research is needed to understand optimized stimulation parameters and operational needs; hence, simultaneous neuroimaging could provide new insights, uncover the effects of neurostimulation in the brain, and provide opportunity to adapt stimulation per participant and in real time.

CONCLUSION AND FUTURE CHALLENGES

Taken together, Neuroergonomics offers conceptual, theoretical, and technical prospects for human factors, neuroscience, and neuroengineering among other disciplines. The neuroergonomic approach has been considerably facilitated by the recent rise of development of portable and wearable neuroimaging devices, including EEG and fNIRS. As use of mobile neuroimaging becomes more practical and widespread, neuroergonomic research is expected to reach its full potential. And the emerging wearable neurostimulation techniques such as tDCS and tACS seem to fuel the growth further. As the Neuroergonomic field grows and moves from lab to routine practice, neuroethics should take a more central role in the design and execution of studies.

Recent trends in Neuroergonomics have established it as a tool to inform design, development, and use of complex interfaces, as well as operational procedures and anywhere human-to-machine and human-to-human interaction is required. In its full capacity, Neuroergonomic approaches are expected to contribute to a diverse array of domains from single participant, product development, and daily procedure design, and hence benefit society.

BOOK ORGANIZATION

This book aims to provide a comprehensive description of state-of–the-art Neuroergonomics research. Following an introduction, the next segment is devoted to methodology, and each subsequent chapter is a tutorial for popular and emerging techniques of interest to Neuroergonomics practitioners. The following three chapters are application areas. Segment three is a collection of the latest neuroadaptive interfaces and operator assessment studies. Segment four is neurostimulation applications. Segment five presents emerging applications in decision-making, usability, trust, and emotions. The last segment contains entries from the Inaugural International Neuroergonomics Conference that took place on October 6 and 7, 2017, in Paris, France.

REFERENCES

1. Hancock PA, Szalma JL. The future of neuroergonomics. *Theoretical Issues in Ergonomics Science* 2003;**4**(1–2):238–49. https://doi.org/10.1080/14 63922021000020927.

1a. Parasuraman R. Neuroergonomics: The study of brain and behavior at work. *Washington, DC: Cognitive Science Laboratory* 1998. Available online at: www.psychology.cua.edu/csl/neuroerg.html.

2. Parasuraman R. Neuroergonomics: research and practice. *Theoretical Issues in Ergonomics Science* 2003;**4**(1–2):5–20. https://doi.org/10.1080/14639220210199753.

3. Sarter N, Sarter M. Neuroergonomics: opportunities and challenges of merging cognitive neuroscience with cognitive ergonomics. *Theoretical Issues in Ergonomics Science* 2003;**4**(1–2):142–50. New York, NY USA: Oxford University Press; 2007 https://doi.org/10.1080/1463922021000020882.

4. Parasuraman R, Rizzo M, Zander TO, Ayaz H. *Neuroergonomics: the brain at work.* New York, NY USA: Oxford University Press; 2007*Frontiers in Human Neuroscience* 2017;**11**(165). https://doi.org/10.3389/fnhum.2017.00165.

5. Gramann K, Fairclough SH, Zander TO, Ayaz H, Watson TD, Pearlson GD. Editorial: trends in neuroergonomics. *Frontiers in Human Neuroscience* 2017;**11**(165):158–67. https://doi.org/10.3389/fnhum.2017.00165.

6. Calhoun VD, Pekar JJ, McGinty VB, Adali T, Watson TD, Pearlson GD, Sato M-A. Different activation dynamics in multiple neural systems during simulated driving. *Hum Brain Mapp* 2002;**16**(3):158–67.

7. Callan DE, Gamez M, Cassel DB, Terzibas C, Callan A, Kawato M, Sato M-A. Dynamic visuomotor transformation involved with remote flying of a plane utilizes the 'Mirror Neuron' system. *PLoS One* 2012;**7**(4):e33873–47. https://doi.org/10.1016/j.neuroimage.2011.06.023.

7a. Durantin G, Dehais F, Gonthier N, Terzibas C, Callan DE. Neural signature of inattentional deafness. *Human brain mapping* 2017;**38**(11):5440–55.

8. Ayaz H, Shewokis PA, Bunce S, Izzetoglu K, Willems B, Onaral B. Optical brain monitoring for operator training and mental workload assessment. *Neuroimage* 2012;**59**(1):36–47. https://doi.org/10.1016/j.neuroimage.2011.06.023.

8a. Liu Y., Ayaz H, Shewokis PA. Multisubject "learning" for mental workload classification using concurrent EEG, fNIRS, and physiological measures. *Frontiers in Human Neuroscience* 2017;**11**(389). https://doi.org/10.3389/fnhum.2017.00389.

9. Mandrick K, Chua Z, Causse M, Perrey S, Dehais F, Kubler A. Why a comprehensive understanding of mental workload through the measurement of neurovascular coupling is a key issue for Neuroergonomics? *Frontiers in Human Neuroscience* 2016;**10**. 14. https://doi.org/10.3389/fnhum.2016.00250.

10. Birbaumer N, Scherer R, Brauneis C, Pfurtscheller G. Brain-computer-interface research: coming of age. *Clinical Neurophysiology* 2006;**117**(3):479–83. https://doi.org/10.1016/j.clinph.2005.11.002.

11. Fazel-Rezai R, Allison BZ, Guger C, Sellers EW, Kleih SC, Kubler A. P300 brain computer interface: current challenges and emerging trends. *Front Neuroeng* 2012;**5**:*Intelligent autonomous systems* 12. Springer; 2013. p. 14–408. https://doi.org/10.3389/fneng.2012.00014.

12. Müller-Putz GR, Scherer R, Brauneis C, Pfurtscheller G. Steady-state visual evoked potential (SSVEP)-based communication: impact of harmonic frequency components. *Journal of Neural Engineering* 2005;**2**(4):Ronzhin A, Rigoll G, Meshcheryakov R, editors. *Interactive collaborative robotics: first international conference, ICR 2016, Budapest, Hungary, August 24–26, 2016, proceedings.* Cham: Springer International Publishing; 2016. p. 123–106.

13. Göhring D, Latotzky D, Wang M, Rojas R. Semi-autonomous car control using brain computer interfaces. *IEEE Transactions on Neural Systems and Rehabilitation Engineering* 2010;**18**(6). In: *Intelligent autonomous systems*, ;**12**. Springer; 2013581–9. p. 393–408. https://doi.org/10.1109/TNSRE.2010.2077654.

14. Stankevich L, Sonkin K, Muller-Putz GR, Scherer R, Slater M, Pfurtscheller G. Human-Robot interaction using brain-computer interface based on EEG signal decoding. *Computational Intelligence and Neuroscience* 2007In: Ronzhin A, Rigoll G, Meshcheryakov R, editors. *Interactive collaborative robotics: first international conference, ICR 2016, Budapest, Hungary, August 24–26, 2016, proceedings.* Cham: Springer International Publishing; 201679642. p. 99–106. https://doi.org/10.1155/2007/79642.

15. Royer AS, Doud AJ, Rose ML, Bin H. EEG control of a virtual helicopter in 3-dimensional space using intelligent control strategies. *IEEE Transactions on Neural Systems and Rehabilitation Engineering* 2010;**18**(6):581–9. https://doi.org/10.1109/TNSRE.2010.2077654.

16. Leeb R, Friedman D, Muller-Putz GR, Scherer R, Slater M, Pfurtscheller G. Self-paced (asynchronous) BCI control of a wheelchair in virtual environments: a case study with a tetraplegic. *Computational Intelligence and Neuroscience* 2007:79642. https://doi.org/10.1155/2007/79642.

17. Zander TO, Kothe C. Towards passive brain–computer interfaces: applying brain–computer interface technology to human–machine systems in general. *Journal of Neural Engineering* 2011;**8**(2):025005–9.

18. Zander TO, Krol LR, Birbaumer NP, Gramann K. Neuroadaptive technology enables implicit cursor control based on medial prefrontal cortex activity. *Proceedings of the National Academy of Sciences* 2016;**85**:889–94. https://doi.org/10.1073/pnas.1605155114.

18a. Gateau T, Ayaz H, Dehais F. In silico versus over the clouds: On-the-fly mental state estimation of aircraft pilots, using a functional near infrared spectroscopy based passive-BCI. *Frontiers in human neuroscience* 2018;**12**(Part 3):187–47.

18b. Gateau T, Durantin G, Lancelot F, Scannella S, Dehais F. Real-time state estimation in a flight simulator using fNIRS. *PLoS one* 2015;**10**(3):e0121279.

18c. Grozea C, Voinescu CD, Fazli S. Bristle-sensors—low-cost flexible passive dry EEG electrodes for neurofeedback and BCI applications. *Journal of neural engineering* 2011;**8**((2):025008.

18d. Dehais F, Causse M, Tremblay S, Qian Z, Onaral B, Wang J, Tong S. Mitigation of conflicts with automation: use of cognitive countermeasures. *Human Factors* 2011;**53**(5):448–60.

18e. Hancock PA, Szalma JL, Wada A, Parasuraman R. Stress and performance. *Frontiers in Human Neuroscience* 2016;**10**. In: Hancock PA, Szalma JL, editors. *Performance under stress.* Hampshire, UK: Ashgate; 2008. p. 1–18.

19. Nitsche M, Paulus W, Bergstedt DT, Ziegler M, Phillips ME. Excitability changes induced in the human motor cortex by weak transcranial direct current stimulation. *The Journal of Physiology* 2000;**527**(3):633–9. https://doi.org/10.3389/fnhum.2016.00034.

20. Clark VP, Parasuraman R. Neuroenhancement: enhancing brain and mind in health and in disease. *Neuroimage* 2014;**85**:889–94. https://doi.org/10.1177/0018720814538815.

21. Flöel A. tDCS-enhanced motor and cognitive function in neurological diseases. *Neuroimage* 2014;**85**(Part 3):934–47. https://doi.org/10.1016/j.neuroimage.2013.05.098.

22. McKendrick R, Parasuraman R, Ayaz H. Wearable functional near infrared spectroscopy (fNIRS) and transcranial direct current stimulation (tDCS): expanding vistas for neurocognitive augmentation. *Frontiers in Systems Neuroscience* 2015;**9**(27). https://doi.org/10.3389/fnsys.2015.00027.

23. Schestatsky P, Morales-Quezada L, Fregni F. Simultaneous EEG monitoring during transcranial direct current stimulation. *Journal of Visualized Experiments* 2013;(76).

24. Curtin A, Sun J, Ayaz H, Qian Z, Onaral B, Wang J, Tong S. Evaluation of evoked responses to pulse-matched high frequency and intermittent theta burst transcranial magnetic stimulation using simultaneous functional near-infrared spectroscopy. *Neurophotonics* 2017;**4**(4):041405. https://doi.org/10.1117/1.NPh.4.4.041405.

25. Callan D, Falcone B, Wada A, Parasuraman R. Simultaneous tDCS-fMRI identifies resting state networks correlated with visual search enhancement. *Frontiers in Human Neuroscience* 2016;**10**. https://doi.org/10.3389/fnhum.2016.00072.

26. Choe J, Coffman BA, Bergstedt DT, Ziegler M, Phillips ME. Transcranial direct current stimulation modulates neuronal activity and learning in pilot training. *Frontiers in Human Neuroscience* 2016;**10**. https://doi.org/10.3389/fnhum.2016.00034.

27. Parasuraman R, McKinley RA. Using noninvasive brain stimulation to accelerate learning and enhance human performance. *Human Factors* 2014. https://doi.org/10.1177/0018720814538815.

Section II

Methods

Chapter 2

The Use of Electroencephalography in Neuroergonomics

Klaus Gramann[1,2], Markus Plank[1]

[1]*Biological Psychology and Neuroergonomic, Technische Universitaet Berlin, Berlin, Germany;* [2]*Center for Advanced Neurological Engineering, University of California San Diego, San Diego, CA, United States*

INTRODUCTION

Compared to other brain-imaging methods, electroencephalogram (EEG) provides three major advantages. First, EEG provides high temporal resolution allowing recording and analyzing brain dynamics at the same millisecond timescale at which cognitive processes take place. This high temporal resolution is an advantage over hemodynamic measures like functional magnetic resonance imaging (fMRI) and functional near infrared spectroscopy (fNIRS), which measure sluggish hemodynamic processes that follow, rather than accompany cognitive processes. The second important advantage of EEG is its portability. Modern amplifier systems are small and lightweight and allow mobile EEG recordings in realistic work settings and everyday life scenarios. The third advantage of EEG is its affordability with relatively low costs as compared to traditional brain imaging approaches like fMRI or magnetoencephalography (MEG). Modern portable systems allow data recordings using mobile phones or other portable devices and allow for EEG experiments with little budget.

The aforementioned advantages of the EEG come with the caveat of a restricted spatial resolution regarding the origin of the surface-measured brain activity. Although hemodynamics measures such as fMRI or fNIRS allow reconstruction of signal-generating structures in the mm range, the reconstructive accuracy of EEG is limited to the cm scale.[1] In addition, the recorded signal is easily mixed with nonbrain activity like eye movements and muscle contractions, which require analytical tools to dissociate brain from nonbrain activity. Such tools usually work best with a relatively large number of electrodes.

PHYSIOLOGICAL FOUNDATION OF THE EEG

The human cortex contains several billion neurons arranged in a specific spatial configuration. Within and across mostly six layers, different kinds of neurons are connected building microcircuits that are grouped into cortical columns. The post-synaptic activity in large populations of pyramidal neurons with their perpendicular orientation toward the scalp is the basis of the EEG signal.[2] Action potentials arriving from intracortical as well as thalamocortical connections lead to excitatory postsynaptic potentials (EPSPs) or inhibitory postsynaptic potentials (IPSPs) in the pyramidal neurons. In case of EPSPs, the rapid influx of positively charged ions into the neurons depolarizes their resting membrane potentials and creates a more negative extracellular voltage around the contact points compared to the voltage along the rest of the neurons. For IPSPs the extracellular voltage becomes more positive due to the influx of negatively or outflux of positively charged ions.

When different regions of the neurons have different charges that are separated by some distance (e.g., the apical dendrite compared to the soma), the overall distribution of charges along the neurons can be described as an electric dipole. The region of the dipole with positive charge is defined as the source. The region with negative charge is defined as the sink of the dipole. It is sufficient that one region of the neuron is charged to a lesser extent to become relatively more positive or negative and thus producing (relatively) opposite charges along the neuron. Dipoles are generated on the level of single neurons and sum up over large neuronal population to become measurable outside the skull. The cortical field potentials produced by all dipoles in the area under the electrode will be measured as one single dipole with the number of summed individual neuronal dipoles determining the overall magnitude of activity.[3]

The field potentials spread through the extracellular fluids that make up a large portion of the human brain. This distribution of charges through the volume of the brain is referred to as *volume conduction*. Outside the brain volume, the field potentials distribute via capacitive conductance through the protective layers of the dura, the skull, and the skin to the electrodes. At the level of the electrode, the signal is picked up and sent to the amplifier to magnify the miniscule signal.

Neuroergonomics. https://doi.org/10.1016/B978-0-12-811926-6.00002-6
Copyright © 2019 Elsevier Inc. All rights reserved.

EEG Amplifiers

Most modern amplifiers are differential amplifiers that amplify the difference between the active electrodes and the reference electrode. The activity measured at each time point at the reference electrode is subtracted from the activity at each time point at the active electrodes. This way, signals that arrive at both the active and the reference electrode in phase (e.g., electrical noise stemming from light, screens or other electrical devices) will be rejected. This is expressed in the *common mode rejection* of the amplifier with higher values indicating improving attenuation of activity that is common at both the reference and the active electrodes. A second important characteristic is the *input impedance* of the amplifier, which describes the tendency of the amplifier to oppose the flow of current from the electrodes through the amplifier. The higher the input impedance (in the range of megaohms) the better the amplifier as it does not affect the signal voltage.

The *number of channels* an amplifier supports determines the spatial density of the recording. More electrodes provide more information about ongoing brain activity and high channel numbers are favorable to allow different analytic approaches to the data. If only very few electrodes are used any description of the signal distribution on the surface or inside the skull is based on interpolation. In addition, most source reconstruction approaches require a high number of equidistant electrodes (100 or more) for reliable estimation of the sources' locations within the signal generating volume.[4] However, higher channel numbers come with increased preparation time and reduced mobility of participants. For ambulatory EEG experiments, mobility of participants might be more important than the range of possible data analyses approaches.

The *sample rate* of the amplifier is the rate at which the analog signal arriving at the electrode is represented as digital signal. With a sample rate of 250 Hz, the analog signal is sampled every 4 ms (1/frequency = 1/250 Hz = 0.004 s). The sample rate also determines the highest frequency that can be analyzed, as described in the Nyquist Theorem.[5] Modern amplifiers usually allow sample rates higher than 250 Hz and thus there is no hardware constraint for a priori restricting the frequency range for data analyses. The sample rate is closely related to the *filter settings* of an amplifier which determine how the analog signal is filtered during amplification. Suitable amplifiers have different filter options and allow to reduce high-frequency aspects in the data or attenuate low-frequency activity or drifts that are irrelevant for the majority of EEG recordings.

The last important technical aspect of an EEG amplifier describes the *resolution* or *bitrate* of the amplifier. The higher the resolution of the amplifier the more steps can be used for displaying the data between the minimum and maximum value of the data range. A 32-bit system, for example, can display 2E32 different values for an EEG signal in the range between 1 and 5000 μV. An amplifier with 8 Bit has only 8 values to display the same data range and thus small differences between subsequent values in the analog signal might not be visible in systems with low bitrates.

EEG Sensors

The EEG sensors or electrodes are the second hardware component besides the amplifier that plays a central role in EEG recordings. To provide the best possible connection between the minimal currents arriving at the skin and the amplifier, surface electrodes are coated with highly conductive materials, mostly silver/silver chloride (Ag/AgCl). To avoid artifacts from distorting the signal at the electrode level or during transmission from the electrode to the amplifier, some electrodes actively amplify the signal directly at the sensor and/or use shielded cables. Actively amplified electrodes need additional electronics in the sensor and are thus bigger than standard passive electrodes. This might play a role in case additional equipment will be placed on the participants' head (e.g., eye tracker, head-mounted VR display).

One of the most important factors influencing the signal quality is the contact of the electrode with the skin. With each minimal movement of the electrodes, the electrode-skin connection and thus the impedance of the electrode-skin system will change. This in turn will impact the signal quality. To secure a stable electrical connection in the electrode-skin system, electrolyte is used after the outmost layer of keratinized dead skin cells is removed to reduce the input impedance between the skin and electrode. Electrolyte is a conductive gel that contains a high amount of saline to allow signal conduction. Electrolyte diffuses into the skin and provides a stable connection in form of an elastic bridge between the electrical currents arriving at the skin and the coating of the electrodes. Thus, even small movements of the electrodes do not lead to changes in conductivity and good data quality can be preserved.

The use of wet electrodes needs time for preparation. An alternative to wet electrodes are dry electrode systems that make direct contact between the conductive layer of the electrode and the skin and thus reduce the time necessary for preparation of participants to a minimum. However, these systems come with the caveat that even minimal movements of the participant lead to changes in the contact of the electrode-skin system and thus impact the signal quality.[6] Thus, dry electrodes are useful only in case participants don't move during the experiment.

Electrodes are placed in elastic caps or nets that come in different sizes to accommodate different head sizes and shapes. In some cases, electrodes can be directly attached to the skin by using adhesive electrode stickers, for example to record the electrooculogram (EOG), i.e., the electrical activity accompanying eye movements. To this end, additional electrodes can be attached under the eyes (infraorbital) and at the outer canthi of the eyes to allow recordings of vertical and horizontal eye movements, respectively. Cap systems provide sockets for the electrodes or have the electrodes already integrated in the cap at specific locations. These locations are arranged according to the international 10–20 system[7] or extended versions thereof.[8,9] The positions are located at percentage steps relative to anatomical landmarks of the skull including the distance from nasion to inion as well as between the left and the right pre-auricular points. The relative positioning of electrodes according to this percent system standardized electrode positioning and allows for comparison and aggregation of recordings across participants with different head shapes and sizes.

SIGNAL PROCESSING

EEG recordings aim at picking up electrical activity generated within the brain. However, due to volume conduction activity of other biological tissues and electronic sources easily interfere with the signal. The human eyes can be considered a dipole with a negatively charged retina and a less negatively charged cornea. The two eye dipoles are located deep in the skull and produce strong voltage fluctuations that will be recorded with the EEG. This nonbrain electrical activity can be reduced using sensor or source based artifact rejection approaches.[10,11] The same is true for muscle activity that is generated by facial muscles and activity of the neck musculature. Both eye and neck muscle activities are biological artifacts that demonstrate higher variability than electrical or mechanical artifacts and thus are more difficult to isolate and reject from functional brain electrical recordings.

The recorded EEG signal consists of three dimensions (voltage over time over electrodes) and is usually preprocessed to emphasize the signals of interest and reduce the impact of what is considered noise. Preprocessing steps depend on the later analytical goals but usually contain data filtered and downsampled to focus on a specific frequency range (e.g., from 0.1 to 40 Hz and downsampling from 500 to 250 Hz for subsequent event-related potential [ERP] analyses). In addition, rereferencing the data (e.g., from an arbitrary reference to linked mastoids) would allow comparison with existing studies that used a specific reference. Downsampling of the data (e.g., from 1 kHz to 250 Hz), and potentially rereferencing the data (e.g., from an arbitrary reference to average mastoids) also would allow comparison with existing studies that used a specific reference electrode. Subsequently, the data are segmented into equidistant windows with partial overlap, or time locked to discrete events of interest. This way, time windows of a defined length relative to the onset of a specific stimulus or a class of stimuli are extracted from the continuous signal.

Data Analysis in the Time Domain—Event-Related Potentials

When analyzing event-related EEG paradigms, the preprocessed and epoched data are averaged sample by sample, resulting in an ERP across the electrode array, which represents the change in mean voltage preceding or following stimuli of interest. Averaging increases the signal-to-noise (SNR) ratio by attenuating any activity that is not time locked to the event while emphasizing functional, stimulus-related electrical processes.[12] ERPs are considered to be composed of underlying components,[13] which modulate the ERP waveform with respect to parameters such as amplitude, latency, and topography. Based on these characteristics, ERP waveforms of different experimental conditions can be statistically compared. For example, an anterior negative component around 200 ms can be detected on a single trial basis whenever a stimulus deviates from the predicted scenario.[14] Thus, this frontal negativity might be used for assessing user expectations during human–machine interaction.

One of the most investigated ERP components is the P300, a positive-going waveform peaking over parietocentral regions around 300 ms poststimulus onset, which is associated with stimulus evaluation and categorization, among other cognitive and motor processes.[15] A typical P300 paradigm is the Oddball task, in which deviant stimuli (*targets*) with low probability are presented randomly interspersed within standard stimuli, which have a much higher probability. Even in the absence of overt behavioral responses, the P300 amplitude is modulated by target probability and participant engagement. In addition, increased cognitive effort has been found to result in diminished P300 differences between targets and standards, which render the P300 an efficient indicator of workload.[16]

Data Analysis in the Frequency Domain—Spectral Variations

As EEG is the result of oscillatory activity of neural assemblies, the data can also be decomposed into weighted sums of sine and cosine functions of different frequencies, phases, and amplitudes.[17] This is accomplished with frequency or time–frequency methods such as FFT (Fast Fourier Analysis) or Wavelets, respectively. Although the former give insights into which frequencies are generally present in the signal, the latter additionally provide information on how these frequencies change over time. Increases or decreases in power over specific cortical sites within discrete frequency bands have been associated with an individual's level of fatigue, attention, task engagement, or mental workload in operational environments.[18] For example, shifts in visual attention are accompanied by (8–12 Hz) alpha modulations in posterior sites, both contralateral and ipsilateral to the attended stimulus position. Therefore, posterior alpha can be used as control signal for applications monitoring EEG correlates of visual attention.[19] Further, Prinzel et al.[20] proposed an EEG-based engagement index based on (13–25 Hz) beta power divided by (8–12 Hz) alpha power plus (4–8 Hz) theta power. The authors monitored this index within a closed-loop system to modulate task allocation in a vigilance task and reported improved operator performance when the EEG engagement index was used to trigger alterations in the stimulus presentation.

APPLICATIONS

In the final part of this chapter, we will briefly describe three recent studies that used different mobile EEG and mobile brain–body imaging[21–23,26]; approaches to investigate the brain dynamics accompanying cognitive processes that are more realistic during interaction with complex environments. The three studies are selected to demonstrate the difference between low- to high-density montages and the possible data analyses approaches used with an increasing number of channels. For a recent review of neuroergonomic studies using EEG in stationary and mobile participants, please see Gramann and colleagues (2017).

The first study by Debener and colleagues[25] describes a new unobtrusive electrode system with 10 channels placed around each ear using printable electrodes. Using a mobile phone to record the data and present an auditory Oddball task to participants, the authors replicated previous P3 amplitude modulations for targets as compared to standard auditory stimuli. Importantly, participants wore the system for a full day without degradation in signal quality, as indicated by comparable classification of P300 differences in the morning and the evening recording sessions. This study demonstrates that concealed EEG recordings are possible for long periods without loss of comfort for the user and without stigmatizing the wearer.

To allow analyses of event-related brain dynamics in realistic working environments, the second study by Wascher et al.,[24] used a workaround. Because realistic working environments seldom provide controllable stimulation that can be used to extract data epochs, Wascher and colleagues resorted to blink activity of participants. Assuming that blinks indicate distinctive points in the information-processing chain, they extracted the onset of blinks in the continuous data and used these time points for epoch extraction. Analyses of the event-related potentials, averaging all epochs with onset of a blink, revealed an increased N2-component as well as an increase in theta activity at frontocentral electrodes. This activity was enhanced only when participants were engaged in the cognitive task but not a physical task or rest. The study used 28 electrodes with a wireless extension to the amplifier allowing for topographical analyses of power changes and event-related activity as compared to the 10-channel setup used in Debener et al.[25]

The final study by Gramann et al.[22] investigated whether the event-related P300 to rare stimuli in a visual Oddball experiment can be observed in four different behavioral states. To this end, high-density EEG was recorded in participants standing, slow walking, fast walking, or running on a treadmill, while responding to visually presented rare targets. The results revealed that it is possible to analyze the P300 with higher amplitudes for targets as compared to nontarget stimuli even when participants perform full-body movements. This result, however, was possible only after data-driven preprocessing of the high-density data and dissociation of nonbrain activity from brain activity using independent component analyses (ICA). Running resulted in large movement-related artifacts that could only be further processed when using movement-related information to compute gait-related artifact templates. This result demonstrates the need for combining high-density EEG with synchronized motion tracking for movements with strong artifact contribution as they can occur in many workplaces (e.g., assembly lines). It could be shown that with increasing movement speed, eye and neck muscle activity explained increasingly more variance in the ERP on the sensor level. This kind of analysis would not be possible without an adequate number of channels. On the other hand, the full setup used in this study would not allow participants to stroll around their natural environment.

SUMMARY

In summary, EEG is one of the most often-used neuroscientific methods with the highest temporal resolution that allows investigation of human brain dynamics in the working environment. Dependent on the research question, stationary or mobile systems can be used to investigate human brain activity during work or other tasks. Increasing mobility of the systems comes with decreasing number of channels and options to analyze the data. However, recent developments already allow for recordings of brain activity in ambulating participants in realistic working environments and during everyday life activities. Future developments in wearable amplifiers and electrode technology will push the range of EEG as research tool even further. This is the starting point of a new and exciting era of mobile brain dynamic recordings that will propel research in Neuroergonomics and beyond.

REFERENCES

1. Nunez PL, Srinivasan R. *Electric fields of the brain: the neurophysics of EEG.* 2nd ed. Oxford (New York): Oxford University Press; 2006.
2. Da Silva FL. *EEG: origin and measurement EEG-fMRI.* Springer; 2009. p. 19–38.
3. Nunez PL. *Electric fields of the brain: the neurophysics of EEG.* Oxford University Press; 1981.
4. Michel CM, Murray MM, Lantz G, Gonzalez S, Spinelli L, Grave de Peralta R. EEG source imaging. *Clinical Neurophysiology* 2004;**115**(10):2195–222.
5. Schomer DL, Da Silva FL. *Niedermeyer's electroencephalography: basic principles, clinical applications, and related fields.* Lippincott Williams & Wilkins; 2012.
6. Saab J, Battes B, Grosse-Wentrup M. *Simultaneous EEG recordings with dry and wet electrodes in motor-imagery: na.* 2011.
7. Jasper HH. The ten twenty electrode system of the international federation. *Electroencephalography and Clinical Neurophysiology* 1958;**10**:371–5.
8. American-Electroencephalographic-Society. Guide lines for standard electrode position nomenclature. *Journal of Clinical Neurophysiology* 1991;**8**(2):200–2.
9. Oostenveld R, Praamstra P. The five percent electrode system for high-resolution EEG and ERP measurements. *Clinical Neurophysiology* 2001;**112**(4):713–9.
10. Gratton G, Coles MGH, Donchin E. A new method for off-line removal of ocular artifact. *Electroencephalography and Clinical Neurophysiology* 1983;**55**(4):468–84. https://doi.org/10.1016/0013-4694(83)90135-9.
11. Jung TP, Makeig S, Humphries C, Lee TW, McKeown MJ, Iragui V, Sejnowski TJ. Removing electroencephalographic artifacts by blind source separation. *Psychophysiology* 2000;**37**(2):163–78.
12. Luck SJ. *An introduction to the event-related potential technique.* MIT Press; 2014.
13. Luck SJ, Kappenman ES. *The Oxford handbook of event-related potential components.* Oxford University Press; 2011.
14. Zander TO, Krol LR, Birbaumer NP, Gramann K. Neuroadaptive technology enables implicit cursor control based on medial prefrontal cortex activity. *Proceedings of the National Academy of Sciences of the United States of America* 2016;**113**(52):14898–903. https://doi.org/10.1073/pnas.1605155114.
15. Polich J. Updating p300: an integrative theory of P3a and P3b. *Clinical Neurophysiology* 2007;**118**(10):2128–48. https://doi.org/10.1016/J.Clinph.2007.04.019.
16. Roy RN, Bonnet S, Charbonnier S, Campagne A. Efficient workload classification based on ignored auditory probes: a proof of concept. *Frontiers in Human Neuroscience* 2016;**10**:519. https://doi.org/10.3389/Fnhum.2016.00519.
17. Freeman W, Quiroga RQ. *Imaging brain function with EEG: advanced temporal and spatial analysis of electroencephalographic signals.* Springer Science & Business Media; 2012.
18. Gevins A, Smith ME. Electroencephalography (EEG) in neuroergonomics. *Neuroergonomics: The Brain at Work* 2006:15–31.
19. van Gerven M, Jensen O. Attention modulations of posterior alpha as a control signal for two-dimensional brain-computer interfaces. *Journal of Neuroscience Methods* 2009;**179**(1):78–84. https://doi.org/10.1016/j.jneumeth.2009.01.016.
20. Prinzel LJ, Freeman FC, Scerbo MW, Mikulka PJ, Pope AT. A closed-loop system for examining psychophysiological measures for adaptive task allocation. *International Journal of Aviation Psychology* 2000;**10**(4):393–410. https://doi.org/10.1207/S15327108ijap1004_6.
21. Gramann K, Ferris DP, Gwin J, Makeig S. Imaging natural cognition in action. *International Journal of Psychophysiology* 2014;**91**(1):22–9. https://doi.org/10.1016/j.ijpsycho.2013.09.003.
22. Gramann K, Gwin JT, Ferris DP, Oie K, Jung TP, Lin CT, Makeig S. Cognition in action: imaging brain/body dynamics in mobile humans. *Reviews in the Neurosciences* 2011;**22**(6):593–608. https://doi.org/10.1515/RNS.2011.047.
23. Makeig S, Gramann K, Jung TP, Sejnowski TJ, Poizner H. Linking brain, mind and behavior. *International Journal of Psychophysiology* 2009;**73**(2):95–100. https://doi.org/10.1016/j.ijpsycho.2008.11.008.
24. Wascher E, Heppner H, Hoffmann S. Towards the measurement of event-related EEG activity in real-life working environments. *International Journal of Psychophysiology* 2014;**91**(1):3–9. https://doi.org/10.1016/j.ijpsycho.2013.10.006.
25. Debener S, Emkes R, De Vos M, Bleichner M. Unobtrusive ambulatory EEG using a smartphone and flexible printed electrodes around the ear. *Scientific Reports* 2015;**5**:16743. https://doi.org/10.1038/Srep16743.
26. Jungnickel E, Gramann K. Mobile brain/body imaging (MoBI) of physical interaction with dynamically moving objects. *Frontiers in Human Neuroscience* 2016;**10**:306. https://doi.org/10.3389/Fnhum.2016.00306.

Chapter 3

The Use of Functional Near-Infrared Spectroscopy in Neuroergonomics

Hasan Ayaz[1,4,5], Meltem Izzetoglu[2,3], Kurtulus Izzetoglu[1], Banu Onaral[1]

[1]Drexel University, Philadelphia, PA, United States; [2]Villanova University, Villanova, PA, United States; [3]Yeshiva University, Bronx, NY, United States; [4]University of Pennsylvania, Philadelphia, PA, United States; [5]Children's Hospital of Philadelphia, Philadelphia, PA, United States

INTRODUCTION

Functional Near Infrared Spectroscopy (fNIRS) is a multiwavelength optical technique: originally developed for the clinical monitoring of tissue oxygenation,[1] it eventually evolved into a useful tool for functional neuroimaging studies.[2–4] The technology has advanced and a variety of fNIRS instruments have been developed to monitor changes in local cerebral oxygenation by measuring the concentration changes of both deoxygenated hemoglobin (deoxy-Hb) and oxygenated hemoglobin (oxy-Hb). Various types of brain activities, such as motor and cognitive activities, have been studied using fNIRS.[5–11]

fNIRS has become increasingly popular for functional neuroimaging studies due to its portability and relatively low cost. It measures hemodynamic changes in the brain in a similar fashion to functional magnetic resonance imaging (fMRI), but fNIRS is quiet (no operating sound), provides higher temporal resolution, does not restrict participants to a confined space, and does not require the participant to lie down. These qualities make fNIRS an ideal candidate for monitoring brain-activity-related hemodynamic changes not only in laboratory settings but also under working conditions and in more ecologically valid environments. Although fNIRS is not immune to all noise caused by motion artifacts, and the failure to remove such noise adequately may lead to biased or false results, fNIRS is more tolerant to motion artifacts induced by the movement of the head compared to fMRI[12] and has even been used in experiments that require participants to exercise. Real-time applications, such as brain–computer interface scenarios, benefit most from algorithms that can both process data in real time and lead to acceptable results using only current and past data points.

This chapter introduces fNIRS principles, signal processing, and analysis techniques as well as representative applications. An introductory overview of these concepts and video tutorial from theory to practice are given in Ayaz et al.[13]

MEASURE

Physiological and Physical Principles

Typically, an optical apparatus consists of a light source by which the tissue is radiated and a light detector that receives light after it has interacted with the tissue. Biological tissues are relatively transparent to light in the near-infrared range between 700 and 900 nm. Photons that enter the tissue undergo two different types of interaction, absorption and scattering. Within the near-infrared range of light, the two primary absorbers are oxy-Hb (HbO2) and deoxy-Hb (Hb).

By measuring optical density (OD) changes at two wavelengths, the relative change of oxy-Hb and deoxy-Hb versus time can be obtained using the modified Beer–Lambert law (MBLL).[14] OD at a specific input wavelength (λ) is the logarithmic ratio of input light intensity (I_{in}) and output (detected) light intensity(I_{out}). OD is also related to the concentration (c) and molar extinction coefficient (ε) of chromophores, the corrected distance (d) of the light source and detector, and a constant attenuation factor (G):

$$OD_\lambda = \log\left(\frac{I_{in}}{I_{out}}\right) \approx \varepsilon_\lambda \cdot c \cdot d + G \qquad (3.1)$$

Neuroergonomics. https://doi.org/10.1016/B978-0-12-811926-6.00003-8
Copyright © 2019 Elsevier Inc. All rights reserved.

Having the same I_{in} at two different time instances and detected light intensity during baseline (I_{rest}) and performance of the task (I_{test}), the difference in OD is:

$$\Delta \, OD_\lambda = \log \left(\frac{I_{rest}}{I_{test}} \right) = \varepsilon_\lambda^{HB} \cdot \Delta \, c^{HB} \cdot d + \varepsilon_\lambda^{HBO_2} \cdot \Delta \, c^{HBO_2} \cdot d \qquad (3.2)$$

Measuring the OD at two different wavelengths gives:

$$\begin{bmatrix} \Delta \, OD_{\lambda 1} \\ \Delta \, OD_{\lambda 2} \end{bmatrix} = \begin{bmatrix} \varepsilon_{\lambda 1}^{HB} d & \varepsilon_{\lambda 1}^{HBO_2} d \\ \varepsilon_{\lambda 2}^{HB} d & \varepsilon_{\lambda 2}^{HBO_2} d \end{bmatrix} \begin{bmatrix} \Delta \, c^{HB} \\ \Delta \, c^{HBO_2} \end{bmatrix} \qquad (3.3)$$

This equation set can be solved for concentrations if the 2×2 matrix is nonsingular. Typically, the two wavelengths are chosen within 700–900 nm where the absorption of oxy-Hb and deoxy-Hb are dominant as compared to other tissue chromophores, and below and above the isosbestic point (~805 nm where absorption spectrums of deoxy- and oxy-Hb cross each other) to focus the changes in absorption to either deoxy-Hb or oxy-Hb, respectively. A historical perspective of the MBLL and its application in fNIRS is available.[4,15–18]

PROCESSING

This subsection introduces the typical signal preprocessing pipeline for raw fNIRS signals before they can be used for analysis. Preprocessing of raw signals is needed to eliminate various types of noise and contaminants that are present in the signals; they can be caused by physiological factors (such as cardiac- and respiration-related signal components), sensor coupling (detector saturation and low light), or motion artifacts. Processing algorithms have been developed for both detection and elimination of contaminated segments prior to analysis, and also for the removal of noise or artifacts to clean the signals. Processing can be applied to light intensity measures as well as hemoglobin signals in both time-series for each channel separately or spatially, or across channels at each time instance. The rest of this subsection discusses each major contamination type and the processing it needs.

Motion Artifacts

When the fNIRS sensors, light sources, and/or detectors slide from their original attached location or lose contact with the skin due to head motion, unexpected sudden bursts or spikes can occur in the fNIRS measurements. Furthermore, if the light source loses coupling with the skin, the detector may record either very low values (since no light may pass through it) or extremely high intensities due to the reflected light from the skin (instead of through the tissue underneath) that can cause momentary saturation. Similar saturation effects may occur if the detector is dislodged and loses contact with the skin, causing the penetration of ambient light. Head movement can further cause changes in the pressure applied to the sensor pad or to the light sources and detectors. These changes may allow more photons to enter the tissue, temporarily varying the detected light intensity.

Other than the visual inspection of data for possible motion artifacts, there are a growing number of motion artifact detection and removal algorithms to automate noise detection and eliminate subjectivity.[19–25] Most of these algorithms are developed and modeled for oxy-Hb, deoxy-Hb, and blood volume (oxy-Hb+deoxy-Hb) data,[20,22,26] while the remaining algorithms perform analysis on raw intensity measurements.[19,21,25] It may be most appropriate to perform the algorithms for motion artifact detection and removal on raw measurements to prevent the propagation of error, bias, and cross-talk between hemodynamic signals. In one thesis[27] a method based on combined temporal independent component analysis (ICA) and principal component analysis (PCA) was proposed and implemented on raw intensity measurements collected at two wavelengths and a dark current condition. A spatial PCA-based filtering algorithm was proposed[21] to exploit the property whereby a motion artifact has an intense and correlated effect on a large area of a near-infrared spectroscopy sensor. In its most simplistic approach, raw data processing is typically limited to low-pass or band-pass filtering to remove high-frequency components or exclude the data altogether if it is an irrecoverable case. However, low-pass or band-pass filtering cannot fully eliminate outliers or spikes that are usually much higher than optical signal levels. It is important to identify these motion-corrupted regions objectively in the time-series data. A recent statistical filtering approach, called sliding-window–motion–artifact–rejection (SMAR) has been developed[19] that uses a covariant of a variation-based approach to identify motion-related artifacts. SMAR does not require a priori knowledge, such as a training dataset from a specific subject or experimental protocol. The algorithm is suitable for use in real time during an experiment since it is causal and simple, and requires practically no computational run-time cost. SMAR is a popular method in the typical signal-processing pipeline to assess signal quality and eliminate contaminated signal segments or channels.

Superficial Layers

Through various Monte Carlo simulations it has been shown that fNIRS is able to measure changes in the hemodynamic response within a banana-shaped volume of underlying tissue between the light source and the detector. The depth of penetration is a fraction of the source–detector distance, and the tissue structure is usually taken as approximately half the source–detector separation for the adult human head.[28,29] Thus within the sampled volume and in addition to the hemodynamic changes originating from the cortical brain layer, there could be additional signal contributions to the overall fNIRS measurements from the superficial head layers of skin/scalp, skull, and cerebrospinal fluid (CSF). Specifically, as compared to the skull and CSF, the skin/scalp layer has more vasculature, and hence hemodynamic changes in this layer can result in additional confounding signals to the targeted cerebral hemodynamic response in fNIRS measurements. However, the amount of signal contribution from the skin layer to the cerebral hemodynamic response is still controversial.[30,31] Study findings[32] suggest that task-related hemodynamic responses disappeared when skin blood flow was occluded. Further evaluation by the authors concluded that a major part of the task-related changes in the oxy-Hb concentration in the forehead was due to task-related changes in the skin blood flow. In a later study,[33] results indicate that skin blood flow influenced both short and long source–detector separation measurements. Contrary to these findings, other studies have shown that measurements obtained from larger source–detector separations (used to guarantee the penetration to cortical brain regions) showed significant task-dependent differences, whereas shorter source–detector separations for measurements obtained only from superficial layers did not show any significant differences.[31,34,35]

Several studies have investigated and proposed appropriate optode configurations and various algorithms to eliminate such potential confounding signals from the skin/scalp layer. As an outcome of these studies, an almost universal agreement has been achieved suggesting the implementation of short and long source–detector separations in hardware configurations. Here, separate measurements originating only from superficial layers can be simultaneously acquired from the short separations. These measurements can then be used as reference signals in the elimination of skin effects from the long separations when extracting cerebral hemodynamic responses. For the elimination of the skin effect with the use of reference signal measurements, algorithms range from signal component elimination techniques (i.e., PCA and ICA) to spatial filtering (i.e., common average rejection).[30,36–38]

Physiological Signals

fNIRS can measure the hemodynamic response related to neuronal activity through the mechanism of neurovascular coupling. In addition to changes related to cognitive activity, fNIRS measurements can also capture hemodynamic signals based on other physiological sources such as heart pulsation, blood pressure, and respiration. Due to their natural rhythm, these almost-periodic signals occupy certain frequency bands that typically arise outside the frequency content of the hemodynamic response related to cognitive activity. In general, separation of such signals is performed using simple frequency selective filters. These additional physiological signals are usually treated like artifacts, and hence removed from fNIRS measurements and eliminated from the overall cognitive activity monitoring study. However, measurement of such signals from different origins using a single sensor can provide insights into different processes in the human body (i.e., information on heart rate variability and respiration rate), forming another powerful aspect of fNIRS monitoring. Similarly, as proposed in various studies, very low-frequency oscillations are sometimes considered as physiological artifacts and frequency selective filters are used to eliminate such effects. Mayer waves, which occur due to arterial blood pressure and have a frequency range centered around ~0.1 Hz, can overlap with the frequency band of hemodynamic response related to cognitive activity and can be a more problematic physiological artifact. Care should be taken if frequency selective filters are to be used for their elimination so that signal content related to cognitive activity is not suppressed in the process. In addition to simple frequency selective filters, more sophisticated methods proposed for the elimination of physiological signals include the curve-fitting method, adaptive, Wiener, and Kalman filtering techniques, least-squares regression algorithms, ICA, and RETROICOR.[39–48]

ANALYZE

There are multiple instrumentation approaches developed for fNIRS, reviewed by Ferrari and Quaresima.[8] Most commercial systems currently available are continuous wave (CW), and for these conversion of fNIRS intensity measurements to relative changes in hemodynamic response in terms of oxy-Hb and deoxy-Hb is usually performed using the previously explained MBLL. There are several parameters given in Eq. (3.3), such as molar extinction coefficients, ε and corrected distance, d for each wavelength used and each chromophore being extracted, that are necessary to be able to perform a matrix inversion. In CW fNIRS applications these parameters are usually taken as constants and must be known a priori.

For molar extinction coefficients there are tabulated values for each chromophore of interest at various wavelengths within the near-infrared range.[14] Corrected distance (d) is linearly related to source–detector separation (sd) scaled by differential pathlength factor (DPF) as $d = \text{sd} \times \text{DPF}$. In addition to individual differences, it has been shown that DPF depends on various factors including source–detector separation, wavelength, head location, age, gender, and even oxygen saturation due to differences in layer thickness and tissue composition.[28,49–52]

Other than the aforementioned constants in CW fNIRS applications, changes in oxy-Hb and deoxy-Hb are obtained relative to a baseline condition. Since comparisons between subject groups, optode locations, or task conditions are made using these relative changes, it is important to design a test protocol with appropriate baseline intervals, usually taken as relaxation/resting periods in between the task periods to allow the cognitive state to return to similar levels. These local baseline regions can then be used in MBLL conversions to obtain changes in the hemodynamic response in the task condition relative the baseline region immediately preceding the task. Note that this operation also corresponds to common baseline correction methods (subtracting the mean of the local baseline region from the following task period data epoch) applied to oxy-Hb and deoxy-Hb data obtained relative to a global baseline region collected at the beginning of the task.

In general, hemodynamic response to neuronal activity is explained through a mechanism called neurovascular coupling, such that neuronal activity causes an increase in oxygen and glucose consumption which then leads to an increase in cerebral blood flow. During the brief period of neuronal activation seen in evoked response studies, this oversupply of oxygen forms the basis of expected changes in hemodynamics as measured by fNIRS, where an increase in oxy-Hb is concomitant with a decrease in the deoxy-Hb time series[6,53] This has been compared to and validated by the blood-oxygenation-level-dependent response obtained by fMRI. Even though the relationship between neural activity and vascular response is usually taken as linear, various nonlinearities have also been noted[54,55] which can be more pronounced in block designs with continuous or rapid presentation of trials. Nevertheless, to reduce dimensionality, statistical comparisons between task conditions, head locations, or subject groups are performed on features extracted from oxy-Hb and deoxy-Hb traces instead of using whole data epochs. Such features involve data values or time components such as the average, maximum or minimum values, time to peak, or full width half maximum. Features extracted from fNIRS measurements are not only used in statistical comparisons but have been recently utilized in machine-learning algorithms for automated classification of healthy and diseased groups or between various conditions.[56–58]

APPLICATIONS

This section highlights select applications of fNIRS in neuroergonomics, listing just a few to represent a growing and diverse array of application areas.[59] The main application domain is related to the human–machine/technology frontier. The efficiency and safety of complex high-precision human–machine systems present in aerospace and robotic surgery are closely related to the cognitive readiness, ability to manage workload, and situational awareness of their operators. Subjective operator reports and physiological and behavioral measures are not sufficiently reliable to monitor the cognitive overload that can lead to adverse outcomes. A key feature of the concept of mental workload (which reflects how hard the brain is working to meet task demands) is that it can be dissociated from behavioral performance data. While experienced human operators can maintain performance at required levels through increased effort and motivation or strategy changes, even in the face of increased task challenge, sustained task demands eventually lead to performance decline unless the upward trend in mental workload can be used to predict subsequent performance breakdown. Consequently, it is important to assess mental workload independent of performance measures during training and operational missions. Neuroergonomic approaches based on measures of human brain hemodynamic activity can provide sensitive and reliable assessments of human mental workload in complex training and work environments. fNIRS is a field-deployable noninvasive optical brain-monitoring technology that provides a measure of cerebral hemodynamics within the prefrontal cortex in response to sensory, motor, or cognitive activation. The following examples examine the relationship of the hemodynamic response in the prefrontal cortex to levels of expertise, mental workload state, and task performance in a variety of application areas.

Aerospace: Cognitive Workload Assessment of Air Traffic Controllers

In a collaborative project, Drexel Optical Brain Imaging Team incorporated fNIRS in a study at the FAA's William J. Hughes Technical Center Human Factors Laboratory where certified controllers were monitored while they managed realistic air traffic control (ATC) scenarios under typical and emergent conditions.[60] The primary objective was to use neurophysiological measures to assess cognitive workload and the usability of new interfaces developed for ATC systems. Throughout the study, certified professional controllers completed ATC tasks with different interface settings and controlled difficulty levels for verification. The results indicated that brain activation as measured by fNIRS provides a valid measure of mental workload in this realistic ATC task[60] (Fig. 3.1).

FIGURE 3.1   Control workstations with high-resolution radarscope, keyboard, trackball, and direct keypad access (left). Operators in the study performing ATC tasks in front of workstations with a fNIRS sensor pad on their foreheads (right).

Aerospace: Expertise Development With Piloting Tasks

Another longitudinal study investigated expertise development with practice in a variety of settings, including of use of complex piloting tasks. fNIRS was used to investigate the relationship of the hemodynamic response in the anterior prefrontal cortex to changes in level of expertise and task performance during learning of simulated unmanned aerial vehicle (UAV) piloting tasks.[60,61] Novice participants with no prior UAV piloting experience participated in a 9-day training program where they used flight simulators to execute real-world maneuvers. Each day, self-reported measures (with NASA TLX), behavioral measures (task performance), and fNIRS measures (prefrontal cortex activity indicating mental effort on task) were recorded. Participants practiced approach and landing scenarios while piloting a virtual UAV (Fig. 3.2). The scenarios were designed to expose novice subjects to realistic and critical tasks for a UAV ground operator directly piloting an aircraft. Results indicate that the level of expertise does appear to influence the hemodynamic response in the dorsolateral/ventrolateral prefrontal cortices. As such, measuring activation in these attentional and control areas relative to task performance can provide an index of level of expertise and illustrate how task-specific practice influences the learning of tasks (Fig. 3.2).

Healthcare: Cognitive Aging

Decline in gait performance is common in aging populations and can result in increased risk of mortality and morbidity, frequent hospitalizations, and in general poorer quality of life.[62–64] Recent epidemiological, cognitive, and neuroimaging studies suggest that gait is influenced by higher-order cognitive and cortical control mechanisms. However, neural underpinnings of gait are still not well understood or studied.

We investigated neural correlates of locomotion in elderly populations by monitoring brain activity in the dorsolateral prefrontal cortex (PFC) using fNIRS (fNIRS Imager 1000, fNIR Devices, Potomac, MD) in a large cohort of elderly participants (age > 65) while they performed real on-the-ground walking tasks with or without a cognitive interference task of letter generation in a longitudinal study for 5 years ("Central Control of Mobility in Aging [CCMA]" project at Albert Einstein College of Medicine, Yeshiva University, Bronx, NY).

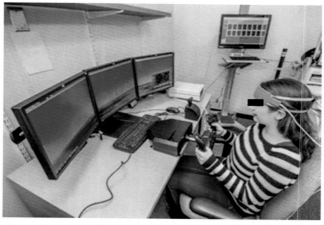

FIGURE 3.2   Low-fidelity flight simulator with unmanned vehicle remote-piloting tasks while operator prefrontal brain activity is monitored with fNIRS.

Our initial pilot study on 11 young and 11 elderly participants indicated that oxygenation levels are increased in the PFC in a dual-task condition (walking while talking—WWT) as compared to a single task (normal walking—NW) in both young and old individuals, but young individuals showed greater increases in PFC oxygenation levels as compared to old participants, suggesting that older adults may underutilize the PFC in attention-demanding locomotion tasks.[65] Reproducible measurements of task-related changes in oxygenation levels found increased PFC activity in WWT as compared to NW in a large cohort of nondemented and ambulatory elderly adults ($n = 348$) in the Central Control of Mobility in Aging (CCMA) study.[66–68] Moreover, elevated PFC oxygenation levels were shown to be maintained throughout the course of WWT but not during NW, since WWT is a dual-task condition and hence more cognitively demanding. In addition, increased oxygenation levels in the PFC were related to better gait and cognitive performance during WWT, consistent with compensatory reallocation models. The individual and combined effects of gender and perceived stress on gait velocity and PFC oxygenation levels during locomotion were studied,[62] and higher levels of perceived task-related stress were found to be associated with more difficulties in negotiating the demands of dual-task walking as well as attenuation of brain oxygenation patterns under attention-demanding walking in older men. Subjective and objective measures of fatigue in the context of the established dual-task walking paradigm[69] indicated that worse perceptions of fatigue were associated with an attenuated increase in oxygenation levels from NW to WWT. In addition to these findings in healthy aging, our studies identified differences in brain activity levels in elderly people with various disease conditions, such as diabeties, neurological gait abnormalities, Parkinsonian syndromes, and MS.[70–74] Findings indicated that higher oxygenation levels during WWT among individuals with peripheral neurological gait abnormalities were associated with worse cognitive performance but faster gait velocity. Patients with Parkinsonian syndromes manifested higher PFC activation to maintain postural stability. Similarly, oxygenated hemoglobin levels were increased in persons with MS compared to controls while walking to compensate for decreased efficiency, whereas presence of diabetes was associated with poorer walking performance and attenuated brain response. With the implementation of fNIRS, our findings have provided more information on neural underpinnings of mobility in aging with or without disease conditions, which can have further implications for risk assessment and interventions in incident mobility impairments and falls.

CONCLUSION

Mobile fNIRS sensors enable monitoring brain activity out of the lab and in ecologically valid settings. Neuroergonomic uses for such fNIRS technology include continuous and ubiquitous measurement of brain function at work and at home. This has the potential to change the way we work and interact with technology and each other. fNIRS technology is still evolving, with ongoing hardware and processing improvements that are making fNIRS systems more portable, more reliable, and more affordable. Ultraportable wearable fNIRS sensors are expected to break the limitations of traditional neuroimaging approaches that previously imposed barriers on experimental protocols, data-collection settings, and task conditions at the expense of ecological validity. As the youngest neuroimaging modality, fNIRS has already reached a level that addresses and benefits real-world problems and industry challenges. As fNIRS technology further matures and enhances its capability, new areas of application will emerge, opening new research directions as well as routine-use cases for industry and home settings.

REFERENCES

1. Jobsis FF. Noninvasive, infrared monitoring of cerebral and myocardial oxygen sufficiency and circulatory parameters. *Science* 1977;**198**(4323):1264–7.
2. Chance B, Zhuang Z, UnAh C, Alter C, Lipton L. Cognition-activated low-frequency modulation of light absorption in human brain. *Proceedings of the National Academy of Sciences* 1993;**90**(8):3770–4.
3. Hoshi Y, Tamura M. Near-infrared optical detection of sequential brain activation in the prefrontal cortex during mental tasks. *NeuroImage* 1997;**5**(4):292–7.
4. Villringer A, Planck J, Hock C, Schleinkofer L, Dirnagl U. Near infrared spectroscopy (NIRS): a new tool to study hemodynamic changes during activation of brain function in human adults. *Neuroscience Letters* 1993;**154**(1–2):101–4.
5. Ayaz H, Onaral B, Izzetoglu K, Shewokis PA, McKendrick R, Parasuraman R. Continuous monitoring of brain dynamics with functional near infrared spectroscopy as a tool for neuroergonomic research: empirical examples and a technological development. *Frontiers in Human Neuroscience* 2013;**7**. https://doi.org/10.3389/fnhum.2013.00871.
6. Batula AM, Mark JA, Kim YE, Ayaz H. Comparison of brain activation during motor imagery and motor movement using fNIRS. *Computational Intelligence and Neuroscience* 2017;**2017**:12. https://doi.org/10.1155/2017/5491296.
7. Cheng L, Ayaz H, Sun J, Tong S, Onaral B. Modulation of functional connectivity and activation during preparation for hand movement. *IIE Transactions on Occupational Ergonomics and Human Factors* 2016;**4**(2–3):175–87. https://doi.org/10.1080/21577323.2016.1191560.
8. Ferrari M, Quaresima V. A brief review on the history of human functional near-infrared spectroscopy (fNIRS) development and fields of application. *NeuroImage* 2012;**63**(2):921–35. https://doi.org/10.1016/j.neuroimage.2012.03.049.

9. Izzetoglu K, Ayaz H, Merzagora A, Izzetoglu M, Shewokis PA, Bunce SC, Onaral B. The evolution of field deployable fNIR spectroscopy from bench to clinical settings. *Journal of Innovative Optical Health Sciences* 2011;**4**(03):239–50.

10. Quaresima V, Ferrari M. Functional near-infrared spectroscopy (fNIRS) for assessing cerebral cortex function during human behavior in natural/social situations: a concise review. *Organizational Research Methods* 2016. https://doi.org/10.1177/1094428116658959.

11. Yücel MA, Selb JJ, Huppert TJ, Franceschini MA, Boas DA. Functional near infrared spectroscopy: enabling routine functional brain imaging. *Current Opinion in Biomedical Engineering* 2017;**4**:78–86. https://doi.org/10.1016/j.cobme.2017.09.011.

12. Okamoto M, Dan H, Shimizu K, Takeo K, Amita T, Oda I, Suzuki T. Multimodal assessment of cortical activation during apple peeling by NIRS and fMRI. *NeuroImage* 2004;**21**(4):1275–88.

13. Ayaz H, Shewokis PA, Curtin A, Izzetoglu M, Izzetoglu K, Onaral B. Using MazeSuite and functional near infrared spectroscopy to study learning in spatial navigation. *Journal of Visualized Experiments* 2011;**56**:e3443. https://doi.org/10.3791/3443.

14. Cope M. *The development of a near infrared spectroscopy system and its application for non invasive monitoring of cerebral blood and tissue oxygenation in the newborn infant* Ph.D. thesis. London: University College London; 1991.

15. Cope M, Delpy D. System for long-term measurement of cerebral blood and tissue oxygenation on newborn infants by near infra-red transillumination. *Medical and Biological Engineering and Computing* 1988;**26**(3):289–94.

16. Cope M, Delpy DT, Reynolds EOR, Wray S, Wyatt J, Van der Zee P. Methods of quantitating cerebral near infrared spectroscopy data. *Advances in Experimental Medicine and Biology* 1988;**222**:183–9.

17. Delpy D, Cope M, Zee P, Arridge S, Wray S, Wyatt J. Estimation of optical pathlength through tissue from direct time of flight measurement. *Physics in Medicine and Biology* 1988;**33**:1433.

18. Wyatt J, Delpy D, Cope M, Wray S, Reynolds E. Quantification of cerebral oxygenation and haemodynamics in sick newborn infants by near infrared spectrophotometry. *The Lancet* 1986;**328**(8515):1063–6.

19. Ayaz H, Izzetoglu M, Shewokis PA, Onaral B. Sliding-window motion artifact rejection for functional near-infrared spectroscopy. In: *Paper presented at the Conf Proc IEEE Eng med Biol Soc, Buenos Aires, Argentina*. 2010.

20. Cui X, Bray S, Reiss AL. Functional near infrared spectroscopy (NIRS) signal improvement based on negative correlation between oxygenated and deoxygenated hemoglobin dynamics. *NeuroImage* 2010;**49**(4):3039–46.

21. Huppert TJ, Diamond SG, Franceschini MA, Boas DA. HomER: a review of time-series analysis methods for near-infrared spectroscopy of the brain. *Applied Optics* 2009;**48**(10):D280–98.

22. Izzetoglu M, Chitrapu P, Bunce S, Onaral B. Motion artifact cancellation in NIR spectroscopy using discrete Kalman filtering. *Biomedical Engineering Online* 2010;**9**(1):16.

23. Izzetoglu M, Devaraj A, Bunce S, Onaral B. Motion artifact cancellation in NIR spectroscopy using Wiener filtering. *IEEE Transactions on Biomedical Engineering* 2005;**52**(5):934–8.

24. Jahani S, Setarehdan SK, Boas DA, Yücel MA. Motion artifact detection and correction in functional near-infrared spectroscopy: a new hybrid method based on spline interpolation method and Savitzky–Golay filtering. *Neurophotonics* 2018;**5**(1):015003.

25. Sweeney KT, Ayaz H, Ward T, Izzetoglu M, McLoone S, Onaral B. A methodology for validating artifact removal techniques for physiological signals. *IEEE Transactions on Information Technology in Biomedicine* 2012;**16**(5):918–26. https://doi.org/10.1109/titb.2012.2207400.

26. Izzetoglu M, Izzetoglu K, Bunce S, Ayaz H, Devaraj A, Onaral B, Pourrezaei K. Functional near-infrared neuroimaging. *IEEE Transactions on Neural Systems and Rehabilitation Engineering* 2005;**13**(2):153–9.

27. Izzetoglu K. *Neural correlates of cognitive workload and anesthetic depth: fNIR spectroscopy investigation in humans* Ph.D. thesis Philadelphia, PA: Drexel University; 2008. Retrieved from: http://hdl.handle.net/1860/2896.

28. Fukui Y, Ajichi Y, Okada E. Monte Carlo prediction of near-infrared light propagation in realistic adult and neonatal head models. *Applied Optics* 2003;**42**(16):2881–7.

29. Okada E, Firbank M, Schweiger M, Arridge SR, Cope M, Delpy DT. Theoretical and experimental investigation of near-infrared light propagation in a model of the adult head. *Applied Optics* 1997;**36**(1):21–31.

30. Herold F, Wiegel P, Scholkmann F, Thiers A, Hamacher D, Schega L. Functional near-infrared spectroscopy in movement science: a systematic review on cortical activity in postural and walking tasks. *Neurophotonics* 2017;**4**(4):041403.

31. Sato H, Yahata N, Funane T, Takizawa R, Katura T, Atsumori H, Koizumi H. A NIRS–fMRI investigation of prefrontal cortex activity during a working memory task. *Neuroimage* 2013;**83**:158–73.

32. Takahashi T, Takikawa Y, Kawagoe R, Shibuya S, Iwano T, Kitazawa S. Influence of skin blood flow on near-infrared spectroscopy signals measured on the forehead during a verbal fluency task. *NeuroImage* 2011;**57**(3):991–1002.

33. Hirasawa A, Yanagisawa S, Tanaka N, Funane T, Kiguchi M, Sørensen H, Ogoh S. Influence of skin blood flow and source-detector distance on near-infrared spectroscopy-determined cerebral oxygenation in humans. *Clinical Physiology and Functional Imaging* 2015;**35**(3):237–44.

34. Barati Z, Shewokis PA, Izzetoglu M, Polikar R, Mychaskiw G, Pourrezaei K. Hemodynamic response to repeated noxious cold pressor tests measured by functional near infrared spectroscopy on forehead. *Annals of Biomedical Engineering* 2013;**41**(2):223–37.

35. Scarapicchia V, Brown C, Mayo C, Gawryluk JR. Functional magnetic resonance imaging and functional near-infrared spectroscopy: insights from combined recording studies. *Frontiers in Human Neuroscience* 2017;**11**:419.

36. Gagnon L, Cooper RJ, Yücel MA, Perdue KL, Greve DN, Boas DA. Short separation channel location impacts the performance of short channel regression in NIRS. *NeuroImage* 2012;**59**(3):2518–28.

37. Haeussinger FB, Dresler T, Heinzel S, Schecklmann M, Fallgatter AJ, Ehlis A-C. Reconstructing functional near-infrared spectroscopy (fNIRS) signals impaired by extra-cranial confounds: an easy-to-use filter method. *NeuroImage* 2014;**95**:69–79.

38. Kohno S, Miyai I, Seiyama A, Oda I, Ishikawa A, Tsuneishi S, Shimizu K. Removal of the skin blood flow artifact in functional near-infrared spectroscopic imaging data through independent component analysis. *Journal of Biomedical Optics* 2007;**12**(6):062111.

39. Devaraj A. *Signal processing for functional near-infrared neuroimaging*. 2005.

40. Diamond SG, Huppert TJ, Kolehmainen V, Franceschini MA, Kaipio JP, Arridge SR, Boas DA. Physiological system identification with the Kalman filter in diffuse optical tomography. In: *Paper presented at the International Conference on Medical image computing and computer-Assisted intervention*. 2005.

41. Glover GH, Li TQ, Ress D. Image-based method for retrospective correction of physiological motion effects in fMRI: RETROICOR. *Magnetic Resonance in Medicine* 2000;**44**(1):162–7.

42. Gratton G, Corballis PM. Removing the heart from the brain: compensation for the pulse artifact in the photon migration signal. *Psychophysiology* 1995;**32**(3):292–9.

43. Mayhew JE, Askew S, Zheng Y, Porrill J, Westby GM, Redgrave P, Harper RM. Cerebral vasomotion: a 0.1-Hz oscillation in reflected light imaging of neural activity. *NeuroImage* 1996;**4**(3):183–93.

44. Morren G, Wolf M, Lemmerling P, Wolf U, Choi JH, Gratton E, Van Huffel S. Detection of fast neuronal signals in the motor cortex from functional near infrared spectroscopy measurements using independent component analysis. *Medical and Biological Engineering and Computing* 2004;**42**(1):92–9.

45. Obrig H, Neufang M, Wenzel R, Kohl M, Steinbrink J, Einhäupl K, Villringer A. Spontaneous low frequency oscillations of cerebral hemodynamics and metabolism in human adults. *NeuroImage* 2000;**12**(6):623–39.

46. Yamada T, Umeyama S, Matsuda K. Multidistance probe arrangement to eliminate artifacts in functional near-infrared spectroscopy. *Journal of Biomedical Optics* 2009;**14**(6):064034.

47. Yücel MA, Selb J, Aasted CM, Lin P-Y, Borsook D, Becerra L, Boas DA. Mayer waves reduce the accuracy of estimated hemodynamic response functions in functional near-infrared spectroscopy. *Biomedical Optics Express* 2016;**7**(8):3078–88.

48. Zhang Q, Strangman GE, Ganis G. Adaptive filtering to reduce global interference in non-invasive NIRS measures of brain activation: how well and when does it work? *NeuroImage* 2009;**45**(3):788–94.

49. Duncan A, Meek JH, Clemence M, Elwell CE, Tyszczuk L, Cope M, Delpy D. Optical pathlength measurements on adult head, calf and forearm and the head of the newborn infant using phase resolved optical spectroscopy. *Physics in Medicine and Biology* 1995;**40**(2):295.

50. Essenpreis M, Elwell C, Cope M, Van der Zee P, Arridge S, Delpy D. Spectral dependence of temporal point spread functions in human tissues. *Applied Optics* 1993;**32**(4):418–25.

51. Scholkmann F, Wolf M. General equation for the differential pathlength factor of the frontal human head depending on wavelength and age. *Journal of Biomedical Optics* 2013;**18**(10):105004.

52. Uludag K, Kohl-Bareis M, Steinbrink J, Obrig H, Villringer A. Crosstalk in the Lambert-Beer calculation for near-infrared wavelengths estimated by Monte simulations. *Journal of Biomedical Optics* 2002;**7**(1):51–60.

53. Abdelnour AF, Huppert T. Real-time imaging of human brain function by near-infrared spectroscopy using an adaptive general linear model. *Neuroimage* 2009;**46**(1):133–43.

54. Devor A, Dunn AK, Andermann ML, Ulbert I, Boas DA, Dale AM. Coupling of total hemoglobin concentration, oxygenation, and neural activity in rat somatosensory cortex. *Neuron* 2003;**39**(2):353–9.

55. Sheth SA, Nemoto M, Guiou M, Walker M, Pouratian N, Toga AW. Linear and nonlinear relationships between neuronal activity, oxygen metabolism, and hemodynamic responses. *Neuron* 2004;**42**(2):347–55. https://doi.org/10.1016/S0896-6273(04)00221-1.

56. Hernandez-Meza G, Izzetoglu M, Osbakken M, Green M, Abubakar H, Izzetoglu K. Investigation of optical neuro-monitoring technique for detection of maintenance and emergence states during general anesthesia. *Journal of Clinical Monitoring and Computing* 2018;**32**(1):147–63.

57. Merzagora AC, Izzetoglu M, Polikar R, Weisser V, Onaral B, Schultheis MT. Functional near-infrared spectroscopy and electroencephalography: a multimodal imaging approach. In: *Paper presented at the International Conference on foundations of augmented cognition*. 2009.

58. Pourshoghi A, Zakeri I, Pourrezaei K. Application of functional data analysis in classification and clustering of functional near-infrared spectroscopy signal in response to noxious stimuli. *Journal of Biomedical Optics* 2016;**21**(10):101411. https://doi.org/10.1117/1.JBO.21.10.101411.

59. Gramann K, Fairclough SH, Zander TO, Ayaz H. Editorial: trends in neuroergonomics. *Frontiers in Human Neuroscience* 2017;**11**(165). https://doi.org/10.3389/fnhum.2017.00165.

60. Ayaz H, Cakir MP, Izzetoglu K, Curtin A, Shewokis PA, Bunce S, Onaral B. Monitoring expertise development during simulated UAV piloting tasks using optical brain imaging. In: *Paper presented at the IEEE aerospace Conference, BigSky, MN, USA*. 2012.

61. Ayaz H, Shewokis PA, Bunce S, Izzetoglu K, Willems B, Onaral B. Optical brain monitoring for operator training and mental workload assessment. *NeuroImage* 2012;**59**(1):36–47.

62. Holtzer R, Yuan J, Verghese J, Mahoney JR, Izzetoglu M, Wang C. Interactions of subjective and objective measures of fatigue defined in the context of brain control of locomotion. *The Journals of Gerontology: Series A* 2017;**72**(3):417–23.

63. Montero-Odasso M, Schapira M, Soriano ER, Varela M, Kaplan R, Camera LA, Mayorga LM. Gait velocity as a single predictor of adverse events in healthy seniors aged 75 years and older. *The Journals of Gerontology Series A: Biological Sciences and Medical Sciences* 2005;**60**(10):1304–9.

64. Verghese J, LeValley A, Hall CB, Katz MJ, Ambrose AF, Lipton RB. Epidemiology of gait disorders in community-residing older adults. *Journal of the American Geriatrics Society* 2006;**54**(2):255–61.

65. Holtzer R, Mahoney JR, Izzetoglu M, Izzetoglu K, Onaral B, Verghese J. fNIRS study of walking and walking while talking in young and old individuals. *The Journals of Gerontology Series A: Biological Sciences and Medical Sciences* 2011;**66A**(8):879–87. https://doi.org/10.1093/gerona/glr068.

66. Chen M, Pillemer S, England S, Izzetoglu M, Mahoney JR, Holtzer R. Neural correlates of obstacle negotiation in older adults: an fNIRS study. *Gait and Posture* 2017;**58**:130–5.

67. Holtzer R, Mahoney JR, Izzetoglu M, Wang C, England S, Verghese J. Online fronto-cortical control of simple and attention-demanding locomotion in humans. *NeuroImage* 2015;**112**:152–9. https://doi.org/10.1016/j.neuroimage.2015.03.002.

68. Verghese J, Wang C, Ayers E, Izzetoglu M, Holtzer R. Brain activation in high-functioning older adults and falls Prospective cohort study. *Neurology* 2017;**88**(2):191–7.
69. Holtzer R, Schoen C, Demetriou E, Mahoney JR, Izzetoglu M, Wang C, Verghese J. Stress and gender effects on prefrontal cortex oxygenation levels assessed during single and dual-task walking conditions. *European Journal of Neuroscience* 2017;**45**(5):660–70.
70. Chaparro G, Balto JM, Sandroff BM, Holtzer R, Izzetoglu M, Motl RW, Hernandez ME. Frontal brain activation changes due to dual-tasking under partial body weight support conditions in older adults with multiple sclerosis. *Journal of NeuroEngineering and Rehabilitation* 2017;**14**(1):65.
71. Hernandez ME, Holtzer R, Chaparro G, Jean K, Balto JM, Sandroff BM, Motl RW. Brain activation changes during locomotion in middle-aged to older adults with multiple sclerosis. *Journal of the Neurological Sciences* 2016;**370**:277–83. https://doi.org/10.1016/j.jns.2016.10.002.
72. Holtzer R, George CJ, Izzetoglu M, Wang C. The effect of diabetes on prefrontal cortex activation patterns during active walking in older adults. *Brain and Cognition* 2018;**125**:14–22.
73. Holtzer R, Verghese J, Allali G, Izzetoglu M, Wang C, Mahoney JR. Neurological gait abnormalities moderate the functional brain signature of the posture first hypothesis. *Brain Topography* 2016;**29**(2):334–43.
74. Mahoney JR, Holtzer R, Izzetoglu M, Zemon V, Verghese J, Allali G. The role of prefrontal cortex during postural control in Parkinsonian syndromes a functional near-infrared spectroscopy study. *Brain Research* 2016;**1633**:126–38. https://doi.org/10.1016/j.brainres.2015.10.053.

Chapter 4

Why is Eye Tracking an Essential Part of Neuroergonomics?

Vsevolod Peysakhovich[1], Frédéric Dehais[1], Andrew T. Duchowski[2]
[1]ISAE-Supaéro, Université de Toulouse, Toulouse, France; [2]School of Computing, Clemson University, Clemson, SC, United States

Neuroergonomics generally promotes the use of brain imaging techniques or electroencephalography to measure the neural mechanisms underpinning human performance in complex real-life situations, so eye tracking is not the first type of neuroergonomics method that comes to mind. Why bring eye movements and pupillary changes into neuroergonomics? Human vision is tightly coupled to a majority of our activities, and vision provides the brain with a wealth of information. It is difficult, though not impossible, to imagine, for example, an aircraft pilot with visual impairment.[1] The eyes are an important mediator between the environment and the brain, facilitating interaction with our everyday world. Our eyes constantly move to direct our foveas (the small region of the central retina that has highest visual acuity) toobjects of interest. Light passes through the pupil to the retina, which then nervates to the brain. Importantly, the retina is a part of the embryonic diencephalon that progressively evolves into a complex connection using several neural pathways to support visual perception and attentional orientation. Hence eye tracking, though an indirect measure of brain activity, is in a way the technique that measures with the closest proximity to the brain–the retina is in fact the only part of the brain visible (e.g., to an optometrist) by the naked eye.

Half a century ago, before the term neuroergonomics was coined, two seminal works were published: Alfred L. Yarbus described the role of eye movements in vision,[2] while Eckhard H. Hess and James M. Polt conducted the first systematic study[3] revealing that pupil size increases with mental activity during problem solving (published in *Science*). These two facets of eye-tracking technology, pupillometry and eye movements, have since gained a solid footing in studying brain functioning in reading and processing of lexical ambiguity,[4] in virtual reality applications,[5] and in complex eye movements in natural behavior.[6] Since the pioneering studies by Paul Fitts and his colleagues,[7] where researchers fastidiously had to process eye movements manually from a film recording of a mirror reflection, eye-tracking technology has made considerable progress. From visual observation through mechanical transmission of eye movements to a recording system,[8] or a scleral search coil,[9] eye-tracking devices have found widespread deployment in the use of video-based technology (i.e., using corneal reflection of an infrared light source[10]). Modern video-based technology has benefited from a century of improvements to its predecessor, Dodge and Cline's machine[11] using corneal reflection of visible light. The result is a large variety of systems with multiple remote (e.g., in-dash or table-mounted) cameras or head-mounted devices. These eye-tracking systems allow the study of human behavior in complex work situations under realistic simulated[12] or ecological conditions[13,13a] without too many constraints. Eye-tracking devices are now ubiquitous in neuroscience, psychology, industrial engineering, marketing/advertising, and computer science,[14] and are now making inroads into the gaming industry.

EYE MOVEMENTS

Since Yarbus,[2] we know that we do not fixate on random points in space just to acquire the whole image but rather extract goal-related relevant features of the scene. Two basic eye movements used in cognitive research are fixations, when the eye is relatively immobile and visual information is extracted (following Just and Carpenter's[15] *eye–mind assumption*), and saccades, when we change the point of fixation and vision is suppressed. The detection of these basic eye movement events for further data analysis is by itself nontrivial.[16] Based on these events, numerous metrics exist to characterize ocular behavior. Among these are dwell time, duration and frequency of fixations, saccade amplitude, matrices of transitions between different areas of interest, etc.[17] Along with quantitative measurements, qualitative approaches using interactive visualization tools are often used.[18] Heat maps[19] and scan-path visualization[20] allow eye-tracking practitioners to gain insights into data without computing any numerical metrics (see Fig. 4.1). While eye movement measurement is an effective tool for

Neuroergonomics. https://doi.org/10.1016/B978-0-12-811926-6.00004-X
Copyright © 2019 Elsevier Inc. All rights reserved.

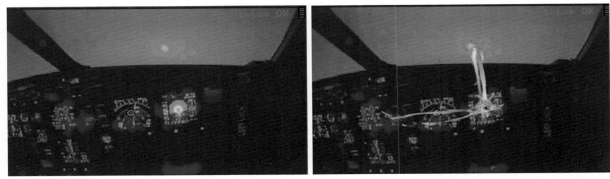

FIGURE 4.1   Two complementary visualization techniques for eye movements. Left: a classical heat map that represents ocular fixation distribution in the cockpit. Right: a scan path indicating the pilot's visual scan of the flight deck, generated by the Attribute-driven edge bundling (ADEB) method,[20,20a] which distinguishes connections between different fixated regions.

evaluating and designing human–machine interfaces,[21,22] eye tracking is rarely used in the three-dimensional (3D) natural world. Future challenges of eye tracking in neuroergonomics include learning to detect 3D eye movements (such as gaze depth[23] and vergence saccades) and studying attention allocation over dynamic volumes of interest. This will allow the design of a neuroergonomic software interface and a neuroergonomic 3D workspace.

PUPIL DIAMETER

While the main function of the pupil is to regulate the amount of the light falling on to the retina, it also reacts to non-visual stimuli. A number of correlations between pupillary changes and attentional effort or cognitive processing have been published since the early 1960s;[3,24,25] see Laeng et al.[26] for a recent review. Pupil diameter is a relevant psychophysiological proxy, but its use has thus far been restricted to controlled laboratory settings. There are many factors that influence pupillary response, among which light conditions are the main challenge for measurement in ecological conditions.[27] Besides the response to ambient illuminance, the pupil also reacts to the luminance of the neighborhood around the fixation point.[28] Recent studies[29] have shown that the luminance impacts the cognitive pupillary response, and thus there is a need for constant measurement of light conditions (both ambient and focal) and the correction of pupillary response according to these conditions. Furthermore, the pupil diameter's tonic and phasic responses appear inversely related.[30,31] Therefore, both tonic and phasic activity should be modeled in complex work situations if pupil diameter is to serve as an index of brain activity. The exact neural mechanisms linking cognitive state and pupillary response remain unclear. Studies suggest the existence of arousal circuits mediated through the locus coeruleus, a small nucleus with a crucial role in cognition,[32] and a parallel pathway from the frontal cortex to pupillary muscles through the superior colliculus,[33] responsible for saccade generation. These two circuits explain pupillary changes as a consequence of attentional shifts or orienting responses.

EYE–COMPUTER INTERFACE

Real-time processing of metrics like pupil diameter, blinks, and eye movements does not require significant computational power and is not as complex as electroencephalography or other brain imaging techniques. This has led some authors since the 1980s to use eye tracking as an online input device to interact directly with the user.[34,35] This gaze-based interactive technique or "eye–computer interface" (ECI) takes advantage of the high velocity and accuracy of eye movements to locate information.[36] Moreover, the eye naturally focuses on targeted items that are intended to be manipulated. Thus ECI supports interaction in a more ecological fashion than the brain–computer interface (BCI), which requires training to learn to use via one's own "brainwave," e.g., when controlling artifacts. The first generations of eye-tracking devices were costly, cumbersome, not very accurate, and involved a long calibration process. Recent technical progress, development of open source software, and the "mass" production of eye-tracking devices are now changing the landscape. These developments have augmented human performance by freeing the hands of surgeons,[37] air traffic controllers,[38] and pilots,[39] and are assisting impaired patients.[40] However, several concerns remain regarding the potential of gaze-based interaction,[38,41] among which are involuntary saccades, the inability of eye movements to produce a smooth trajectory, and the need for a long dwell time to select an area/item of interest. Indeed, efficient interaction to select the desired item with gaze is still an open research issue. Some authors have suggested an elegant solution combining the potential of ECI to identify the relevant items combined with an electroencephalography-based BCI to trigger their selection.[40,42,42a] Eventually, ECI will also support "implicit" or "passive" interaction to monitor the user's scanning performance or infer degraded attentional states such as attentional tunneling.[43] Meanwhile, "active" ECIs can benefit from their hybridization with electroencephalography to obtain a better assessment of visual scene processing along with their associated covert attentional mechanisms.[44] As an example, this approach can pave the way for online human–system interaction in the design of an adaptive cockpit that overcomes the cognitive bottleneck wherein the user's cognitive state is monitored to detect, for example, drowsiness. Such drowsiness vigilance systems have been integrated into vocational trucks for many years, and illustrate the potential of monitoring technology to support safer operations in real-life situations.

REFERENCES

1. Valéry B, Scannella S, Peysakhovich V, Barone P, Causse M. Can an aircraft be piloted via sonification with an acceptable attentional cost? A comparison of blind and sighted pilots. *Applied Ergonomics* 2017;**62**:227–36.
2. Yarbus AL. *Eye movements and vision*. New York: Plenum Press; 1967.
3. Hess EH, Polt JM. Pupil size in relation to mental activity during simple problem-solving. *Science* 1964;**143**(3611):1190–2.
4. Rayner K. Eye movements and attention in reading, scene perception, and visual search. *The Quarterly Journal of Experimental Psychology* 2009;**62**(8):1457–506.
5. Skulmowski A, Bunge A, Kaspar K, Pipa G. Forced-choice decision-making in modified trolley dilemma situations: a virtual reality and eye tracking study. *Frontiers in Behavioral Neuroscience* 2014;**8**:426.
6. Hayhoe M, Ballard D. Eye movements in natural behavior. *Trends in Cognitive Sciences* 2005;**9**(4):188–94.
7. Fitts PM, Jones RE, Milton JL. Eye movements of aircraft pilots during instrument-landing approaches. *Aeronautical Engineering Review* 1950;**9**(2):24–9.
8. Huey EB. Preliminary experiments in the physiology and psychology of reading. *The American Journal of Psychology* 1898;**9**(4):575–86.
9. Robinson DA. A method of measuring eye movement using a scleral search coil in a magnetic field. *IEEE Transactions on Bio-medical Electronics* 1963;**10**(4):137–45.
10. Duchowski A. Eye tracking methodology: theory and practice. Vol. 373. Springer Science & Business Media; 2007.
11. Dodge R, Cline TS. The angle velocity of eye movements. *Psychological Review* 1901;**8**(2):145.
12. Dehais F, Peysakhovich V, Scannella S, Fongue J, Gateau T. Automation surprise in aviation: real-time solutions. In: *Proceedings of the 33rd annual ACM conference on human factors in computing systems*. ACM; April 2015. p. 2525–34.
13. Dehais F, Causse M, Pastor J. Embedded eye tracker in a real aircraft: new perspectives on pilot/aircraft interaction monitoring. In: *Proceedings from the 3rd international conference on research in air transportation*. Fairfax, USA: Federal Aviation Administration; March 2008.
13a. Scannella S, Peysakhovich V, Ehrig F, Lepron F, Dehais F. (in Press). Assessment of ocular and physiological metrics to discriminate flight phases in real light aircraft. *Human factors*.
14. Duchowski AT. A breadth-first survey of eye-tracking applications. *Behavior Research Methods, Instruments, & Computers* 2002;**34**(4):455–70.
15. Just MA, Carpenter PA. Eye fixations and cognitive processes. *Cognitive Psychology* 1976;**8**(4):441–80.
16. Andersson R, Larsson L, Holmqvist K, Stridh M, Nyström M. One algorithm to rule them all? An evaluation and discussion of ten eye movement event-detection algorithms. *Behavior Research Methods* 2016:1–22.
17. Holmqvist K, Nyström M, Andersson R, Dewhurst R, Jarodzka H, Van de Weijer J. *Eye tracking: a comprehensive guide to methods and measures*. OUP Oxford; 2011.
18. Blascheck T, Kurzhals K, Raschke M, Burch M, Weiskopf D, Ertl T. State-of-the-art of visualization for eye tracking data. In: *Proceedings of EuroVis. Proceedings of EuroVis*Vol. 2014, Vol. 2014. June 2014.
19. Wooding DS. Fixation maps: quantifying eye-movement traces. In: *Proceedings of the 2002 symposium on Eye tracking research & applications*. ACM; March 2002. p. 31–6.
20. Peysakhovich V, Hurter C, Telea A. Attribute-driven edge bundling for general graphs with applications in trail analysis. In: *Visualization symposium (PacificVis), 2015 IEEE Pacific*. IEEE; 2015a. p. 39–46.
20a. Peysakhovich V, Hurter C. Scanpath visualization and comparison using visual aggregation techniques. *Journal of Eye Movement Research* 2018;**10**(5):1–14.
21. Goldberg JH, Kotval XP. Computer interface evaluation using eye movements: methods and constructs. *International Journal of Industrial Ergonomics* 1999;**24**(6):631–45.
22. Poole A, Ball LJ. Eye tracking in HCI and usability research. *Encyclopedia of Human Computer Interaction* 2006;**1**:211–9.
23. Duchowski AT, House DH, Gestring J, Congdon R, Świrski L, Dodgson NA, Krejtz K, Krejtz I. Comparing estimated gaze depth in virtual and physical environments. In: *Proceedings of the symposium on Eye Tracking Research & Applications (ETRA), March 26–28, 2014*. Safety Harbor, FL: ACM; 2014.
24. Hess EH, Polt JM. Pupil size as related to interest value of visual stimuli. *Science* 1960;**132**(3423):349–50.
25. Kahneman D, Beatty J. Pupil diameter and load on memory. *Science* 1966;**154**(3756):1583–5.
26. Laeng B, Sirois S, Gredebäck G. Pupillometry a window to the preconscious? *Perspectives on Psychological Science* 2012;**7**(1):18–27.

27. Peysakhovich V, Causse M, Scannella S, Dehais F. Frequency analysis of a task-evoked pupillary response: luminance-independent measure of mental effort. *International Journal of Psychophysiology* 2015b;**97**(1):30–7.

28. Pereverzeva M, Binda P, Murray SO. Covert attention to bright and dark surfaces drives pupillary responses. *Journal of Vision* 2012;**12**(9):661.

29. Peysakhovich V, Vachon F, Dehais F. The impact of luminance on tonic and phasic pupillary responses to sustained cognitive load. *International Journal of Psychophysiology* 2017;**112**:40–5.

30. Gilzenrat MS, Nieuwenhuis S, Jepma M, Cohen JD. Pupil diameter tracks changes in control state predicted by the adaptive gain theory of locus coeruleus function. *Cognitive, Affective & Behavioral Neuroscience* 2010;**10**(2):252–69.

31. Mandrick K, Peysakhovich V, Rémy F, Lepron E, Causse M. Neural and psychophysiological correlates of human performance under stress and high mental workload. *Biological Psychology* 2016;**121**:62–73.

32. Joshi S, Li Y, Kalwani RM, Gold JI. Relationships between pupil diameter and neuronal activity in the locus coeruleus, colliculi, and cingulate cortex. *Neuron* 2016;**89**(1):221–34.

33. Wang CA, Munoz DP. A circuit for pupil orienting responses: implications for cognitive modulation of pupil size. *Current Opinion in Neurobiology* 2015;**33**:134–40.

34. Bolt RA. Gaze-orchestrated dynamic windows. In: *ACM SIGGRAPH computer graphics. ACM SIGGRAPH computer graphics*Vol. 15, Vol. 15. ACM; 1981. p. 109–19.

35. Jacob RJ. What you look at is what you get: eye movement-based interaction techniques. In: *Proceedings of the SIGCHI conference on human factors in computing systems*. ACM; 1990. p. 11–8.

36. Sibert LE, Jacob RJ. Evaluation of eye gaze interaction. In: *Proceedings of the SIGCHI conference on human factors in computing systems*. ACM; 2000. p. 281–8.

37. Noonan DP, Mylonas GP, Shang J, Payne CJ, Darzi A, Yang G-Z. Gaze contingent control for an articulated mechatronic laparoscope. In: *Biomedical robotics and biomechatronics (BioRob), 2010 3rd IEEE RAS and EMBS international conference on*. IEEE; 2010. p. 759–64.

38. Alonso R, Causse M, Vachon F, Parise R, Dehais F, Terrier P. Evaluation of head-free eye tracking as an input device for air traffic control. *Ergonomics* 2013;**56**(2):246–55.

39. Merchant S, Schnell T. Applying eye tracking as an alternative approach for activation of controls and functions in aircraft. In: *Presented at the 19th digital avionics systems conference*, Vol. 2. IEEE; 2000. 5A5/1-5A5/9.

40. Shishkin SL, Nuzhdin YO, Svirin EP, Trofimov AG, Fedorova AA, Kozyrskiy BL, Velichkovsky BM. EEG negativity in fixations used for gaze-based control: toward converting intentions into actions with an eye-brain-computer interface. *Frontiers in Neuroscience* 2016;**10**.

41. Jacob R, Stellmach S. What you look at is what you get: gaze-based user interfaces. *Interactions* 2016;**23**(5):62–5.

42. Zander TO, Gaertner M, Kothe C, Vilimek R. Combining eye gaze input with a brain–computer interface for touchless human–computer interaction. *International Journal of Human–Computer Interaction* 2010;**27**(1):38–51.

42a. Peysakhovich V, Lefrançois O, Dehais F, Causse M. The neuroergonomics of aircraft cockpits: the four stages of eye-tracking integration to enhance flight safety. *Safety* 2018;**4**(1):8.

43. Regis N, Dehais F, Rachelson E, Thooris C, Pizzio S, Causse M, Tessier C. Formal detection of attentional tunneling in human operator automation interactions. *IEEE Transaction on Human Factors* 2014;**44**(3):326–336.

44. Baccino T, Manunta Y. Eye-fixation-related potentials: insight into parafoveal processing. *Journal of Psychophysiology* 2005;**19**(3):204–15.

Chapter 5

The Use of tDCS and rTMS Methods in Neuroergonomics

Daniel Callan[1,2], Stephane Perrey[3]

[1]*Center for Information and Neural Networks (CiNet), National Institute of Information and Communications Technology (NICT), Osaka University, Osaka, Japan;* [2]*ISAE-SUPAERO, Université de Toulouse, Toulouse, France;* [3]*EuroMov, University of Montpellier, Montpellier, France*

INTRODUCTION

The possibility of influencing brain activity and enhancing behavioral performance through external intervention has long fascinated. The past decade has seen a rapid development of noninvasive brain stimulation (NIBS) techniques such as transcranial magnetic stimulation (TMS) and transcranial direct-current stimulation (tDCS) that allow interaction with brain function. The term "noninvasive" brain stimulation refers to those techniques that act on brain physiology without the need for surgical procedures involving electrode implantation. NIBS techniques play a pivotal role in human systems and cognitive neuroscience as they can reveal the relevance of certain brain structures or neuronal activity patterns for a given cognitive or motor function, especially when used in conjunction with neuroimaging and electrophysiology methods. The main NIBS techniques affect brain function via electrical or magnetic impulses. Stimulation techniques (such as TMS) that primarily induce activity of neurons (suprathreshold stimulation) are distinguished from those (such as tDCS) that primarily exert modulatory effects on ongoing neuronal activity and excitability (subthreshold). This chapter will provide an overview regarding TMS and tDCS techniques that are able to produce functionally relevant changes in human brain functions.

TMS and tDCS have considerable implications for enhancing human performance with regard to neuroergonomic applications (See Ref. 1 for a more extensive review). TMS has been found to improve perceptual, motor, and cognitive performance on a variety of tasks in laboratory settings (See Ref. 2, 3; for reviews). An understanding of the neural processes mediating TMS- and tDCS-induced enhancement of behavioral performance, including the various mechanisms and associated stimulation parameters, may provide insight into paradigms by which real-world neuroergonomic applications can be implemented.

TMS PRINCIPLES

TMS involves delivering a magnetic pulse from a coil placed on the head over the desired cortical site of stimulation. The rapid pulse of the magnetic field induces electrical currents that depolarize large groups of neurons (that can lead to action potentials depending on the stimulation intensity) primarily in cortical regions just under the stimulator.[4,5] Distributed networks connected to the region of stimulation are also affected by TMS,[6] allowing modulation of subcortical and distal cortical regions far from the site of stimulation. Depending on the type of stimulation method, TMS can be used to act, for example, as a short-term virtual lesion of a specific brain region or can act to enhance longer-term processing and learning in brain regions and networks associated with the site of stimulation.[6] These longer-term effects are often a result of a sequence of repetitive pulses of TMS (rTMS). The effect that rTMS has on cortical processes modulating behavioral and cognitive performance is mediated by many factors including the number of stimulation pulses, the rate of stimulation, and the pattern of stimulation, as well as the intensity of the stimulation.[4] Long-term potentiation (LTP) and long-term depression (LTD) of synaptic connections are thought to be, in part, responsible for the neural plastic alterations associated with enhanced learning and performance induced by rTMS.[7] Similar to the firing pattern of some neurons in the brain known to induce LTP and LTD, Theta burst stimulation (TBS) paradigms for TMS involve short bursts of high-frequency (50 Hz) pulses presented at intervals of, for example, 200 ms (5 Hz Theta frequency).[2] TBS paradigms for TMS are thought to produce robust longer-term effects on cognitive learning and performance.[2,3]

There are two primary means by which TMS is thought to enhance behavioral performance. The first involves direct facilitative modulation of task-relevant cortical networks; the second involves the principle of "addition-by-subtraction," which enhances task-relevant behavioral performance by disrupting competing or distracting processes that may normally interfere with the primary task.[3] Many factors determine whether enhancement in performance is mediated by facilitative modulation of task-relevant cortical networks or disruption of competing processes including the pattern of stimulation, the site of stimulation, and the interaction between TMS and ongoing (as well as prior) task implementation.[1-4] The enhancement/suppression of cortical excitability (more or less responsive to produce activity) by TMS is generally thought to facilitate/impair functions mediated by underlying cortical networks being stimulated. High-frequency (above 5 Hz) rTMS and intermittent iTBS (stimulation pattern, for example, of 2 s of TBS and 8 s off repeated for a duration of 190 s) are associated with enhancement of greater cortical excitability,[2,3] whereas low-frequency (around 1 Hz) rTMS and continuous cTBS (continuous pattern of TBS) are associated with suppression of cortical excitability.[2,3] The relationship between enhancement and suppression of cortical excitability induced by TMS on facilitating behavioral performance is dependent on the task at hand and its relationship to the functions of the cortical networks being stimulated. Task-relevant cortical networks stimulated by high-frequency rTMS or iTBS may facilitate behavioral performance, whereas low-frequency rTMS and cTBS to the same cortical region may degrade behavioral performance. It also follows that high-frequency rTMS or iTBS to a cortical region that may instantiate a competing or distracting process to the task may degrade behavioral performance, whereas low-frequency rTMS or cTBS to the same region facilitates behavioral performance (addition by subtraction).[3] Even with similar types of stimulation (e.g., high-frequency rTMS), it has been shown that behavioral enhancement is dependent on the task performed (e.g., attention to local or global aspects of a visual stimulus, both of which are functionally related to the site of stimulation) and the specific frequency of stimulation (e.g., 20 Hz or 5 Hz).[3,8] A considerable number of factors determine whether TMS will enhance perceptual, motor, and cognitive performance, in addition to the ones discussed earlier including individual differences in susceptibility, time of day, and duration of stimulation, as well as the relationship of the timing of the stimulation to the task.[1-4]

tDCS PRINCIPLES

tDCS involves delivering prolonged (~10–20 min), low-intensity electric current (1–2 mA) directly through a pair of electrodes placed on the scalp, and is thought to shift the neuronal resting membrane potential causing a slight depolarization (excitation) or hyperpolarization (inhibition) of the cortical neurons.[9] Although the effects of weak direct-current stimulation on the excitability of the central nervous system were reported decades ago, tDCS receives nowadays a growing interest as a portable, simplistic natural and noninvasive tool for modulating behavioral and neurophysiological activity. Generally, the effects of tDCS are polarity specific. Anodal tDCS (a-tDCS) increases neuronal excitability and promotes mechanisms involved in LTP, whereas cathodal tDCS will induce the opposite effects,[9] with changes remaining above baseline for up to an hour following a single tDCS session.[10] Prolonged adaptations appear to be modulated by strengthening or weakening of synaptic activity, which is dependent on the activation of the glutamatergic NMDA (N-methyl-D-aspartate) receptor.[11] Depending on stimulation parameters, tDCS can induce changes at distant points through effective modulations of remote, interconnected networks. By using functional neuroimaging, tDCS-induced changes in regional cerebral blood flow have been observed in the human brain at rest, during a motor task over the stimulated cortical area, as well as distant functionally related regions.[12]

Traditionally, the positive (anode) electrode is placed on the area of the brain to be stimulated, and the negative (cathode) electrode is put over another cranial or extracranial location. Typically, current intensities of 0.5–2 mA are applied for durations of 5–20 min. For the conventional tDCS, a direct current is delivered through two large (~30 cm²) dampened-surface electrodes for maintaining a low current intensity such that the skin sensation of the electrical stimulation is bearable. tDCS with electrodes smaller than approximately 2 cm² is called high-definition (HD)-tDCS and often uses array of electrodes to guide current through the brain. This is one solution to improve optimization of the tDCS technique due to the expected focality of the induced current[13] and the persistence of the aftereffects on cortical excitability.[14] The electrode montage (electrode position and size) along with the applied current determine the generated electrical field strength in the brain, which, in turn, determines the efficacy of tDCS. One of the advantages of tDCS compared to other NIBS techniques is to use an initial and brief stimulation as a reliable method of placebo. Because the arising sensations resulted from the anode tend to occur only at early stages of application, a sham-tDCS applied for 30 s makes it difficult for the individual to distinguish the placebo from a real tDCS application.

Although the optimal timing of tDCS delivery remains unclear, there is emerging evidence to suggest that a-tDCS applied concurrently with the cognitive/motor task produces larger performance gains and learning compared with tDCS applied independently or prior to the task.[15-17] Furthermore, with session duration of 20 min or more and with multiple sessions over consecutive days (3–5), the aftereffects of tDCS will last longer. A large number of studies provide evidence for efficacy of multiple-session a-tDCS (see for a review Ref. 18 and seminal work of Reis et al.[17]). Altogether, current findings suggest a time-dependent development of consecutive aftereffects gains in the first few hours after each concurrent a-tDCS and training session.

CONCLUSION

The utilization of TMS and tDCS in neuroergonomic applications is dependent on the nature and time-sensitive goals of the task to be enhanced and is constrained by aspects of TMS/tDCS technology. For tasks that involve online improvement in a specific aspect such as faster response time in a motor task or increased accuracy in a visual search task, TMS and tDCS can be applied interactively utilizing the proper stimulation parameters to induce cortical excitability leading to enhanced performance. One could possibly utilize this application to enhance operator performance in such tasks, for example, as search and rescue and radar detection, in which online modulation of specific task-relevant cortical processes are facilitated. In many cases, it is desired that the enhancement in performance induced by TMS or tDCS has effects that are perpetuated for a long period after stimulation. This is particularly important in neuroergonomic applications involving acquisition of new skills. Prolonging the enhancing effects of TMS or tDCS to allow for accelerated learning is facilitated by utilizing either TBS[3] or a-tDCS[18] over multiple sessions in conjunction with the task to be acquired. A large constraint of NIBS techniques, for them to be used most effectively, is that they require neuroimaging techniques to localize the site of the task-relevant cortical regions to be stimulated[3] as well as the temporal properties of neural processing.[19] This may limit the scope by which both TMS and tDCS can be used practically in neuroergonomic applications. With the advancement of technology including combination of techniques, multichannel NIBS (having the ability to stimulate multiple brain areas simultaneously), and greater portability of the whole equipment, the potential exists for considerable neuroergonomic applications to enhance human performance.

REFERENCES

1. McKinley RA, Bridges N, Walters CM, Nelson J. Modulating the brain at work using noninvasive transcranial stimulation. *Neuroimage* 2012;**59**:129–37.
2. Demeter E. Enhancing cognition with theta burst stimulation. *Current Behavioral Neuroscience Reports* 2016;**3**:87–94.
3. Luber B, Lisanby SH. Enhancement of human cognitive performance using transcranial magnetic stimulation (TMS). *Neuroimage* 2014;**85**:961–70.
4. Ridding MC, Rothwell JC. Is there a future for therapeutic use of transcranial magnetic stimulation? *Nature Reviews: Neuroscience* 2007;**8**:559–67.
5. Wagner T, Valero-Cabre A, Pascual-Leone A. Noninvasive human brain stimulation. *Annual Review of Biomedical Engineering* 2007;**9**(19):1–19. 39.
6. Reithler J, Peters JC, Sack AT. Multimodal transcranial magnetic stimulation: using concurrent neuroimaging to reveal the neural network dynamics of noninvasive brain stimulation. *Progress in Neurobiology* 2011;**94**:149–65.
7. Lenz M, Vlachos A. Releasing the cortical brake by non-invasive electromagnetic stimulation? rTMS induces LTD of GABAergic neurotransmission. *Frontiers in Neural Circuits* 2016;**10**:96.
8. Romei V, Driver J, Schyns PG, Thut G. Rhythmic TMS over parietal cortex links distinct brain frequencies to global versus local visual processing. *Current Biology* 2011;**21**:334–7.
9. Nitsche MA, Paulus W. Excitability changes induced in the human motor cortex by weak transcranial direct current stimulation. *Journal of Physiology* 2000;**527**:633–9.
10. Frickle K, Seeber AA, Thirugnanasambandam N, Paulus W, Nitsche MA, Rothwell JC. Time course of the induction of homeostatic plasticity generated by repeated transcranial direct current stimulation of the human motor cortex. *Journal of Neurophysiology* 2011;**105**:1141–9.
11. Liebetanz D, Nitsche MA, Tergau F, Paulus W. Pharmacological approach to the mechanisms of transcranial DC-stimulated-induced after-effects of human cortex excitability. *Brain* 2002;**125**:2238–47.
12. Lang N, Siebner HR, Ward NS, Lee L, Nitsche MA, Paulus W, Rothwell JC, Lemon RN, Frackowiak RS. How does transcranial DC stimulation of the primary motor cortex alter regional neuronal activity in the human brain? *European Journal of Neuroscience* 2005;**22**:495–504.
13. Edwards D, Cortes M, Datta A, Minhas P, Wassermann EM, Bikson M. Physiological and modeling evidence for focal transcranial electrical brain stimulation in humans: a basis for high-definition tDCS. *Neuroimage* 2013;**74**:266–75.
14. Kuo H-I, Bikson M, Datta A, Minhas P, Paulus W, Kuo M-F, Nitsche MA. Comparing cortical plasticity induced by conventional and high-definition 4 × 1 ring tDCS: a neurophysiological study. *Brain Stimulation* 2013;**6**:644–8.
15. Cohen Kadosh R, Soskic S, Iuculano T, Kanai R, Walsh V. Modulating neuronal activity produces specific and long-lasting changes in numerical competence. *Current Biology* 2010;**20**:2016–20.
16. Parikh PJ, Cole KJ. Effects of transcranial direct current stimulation in combination with motor practive on dexterous grasping and manipulation in healthy older adults. *Physiological Reports* 2014;**2**:e00255.
17. Reis J, Schambra HM, Cohen LG, Buch ER, Fritsch B, Zarahn E, Celnik PA, Krakauer JW. Noninvasive cortical stimulation enhances motor skill acquisition over multiple days through an effect on consolidation. *Proceedings of the National Academy of Sciences* 2009;**106**:1590–5.
18. Hashemirad F, Zoghi M, Fitzgerald PB, Jaberzadeh S. The effect of anodal transcranial direct current stimulation on motor sequence learning in healthy individuals: a systematic review and meta-analysis. *Brain and Cognition* 2016;**102**:1–12.
19. Bergmann TO, Karabanov A, Hartwigens G, Thielscher A, Roman Siebner H. Combining non-invasive transcranial brain stimulation with neuroimaging and electrophysiology: current approaches and future perspectives. *Neuroimage* 2016;**140**:4–19.

Chapter 6

Transcranial Doppler Sonography in Neuroergonomics

Tyler H. Shaw[1], Amanda E. Harwood[1], Kelly Satterfield[2], Victor S. Finomore[3]

[1]George Mason University, Fairfax, VA, United States; [2]Oak Ridge Institute for Science and Education, Wright-Patterson Air Force Base, OH, United States; [3]West Virginia University, Morgantown, WV, United States

Transcranial Doppler sonography (TCD) was used as a tool in human factors and neuroergonomics research as early as the 1990s.[1,2] Prior to that it had been used in medicine for many years to assess and diagnose various aspects of cardiovascular dysfunction.[3] Since TCD's adoption in psychological research it has been used in studies examining visual and auditory processing, spatial processes, working memory, gender and age, language processes, and general attention (see reviews by Ref. 4; Dusheck and Schandry, 2003). It has primarily been used in neuroergonomics research to examine workload and effort during vigilance performance. This chapter serves as a selective review of the literature from the past decade. It begins with a brief overview of the technical elements and the logic underlying the use of the technique, and continues with a discussion of TCD's use within neuroergonomics to study various tasks, processes, and individual differences.

TCD INSTRUMENTATION AND APPLICATION

TCD is an ultrasound device used to examine cerebral blood flow velocity (CBFV) in the mainstem intracranial arteries. The three major branched arteries are most frequently insonated for measurement: the anterior cerebral arteries, the posterior cerebral arteries, and the middle cerebral arteries (MCAs). While most blood vessels constrict and dilate rapidly, the diameter of the three major cerebral arteries remains relatively stable over time,[5] hence any changes in CBFV can be attributed to the demand of metabolically active cortical regions for oxygenated hemoglobin. Unlike other hemodynamic instruments, such as functional magnetic resonance imaging (fMRI) and functional near-infrared spectroscopy (fNIRS) that measure oxyhemoglobin and deoxyhemoglobin directly, TCD measures the speed at which blood carrying oxygen is perfused to relevant cortical areas. CBFV and overall cerebral blood flow are highly correlated; validation studies demonstrate that TCD and fMRI can be used interchangeably to examine task and hemispheric lateralization (Duschek and Schandry, 2002).[6]

TCD uses a small 2 MHz pulsed Doppler transducer to monitor arterial blood flow velocity. The transducer, worn in a headband, is placed just above the zygomatic arch along the temporal bone, a part of the skull that is functionally transparent to ultrasound (see Fig. 6.1). The logic underlying the technique is that when an area of the brain becomes active, the area consumes oxygen. This increases the speed at which oxygenated blood is delivered to these brain areas to replace the oxygen-depleted blood cells.[7] While TCD does not have good spatial resolution, it has high temporal resolution; the low weight and small size of the transducers permit real-time measurement of CBFV while not restricting or interfering with body motion. Moreover, TCD units are low cost and portable, and the acquisition of an ultrasonic signal can be achieved with relative ease. These practical advantages make TCD a good candidate for use in human performance studies. More details on the fundamentals and technical aspects of TCD are given elsewhere.[4,8]

The general trend in psychological research findings is that CBFV accelerates during task performance over a resting baseline (relaxed wakefulness), and these changes covary with the cognitive demand imposed by these tasks.[9] For that reason it is assumed that TCD is a measure of workload and effort. Traditionally, workload and effort have been measured with self-report scales (e.g., the NASA-TLX index[10]), but there are several advantages to using physiological measures rather than self-report measures. First, administration of self-report scales often interrupts task performance,[11] whereas workload information from physiological measures can be obtained in a relatively unobtrusive manner. Second, subjective measures usually provide workload information only after the task has been completed, while physiological measures give continuous workload assessment throughout the entire duration of the task. Thus measures should be sought that can provide reliable assessments of workload without the drawbacks of intrusion and post hoc subjective assessment.

Neuroergonomics. https://doi.org/10.1016/B978-0-12-811926-6.00006-3
Copyright © 2019 Elsevier Inc. All rights reserved.

FIGURE 6.1   Dr. Joel Warm wearing the TCD headset.

TCD AND TASK CHARACTERISTICS

Most of the work examining TCD in human performance research in the past decade has focused on vigilance or sustained attention. Vigilance remains a high priority in human factors research because it is required in high-risk public safety jobs (e.g., air traffic control) and is a crucial component of the way humans interact with technology (e.g., autonomous vehicle monitoring). Vigilance tasks require individuals to monitor some aspect of their environment for a rare critical signal and make the appropriate response to indicate the presence of that signal. Often, performance on vigilance tasks is initially high but after a short time it significantly declines. This decline in performance is known as the vigilance decrement, and it is typical in most vigilance studies. The conceptual framework most widely used by vigilance researchers to explain the vigilance decrement is resource theory.[12–14] According to resource theory, the decrement is the result of the depletion of information-processing resources that are not replenished over time[15] until a period of rest. Some have considered resource theory to be a circular explanation for vigilance performance,[16] but neuroergonomics measures can strengthen the claim made by advocates of resource theory by quantifying resource availability and utilization.

Several vigilance studies have used TCD to examine task factors that are theoretically and empirically linked to sustained attention, including a manipulation of working memory demands imposed by the task,[2] visual search,[17] providing cues for the arrival of critical signals (Hitchcock et. al, 2003), and the event rate effect.[17] More recent studies have demonstrated that TCD is sensitive to the effect of providing observers with knowledge of results,[18] the sensory modality of signals,[19] irregular event schedules,[20] and spatial uncertainty.[21] These studies examining task characteristics have revealed a consistent pattern of results. The absolute level of blood flow velocity is directly related to increases in task difficulty.[22–25] Furthermore, a near-universal finding is that as vigilance decreases, this decrement is accompanied by a temporal decline in CBFV[1]; Shaw et al., 2009).[25a,25b] In addition, the CBFV effects are generally lateralized to the right cerebral hemisphere, consistent with fMRI studies that point to a right-hemispheric system in the functional control of vigilance performance.[26] Finally, several studies have shown that the temporal CBFV effect only occurs when participants are actively engaged in the task. When compared to participants performing the task, there is no decrement in CBFV in control participants who were exposed to an identical vigilance task, but with no work imperative (Shaw et al., 2009).[22,23] It is also important to note that all of these findings occurred in the MCA, which is the blood vessel that carries 80% of blood to the brain and has the largest perfusion territory.[8]

TCD and Supervisory Control

In addition to the findings regarding TCD and vigilance, CBFV is sensitive to dynamic transitions in task demand in complex decision-making tasks.[24,27] Resource theory suggests that when adapting to changing task demands, performance is

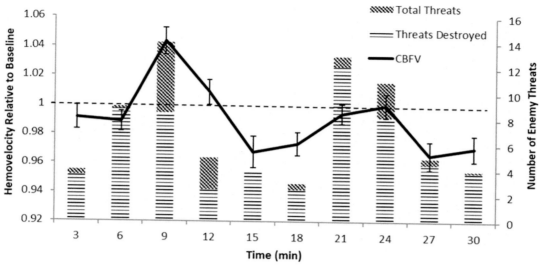

FIGURE 6.2   Cerebral hemovelocity relative to baseline (left vertical axis) overlaid on number of enemy threats and number of threats destroyed (right vertical) across the 30 min task. *Dashed line* represents baseline hemovelocity scores. *Error bars* are standard error.

related to the amount of attentional resources that can be applied to the task,[28] and workload represents the proportion of resources required to meet the demands of the task.[29] Thus in addition to indexing resource utilization in vigilance, TCD might be used to index resources during transitions in task load in more complex dynamic tasks.

This idea was tested in one study[27] in which participants performed a supervisory task where they controlled multiple unmanned aerial vehicles (UAVs). Participants had many tasks during the operation of the UAVs—they had to defend a "no-fly" zone from the incursion of enemy threats, destroy as many enemy threats as possible, and protect their own assets from destruction. In one condition the number of enemy threats (which was generally held constant) increased by 200% at two instances across the 30-min task. The schedule of enemy events, the number of threats destroyed, and the associated CBFV levels are shown in Fig. 6.2. During the performance of the task, CBFV was recorded from the left and right MCAs. Results demonstrated that CBFV was sensitive to moment-to-moment transitions in task load, as the hemovelocity scores very closely mirrored the event schedule of enemy threats (see Fig. 6.2 for an illustration). This finding also held true for a condition in which there was only a single task load transition, and a condition when there was no task load transition (more details are available elsewhere[27]). These results are consistent with the expectations derived from resource theory, and demonstrate that the TCD measure can be used to index moment-to-moment changes in workload even in dynamic tasks.

TCD AND OPERATOR CHARACTERISTICS

In addition to the numerous task characteristics discussed in the previous section, more recent research has shown that the TCD measure is sensitive to a variety of operator characteristics. Individual differences have a long history in vigilance, and a great deal of work has been conducted in identifying individual differences and operator characteristics that might be related to vigilance (reviewed elsewhere[30]). Despite these research efforts, the results pertaining to individual differences and their relation to vigilance have been mixed. Some evidence has been found to support the position that individual difference profiles can be useful in predicting vigilance performance.[31,32] Other evidence suggests that the specific personality dimensions investigated often appear as weak and inconsistent predictors of vigilance.[33] Indeed, some researchers have suggested that a multivariate assessment strategy is necessary for individual differences to predict vigilance reliably.[24,30]

Exploring neurobiological correlates of performance at the individual level may help resolve some of the inconsistency that exists in the literature. Researchers have argued for several years that exploring the link between individual differences and neurophysiological measures may provide better insight as to how attentional resources are allocated during vigilance.[24,34–37] Moreover, one possible explanation for the low effect sizes often seen with unitary predictors is that most studies have focused upon performance *outcomes*, whereas an understanding of performance *process* may be just as important.[38] Others have made similar claims, referring to the difference between performance *effectiveness* (i.e., quality of task performance) and performance *efficiency* (i.e., how much effort is necessary to attain high-quality performance). To that end, recent research has examined individual differences through the lens of TCD to examine the underlying efficiency associated with performance. The approach is to examine individual differences that are theoretically linked to vigilance

FIGURE 6.3 Cerebral hemovelcity scores relative to baseline as a function of periods of watch for the experienced and novice groups. The *dashed line* represents baseline hemovelocity score. *Error bars* are standard error.

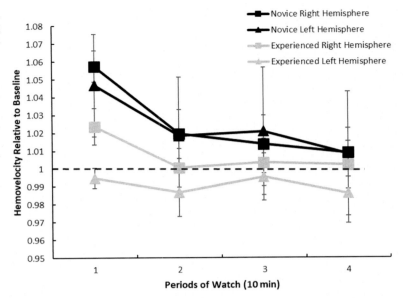

and use TCD to measure how resources are allocated to task performance. Some of these personality dimensions include characteristics such as neuroticism,[39] impulsivity,[24] and effects of practice and experience.[40] The following sections highlight two examples of this approach.

TCD and the Effects of Operator Experience

Studies conducted on the acquisition of skill in vigilance tasks have shown that improvements in detection accuracy and reductions in false alarms in vigilance occur with increased practice.[41–43] Several pieces of evidence point to *reductions* in brain activity for regions involved in executive control and higher-order cognitive functions with repeated practice (Kelly and Garavan, 2004).[44–46] These findings suggest a form of neural processing efficiency (Kelly and Garavan, 2004)[46] in which a more precise neural circuit develops, governing performance after a task is well practiced. The processing efficiency hypothesis suggests that fewer information-processing resources should be required for task performance in well-practiced observers, and specifically those resources stemming from the regions underlying the top-down, fronto-parietal network of vigilance.[26] Since the CBFV measure is an index of these processing resources, one possibility is to use the CBFV measure to assess, and therefore serve as a marker for, the amount of resources required for task performance in both novice and well-practiced operators.

In one study[40] operators performed a 40-min communication-monitoring vigilance task in which they were required to monitor six radio channels simultaneously for the occurrence of critical phrases. Some of the operators had extensive experience with the task—they engaged in approximately 14 h of practice on this vigilance task over a 4–5-week period. They were compared to participants who were considered novices at the task, as they had only practiced the task for 5 min before performing the experimental trial. CBFV measurements were taken bilaterally from the right and left MCAs. While performance was comparable across the two experience levels, results related to the TCD measure revealed that CBFV was only elevated in the novice group, and the temporal decline in CBFV only occurred in the novice group (see Fig. 6.3). This suggests that the experienced group was able to achieve the same level of performance as the novice group while utilizing fewer information-processing resources—a finding that supports the processing efficiency hypothesis[46,47] and the resource account of vigilance (Shaw et al., 2009).[25]

TCD and the Effects of Cognitive Aging

Vigilance studies on cognitive aging have yielded mixed results. Some research points to worse performance with age,[48] while other results suggest no age-related performance deficits.[42] An examination of the underlying neural mechanisms may shed light on the basic process involved, which could help to resolve some of the inconsistency. Research has shown that for tasks during which young people show largely unilateral signals dependent on blood oxygen level, healthy older people show bilateral activation.[49] This research was incorporated in the "hemispheric asymmetry reduction in older adults"

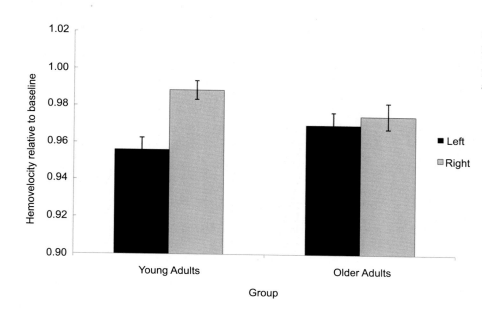

FIGURE 6.4 CBFV relative to baseline plotted as a function of young adults (ages 18–24) and older adults (ages 66–77) for the left and right cerebral hemisphere. *Error bars* are standard error.

(HAROLD) hypothesis of functional brain organization,[50] which suggests that older adults might use bilateral activation in the prefrontal cortex to compensate for cognitive decline.

Other researchers used TCD to monitor CBFV in a vigilance task to determine if the TCD measure reveals reduced hemispheric asymmetry during vigilance.[51] Vigilance operators performed a 60-min traditional vigilance task and CBFV was recorded from the right and left MCAs. A young adult group (ages 18–24, mean=20) was compared to a healthy older adult group (ages 66–77, mean=69). Results showed that older adults performed significantly worse on the vigilance task than young adults. The CBFV data revealed that the hemispheric lateralization often observed in vigilance studies was present for younger adults, but disappeared for older adults (see Fig. 6.4). This bilateral activation in CBFV is consistent with the compensatory mechanism seen in the HAROLD hypothesis,[50] and also with evidence from vigilance research that more difficult vigilance assignments are associated with reduced hemispheric asymmetry.[52] Thus it can be concluded that older adults were recruiting additional resources from the left hemisphere to complete the vigilance task, despite performing worse than the young adults.

CONCLUSION

TCD has shown much promise as a useful tool in human performance research. The last 10 years of research have revealed several key insights regarding the technique. While it has been well documented that the CBFV index was sensitive to the workload associated with task characteristics,[53] newer findings indicate that TCD is also sensitive to individual differences in resource allocation/utilization. These findings suggest that the TCD measure can aid in distinguishing those traits that are associated with superior performance in vigilance. This demonstration has clear implications for the longstanding selection issue in vigilance. TCD could potentially be used to identify individuals who have an inherent disposition to perform better at vigilance tasks. Equally important, the TCD measure appears diagnostic in its ability to identify those operators who could benefit from additional training.

Another research insight worth mentioning concerns the hemisphere effect that is often found in vigilance studies. Vigilance tasks tend to be right lateralized, and this is corroborated by positron-emission tomography (PET) and fMRI vigilance studies.[26,54] However, recent CBFV research has shown that the right-hemisphere effect is not as pervasive as originally thought. For example, one study[52] suggests that it is likely that task difficulty serves as a moderator for the hemisphere effect, so more difficult tasks elicit responses from both hemispheres. Moreover, it appears that bilateral activation also serves a compensatory function in vigilance; in other words, when operators are applying more effort to the task to meet task demand, it is likely that bilateral activation as opposed to exclusively right-hemisphere activation will occur (Harwood et al., 2017; Shaw et al., 2016).[40]

Although TCD is an imaging tool that has been used successfully in human performance studies, it is not without its limitations. Perhaps the most pronounced limitation concerns the lack of spatial resolution of the technique. It is common knowledge that vigilance performance is subserved by many cortical areas,[26] and the lack of spatial specificity of TCD does not allow us to examine the individual contributions of each cortical region. One advantage of TCD is that it has excellent temporal resolution, and several studies have noted that it is comparable in temporal resolution to other hemodynamic measures.[55] In addition, it has been demonstrated that the CBFV response can be time-locked to critical signal presentation.[40] In that study it was shown that the amplitude of the event-related CBFV response was greater during the time phase in which signals were correctly detected, but not during time phases where a critical signal was missed. This finding is consistent with the results of vigilance studies that have examined the P300 component of the event-related potential and the magnitude of pupil dilation.[56] Another advantage of TCD is its relatively low cost and unobtrusive nature. Hence if a researcher is interested in an imaging method that is sensitive to changes in workload and questions related to hemispheric lateralization, TCD is a very suitable candidate.

REFERENCES

1. Schnittger C, Johannes S, Arnavaz A, Münte TF. Relation of cerebral blood flow velocity and level of vigilance in humans. *NeuroReport* 1997;**8**:1637–9.
2. Mayleben DW, Warm JS, Dember WN, Rosa RR, Shear PK, Temple J. Cerebral bloodflow velocity and vigilance. In: *Third automation technology and human performance Conference, Norfolk, VA*. 1998.
3. Caplan LR, Brass LM, DeWitt LD, Adams RJ, Gomex C, Otis S, Weschler LR, von Reutern GM. Transcranial Doppler ultrasound: present status. *Neurology* 1990;**40**:696–700.
4. Babikian VL, Wechsler LR, editors. *Transcranial Doppler ultrasonography.*. Butterworth-Heinemann Medical; 1999.
5. Giller CA, Bowman G, Dyer H, Mootz L, Krippner W. Cerebral arterial diameters during changes in blood pressure and carbon dioxide during craniotomy. *Neurosurgery* 1993;**32**:737–41.
6. Stroobant N, Vingerhoets G. Transcranial Doppler ultrasonography monitoring of cerebral hemodynamics during performance of cognitive tasks: a review. *Neuropsychology Review* 2000;**10**:213–31.
7. Aaslid R. Transcranial Doppler examination techniques. In: *Transcranial Doppler sonography*. Vienna: Springer; 1986. p. 39–59.
8. Tripp LD, Warm JS. In: Parasuraman R, Rizzo M, editors. *Transcranial Doppler sonography. Neuroergonomics: the brain at work*. 2006. p. 82–94.
9. Duschek S, Schandry R. Functional transcranial Doppler sonography as a tool in psychophysiological research. *Psychophysiology* 2003;**40**:436–54.
10. Hart SG, Staveland LE. Development of NASA-TLX (task load index): results of empirical and theoretical research. In: Hancock PA, Meshkati N, editors. *Human mental work-load*. Amsterdam (Netherlands): North-Holland; 1988. p. 139–83.
11. Moroney WF, Biers DW, Eggemeier FT. Some measurement and methodological considerations in the application of subjective workload measurement techniques. *The International Journal of Aviation Psychology* 1995;**5**:87–106.
12. Kahneman D. *Attention and effort*, vol. 1063. Englewood Cliffs (NJ): Prentice-Hall; 1973.
13. Wickens CD. The structure of attentional resources. *Attention and Performance VIII* 1980;**8**:239–57.
14. Parasuraman R, Davies DR, editors. Parasuraman R, Davies DR, editors. *Varieties of attention*, vol. 40. Orlando: Academic Press; 1984. p. 47–52.
15. Parasuraman R, Warm JS, Dember WN. Vigilance: taxonomy and utility. In: *Ergonomics and human factors*. New York (NY): Springer; 1987. p. 11–32.
16. Navon D. Resources—a theoretical soup stone? *Psychological Review* 1984;**91**:216.
17. Hollander TD, Warm JS, Matthews G, Shockley K, Dember WN, Weiler E, Tripp LD, Scerbo MW. Feature presence/absence modifies the event rate effect and cerebral hemovelocity in vigilance performance. In: *Proceedings of the human factors and ergonomics Society Annual meeting*, **48**. Los Angeles (CA): SAGE Publications; 2004. p. 1943–7.
18. Shaw TH, Parasuraman R, Sikdar S, Warm J. Knowledge of results and signal salience modify vigilance performance and cerebral hemovelocity. In: *Proceedings of the human factors and ergonomics Society Annual meeting*, **53**. Los Angeles (CA): Sage Publications; 2009a. p. 1062–5.
19. Shaw TH, Warm JS, Finomore VS., Tripp L, Matthews G, Weiler E, Parasuraman R. Effects of sensory modality on cerebral blood flow velocity during vigilance. *Neuroscience Letters* 2009b;**461**:207–11.
20. Shaw TH, Funke ME, Dillard M, Funke GJ, Warm JS, Parasuraman R. Event-related cerebral hemodynamics reveal target-specific resource allocation for both "go" and "no-go" response-based vigilance tasks. *Brain and Cognition* 2013a;**82**:265–73.
21. Funke ME, Warm JS, Matthews G, Funke GJ, Chiu PY, Shaw TH, Greenlee ET. The neuroergonomics of vigilance: effects of spatial uncertainty on cerebral blood flow velocity and oculomotor fatigue. *Human Factors* 2017;**59**:62–75.
22. Hitchcock EM, Warm JS, Matthews G, Dember WN, Shear PK, Tripp LD, Mayleben DW, Parasuraman R. Automation cueing modulates cerebral blood flow and vigilance in a simulated air traffic control task. *Theoretical Issues in Ergonomics Science* 2003;**4**:89–112.
23. Shaw T, Finomore V, Warm J, Matthews G. Effects of regular or irregular event schedules on cerebral hemovelocity during a sustained attention task. *Journal of Clinical and Experimental Neuropsychology* 2012;**34**:57–66.
24. Shaw TH, Matthews G, Warm JS, Finomore VS, Silverman L, Costa Jr PT. Individual differences in vigilance: personality, ability and states of stress. *Journal of Research in Personality* 2010;**44**:297–308.
25. Warm JS, Matthews G, Parasuraman R. Cerebral hemodynamics and vigilance performance. *Military Psychology* 2009;**21**(S1):S75.
25a. Shaw TH, Finomore VS, Warm JS, Matthews G, Weiler E, Parasuraman R. Effects of sensory modality on cerebral blood flow velocity during vigilance, *Neuroscience Letters* 2009;**461**:207–211.
25b. Shaw TH, Nguyen C, Satterfield K, Ramirez R, McKnight PE. Cerebral hemovelocity reveals differential resource allocation strategies for extraverts and introverts during vigilance. *Experimental brain research* 2016;**244**(2):577–585.
26. Parasuraman R, Warm JS, See JE. Brain systems of vigilance. In: Parasuraman R, editor. *The attentive brain*. Cambridge (MA): MIT Press; 1998. p. 221–56.

27. Satterfield K, Ramirez R, Shaw T, Parasuraman R. Measuring workload during a dynamic supervisory control task using cerebral blood flow velocity and the NASA-TLX. In: *Proceedings of the human factors and ergonomics Society Annual meeting*, **56**. Los Angeles (CA): Sage Publications; 2012. p. 163–7.

28. Norman DA, Bobrow DG. On data-limited and resource-limited processes. *Cognitive Psychology* 1975;**7**(1):44–64.

29. Welford AT. Mental work-load as a function of demand, capacity, strategy and skill. *Ergonomics* 1978;**21**:151–67.

30. Finomore V, Matthews G, Shaw T, Warm J. Predicting vigilance: a fresh look at an old problem. *Ergonomics* 2009;**52**:791–808.

31. Helton WS, Dember WN, Warm JS, Matthews G. Optimism, pessimism, and false failure feedback: effects on vigilance performance. *Current Psychology* 1999;**18**(4):311–25.

32. Rose CL, Murphy LB, Byard L, Nikzad K. The role of the Big Five personality factors in vigilance performance and workload. *European Journal of Personality* 2002;**16**:185–200.

33. Koelega HS. Extraversion and vigilance performance: 30 years of inconsistencies. *Psychological Bulletin* 1992;**112**:239.

34. Matthews G, Warm JS, Reinerman-Jones LE, Langheim LK, Guznov S, Shaw TH, Finomore VS. The functional fidelity of individual differences research: the case for context-matching. *Theoretical Issues in Ergonomics Science* 2011;**12**:435–50.

35. Matthews G, Warm JS, Shaw TH, Finomore VS. Predicting battlefield vigilance: a multivariate approach to assessment of attentional resources. *Ergonomics* 2014;**57**:856–75.

36. Parasuraman R. Assaying individual differences in cognition with molecular genetics: theory and application. *Theoretical Issues in Ergonomics Science* 2009;**10**:399–416.

37. Shingledecker C, Weldon DE, Behymer K, Simpkins B, Lerner E, Warm J, Matthews G, Finomore V, Shaw T, Murphy JS. Measuring vigilance abilities to enhance combat identification performance. *Human Factors Issues in Combat Identification* 2010:47–66.

38. Eysenck MW, Calvo MG. Anxiety and performance: the processing efficiency theory. *Cognition and Emotion* 1992;**6**:409–34.

39. Mandell AR, Becker A, VanAndel A, Nelson A, Shaw TH. Neuroticism and vigilance revisited: a transcranial Doppler investigation. *Consciousness and Cognition* 2015;**36**:19–26.

40. Shaw TH, Satterfield K, Ramirez R, Finomore V. Using cerebral hemovelocity to measure workload during a spatialised auditory vigilance task in novice and experienced observers. *Ergonomics* 2013b;**56**:1251–63.

41. Fisk AD, Schneider W. Control and automatic processing during tasks requiring sustained attention: a new approach to vigilance. *Human Factors* 1981;**23**:737–50.

42. Parasuraman R, Giambra L. Skill development in vigilance: effects of event rate and age. *Psychology and Aging* 1991;**6**(2):155.

43. Schneider W, Shiffrin RM. Controlled and automatic human information processing: I. Detection, search, and attention. *Psychological Review* 1977;**84**:1.

44. Andreasen NC, O'Leary DS, Cizadlo T, Arndt S, Rezai K, Watkins GL, Ponto LL, Hichwa RD. II. PET studies of memory: novel versus practiced free recall of word lists. *Neuroimage* 1995;**2**:296–305.

45. Hempel A, Giesel FL, Caraballo GNM, Amann M, Meyer H, Wüstenberg T, Essig M, Schröder J. Plasticity of cortical activation related to working memory during training. *American Journal of Psychiatry* 2004;**161**:745–7.

46. Hill NM, Schneider W. Brain changes in the development of expertise: neuroanatomical and neurophysiological evidence about skill-based adaptations. *The Cambridge Handbook of Expertise and Expert Performance* 2006:653–82.

47. Kelly AC, Garavan H. Human functional neuroimaging of brain changes associated with practice. *Cerebral Cortex* 2004;**15**:1089–102.

48. Mouloua M, Parasuraman R. Aging and cognitive vigilance: effects of spatial uncertainty and event rate. *Experimental Aging Research* 1995;**21**:17–32.

49. Grady CL, McIntosh AR, Craik FI. Task-related activity in prefrontal cortex and its relation to recognition memory performance in young and old adults. *Neuropsychologia* 2005;**43**:1466–81.

50. Cabeza R, Anderson ND, Locantore JK, McIntosh AR. Aging gracefully: compensatory brain activity in high-performing older adults. *Neuroimage* 2002;**17**:1394–402.

51. Harwood AE, Greenwood PM, Shaw TH. Transcranial Doppler sonography reveals reductions in hemispheric asymmetry in healthy older adults during vigilance. *Frontiers in Aging Neuroscience* 2017;**9**:21–9.

52. Helton WS, Warm JS, Tripp LD, Matthews G, Parasuraman R, Hancock PA. Cerebral lateralization of vigilance: a function of task difficulty. *Neuropsychologia* 2010;**48**:1683–8.

53. Warm JS, Parasuraman R. Cerebral hemodynamics and vigilance. Neuroergonomics. *The Brain at Work* 2007:146–58.

54. Coull JT, Frith CD, Frackowiak RSJ, Grasby PM. A fronto-parietal network for rapid visual information processing: a PET study of sustained attention and working memory. *Neuropsychologia* 1996;**34**:1085–95.

55. Deppe M, Knecht S, Papke K, Lohmann H, Fleischer H, Heindel W, Ringelstein EB, Henningsen H. Assessment of hemispheric language lateralization: a comparison between fMRI and fTCD. *Journal of Cerebral Blood Flow and Metabolism* 2000;**20**:263–8.

56. Murphy PR, Robertson IH, Balsters JH, O'Connell RG. Pupillometry and P3 index the locus-coeruleus-noradrenergic arousal function in humans. *Psychophysiology* 2011;**48**:1532–43.

FURTHER READING

1. Alcañiz M, Rey B, Tembl J, Parkhutik V. A neuroscience approach to virtual reality experience using transcranial Doppler monitoring. *Presence* 2009;**18**:97–111.

2. Droste DW, Harders AG, Rastogi E. A transcranial Doppler study of blood flow velocity in the middle cerebral arteries performed at rest and during mental activities. *Stroke* 1989;**20**(8):1005–11.

3. Schuepbach D, Goenner F, Staikov I, Mattle HP, Hell D, Brenner HD. Temporal modulation of cerebral hemodynamics under prefrontal challenge in schizophrenia: a transcranial Doppler sonography study. *Psychiatry Research: Neuroimaging* 2002;**115**:155–70.
4. Sturzenegger M, Newell DW, Aaslid R. Visually evoked blood flow response assessed by simultaneous two-channel transcranial Doppler using flow velocity averaging. *Stroke* 1996;**27**:2256–61.
5. Vingerhoets G, Luppens E. Cerebral blood flow velocity changes during dichotic listening with directed or divided attention: a transcranial Doppler ultrasonography study. *Neuropsychologia* 2001;**39**:1105–11.

Chapter 7

Brain–Computer Interface Contributions to Neuroergonomics

Fabien Lotte[1,a], Raphaëlle N. Roy[2,a]
[1]INRIA Bordeaux Sud-Ouest/LaBRI, Talence, France; [2]ISAE-SUPAREO, Université de Toulouse, Toulouse, France

INTRODUCTION

Brain–Computer Interfaces (BCIs) are communication and control systems that enable their users to send commands and messages to a computer application by using only their brain activity, this activity being measured and processed by the system.[1] A typical example of a BCI would be an application in which the user can move a cursor on a computer screen toward the left or towards the right, by imagining left- or right-hand movements respectively. Although there are various ways to measure brain activity in BCIs,[2] portable brain-imaging techniques are typically used for practical applications. In particular, Electroencephalography (EEG) and functional Near InfraRed Spectroscopy (fNIRS) have been used for practical BCI applications, such as outside laboratories. Nonetheless, EEG remains by far the most-used measure of brain activity for BCI design, in both laboratories and real-life applications. Therefore, in the rest of this chapter, we are going to focus only on EEG-based BCIs.

BCIs can be divided into three categories: active, reactive, and passive BCIs.[3] With an **active BCI**, the user voluntarily imagines some specific mental tasks (e.g., imagining left- or right-hand movements), the resulting EEG patterns for which are translated into specific commands, e.g., moving the cursor left when the BCI recognizes an imagine left-hand movement in EEG signals.

With **reactive BCIs**, various stimuli (often visual ones) are presented to the user, each one associated to a different command. Each stimulus is designed to evoke a different brain response (Event-Related Potential [ERP] or Evoked-Response Potential) when the user pays attention to it. This brain response can be detected in EEG signals and thus translated into the command associated with this stimulus. The most iconic example of reactive BCIs is the P300-speller, in which the user is presented with a matrix containing all letters of the alphabet, these letters randomly flashing.[4] The user is asked to pay attention to flashes on the letter he wants to spell (the target letter), which will give rise to a P300 ERP (a positive increase in EEG signal amplitude appearing about 300 ms after a rare and relevant stimulus) in the user's EEG signals when the target letter is flashed. No such P300 will appear when other letters are flashed. This enables identification of the target letter by finding out which letter evokes a P300 when flashed.

Another very widespread type of reactive BCIs are Steady-State Visual Evoked Potential (SSVEP)-based BCIs. With such BCIs, the user is presented with various flickering visual objects (e.g., buttons on screen), each object flickering at a different frequency and being associated with a different command. If the user pays attention to one of these flickering objects, it will give rise to an SSVEP in his/her EEG signals, i.e., in an increase of occipital EEG signal power at the same frequency as the flickering frequency of the object, and at its harmonics. For instance, paying attention to a button flickering at 10 Hz, would lead to an increase in 10-Hz EEG power, and possibly 20-Hz EEG power as well. Detecting the SSVEP enables identification of the object the user is paying attention to, and thus sending the corresponding command.

Finally, **passive BCIs** are used to monitor the user's mental states to adapt the application accordingly, without the user sending any voluntary command through EEG signals. For instance, a passive BCI can be used to continuously estimate mental workload levels in EEG signals to present the user with a human–computer interface, e.g., a plane cockpit interface, that is not too cognitively difficult to use, nor too boring.

Passive BCIs are typically the kind of BCIs that can be used for neuroergonomics research and applications. Nonetheless, many of the tools developed for active and reactive BCI, and, in particular, EEG signal-processing tools, are the same as

---

a. Denotes equal contribution.

Neuroergonomics. https://doi.org/10.1016/B978-0-12-811926-6.00007-5
Copyright © 2019 Elsevier Inc. All rights reserved.

the ones used for passive BCIs, and thus can be used for neuroergonomics as well. Therefore, in this chapter, we present a short overview of the tools developed for BCI research that can contribute to neuroergonomics. In particular, we will first briefly present tools to process and classify EEG signals online, to estimate the user mental state. Then we will show how passive EEG-based BCIs can be used for neuroergonomics and illustrate this with existing works. We then present some brief perspectives for the field.

SIGNAL PROCESSING

In BCI research, various signal-processing tools were developed to estimate in real time the users' mental states from their EEG signals, and this despite the noisy, nonstationary and data-scarce nature of those signals. Typically, EEG signal processing in BCI follows a pattern-recognition pipeline, which consists in:

1. Preprocessing EEG signals, which mostly consists in filtering them to increase their Signal-to-Noise Ratio (SNR);
2. Extracting features to describe EEG data in a compact way;
3. Classifying these features.[5–7]

We later describe the main approaches available to perform these different steps, with a focus on approaches that can be used online. In addition, all these steps can be dynamically adapted online as well, so we briefly mention how this can be done.

Preprocessing

Preprocessing EEG signals typically consists in filtering the signal in various ways, to reduce the influence of artifacts such as eye movements (Electrooculography [EOG]) or muscle tension (Electromyography [EMG]),[8] and to highlight the EEG patterns representative of the mental state of interest. The most basic filtering is spectral filtering, i.e., restricting the EEG signals to some specific oscillatory components, e.g., only the alpha (8–12 Hz) and theta (4–7 Hz) rhythms to estimate mental workload. Interestingly enough, some algorithms were developed to automatically identify the best frequency band for each subject, see, e.g., Pregenzer and Pfurtscheller.[9]

Another essential preprocessing step is spatial filtering, i.e., combining the signals from multiple EEG channels to obtain a new signal with higher SNR. Different algorithms were developed to optimize spatial filters from examples of EEG data, to obtain EEG features that are maximally different between mental states to recognize them as well as possible.[10] For online applications, we can notably cite the Common Spatial Patterns (CSP) algorithm to estimate mental states based on oscillatory activity,[10] or xDAWN to estimate states based on Event-Related Potentials.[11,12] The Source Power Comodulation (SPOC) algorithm also enables one to find spatial filters such that the power of the spatially filtered signals maximally covary with a continuous target variable.[13] As such, it can be used to estimate continuous mental states, such as attention or workload levels. Some extensions of such algorithms were also proposed to be more robust to noise or limited training data.[14,15]

Finally, reducing the influence of EOG or EMG artifacts is also desirable. Although there are many effective algorithms to remove artifacts offline based on Independent Component Analysis (ICA),[16] these algorithms typically cannot be used online as they are not computationally efficient enough. For online application, simpler and faster, but nonetheless useful, algorithms are used. Notably, to remove EOG, regression-based algorithms based on explicit measures of EOG are often used.[17] For removing types of artifacts online that are more general, an interesting recent development is the FORCe algorithm, which combines wavelet decompositions and heuristics to remove artifactual wavelet components.[18]

Feature Extraction

Once the EEG is processed, they can be described by features.[6] For BCIs based on oscillatory activity, the typically extracted features are the Band Power (BP) of the EEG signals in various frequency bands and channels. For ERP-based BCIs, the used features are typically the amplitude of the preprocessed EEG time points, for each channel, after downsampling.[19] These two types of features are by far the most used and give good results. It should be mentioned, though, that other types of features are being explored, such as complexity features, describing signal regularities, or connectivity features, quantifying how synchronized signals are from different channels or frequency bands.[20] Although these are not as efficient as BP or time points, they can improve the overall accuracy when used in combination with them.

Classification

Classifiers learn from data that feature values correspond to which class, i.e., to which mental state here. Multiple variants of classifiers have been explored for BCIs (see Lotte et al.[5] for a review). When it comes to online use though, only a few classifiers are typically used, the main ones being Linear Discriminant Analysis (LDA; and its variants such as shrinkage LDA[19] or Stepwise LDA[5]) and Support Vector Machine (SVM). Both classifiers are linear, and are fast to train and use. They can also be trained from rather little training data, which makes them ideal for practical online BCI use. In recent developments, Riemannian geometry-based classifiers, which classify covariance matrices rather than vectors of features, also prove very promising, including online.[21,22]

Adaptation

As previously mentioned, EEG signals tend to be nonstationary, and the environment in which the BCI is used also leads to varying amount of external noise. As such, to reach optimal performances, it is worth considering adaptive signal-processing algorithms, the parameters for which are dynamically changed and optimized during online use.[23] A number of variants of the aforementioned algorithms were thus designed to optimize online, in an incremental way, the spectral filters, spatial filters, features, and classifiers, as new EEG data become available (see Mladenovic et al.[24] for a review). Although most of these algorithms remain to be tested in ecological conditions in outside laboratories, they seem promising to deal with EEG fluctuations due to variation in context, noise, and recording conditions that are typically encountered in real-life applications.

CONTRIBUTIONS TO NEUROERGONOMICS

As explained earlier, since the beginning of the 20th century, Brain–Computer Interface technology has been transposed to monitoring mental states of users. Systems that take into account information about mental states extracted from neurophysiological measures have been called biocybernetic systems or passive BCIs.[3,25,26] Such systems can be used either offline or online, with different applicative goals and for various mental states. Definitions of mental states critical for the neuroergonomics field are given later, as well as examples of applications.

Mental States

A diversity of mental states are relevant for characterizing a user/operator's state. The mental states of interest can be separated into two categories, the mental states linked to the main characteristics of the task performed by the user/operator, e.g., fatigue, and mental states linked to critical states of the system the operator interacts with, e.g., inattentional blindness. Those two types of mental states globally generalize to the level to which the subject/operator has recruited and engaged cognitivo-attentional resources. Therefore, one can consider that the mental states linked to the main characteristics of the task relate to **global** resource engagement, whereas mental states linked to critical states of the system relate to **local** resource engagement. Examples of mental states that fall into these categories and that are classically estimated in neuroergonomics applications are listed in Table 7.1 later. For more details on each of these mental states, please refer to Roy and Frey.[26]

It is worth noting that in addition to the mental states listed previously, which mainly rely on cognitive processes, emotional/affective states are of tremendous importance, as they are inseparable from cognitive states. For instance, workload can induce stress and frustration. It is, therefore, important to realize that mental states are never measured separately, which may be why systems trained on a particular set of data acquired in a specific setting are generally difficult to apply on another set acquired in a different setting. What is more, all these mental states interact in real-life settings and thus decrease the system's performance if the latter is not conceived accordingly.[27] For a review on affective BCIs, please refer to Mühl et al.[28]

TABLE 7.1 Mental States Classically Estimated in Neuroergonomics Applications and Generated in Response to Either Main Characteristics of the Task or Critical States of the System

Main Characteristics of the Task	Linked to time-on-task: fatigue, vigilance, boredom and mind wandering; Linked to mental workload: load in working memory, divided attention, social or temporal stress.
Critical states of the system	Inattentional blindness or deafness phenomena, automation surprise/confusion.

Offline Use: Evaluation

Passive BCI technologies can be used offline, and in fact to this day they mostly are. More particularly, they are used for different purposes, the primary one being the **evaluation of a product**, a work setting, or a work task, to determine their usability, performance, and generally their impact on the user. Hence, passive BCIs can be used for the evaluation of the comfort of stereoscopic displays,[29] for the evaluation of the difficulty of a game,[30] a multitasking environment,[31] a flying task,[32] or a surgical training procedure,[33] and also for the evaluation of prolonged and monotonous tasks such as driving.[34] Another way of using these systems is to perform an **evaluation of the user** him/herself, for instance to determine his/her fitness to perform a coming task, or to determine his/her learning type. This could be promising, and to our knowledge has not yet been done.

Online Use: Adaptation

Passive BCIs aim at being used online, and ideally should be so. Although to this day the scientific literature on the subject is mostly speculative, it seems to be the goal of most researchers in the field. The online use of passive BCIs allows to "close the loop" between a user and the system, and also to include the user in a more global system and regard him/her as a subsystem him/herself. To do so, the system has to adapt to the measured and inferred mental states of the user using countermeasures—if the detected state has a negative impact on performance—or more generally, implicit modifications of the system. Most studies developed and presented in the offline use section actually intend to progress toward an online evaluation of the user's mental states.

Recently, the technological developments and the increase in variety of origin of the community members have allowed the implementation of systems that perform the measurements online, with for instance using fNIRS in online inflight workload monitoring,[35] and using EEG during classical human–computer interaction tasks[36] and numerical learning tasks.[37]

In addition, a few studies have recently been published with an actual **adaptation** of the system to the user's mental state as inferred from neurophysiological measures. For instance, workload-level detection through EEG can be used to adapt the level of difficulty of a multitasking environment (i.e., the MultiAttribute Task Battery)[38] and more recently that of a *Tetris* video game[39] and of an air-traffic controller display.[40] Additionally, passive BCIs can also be used to improve active BCIs, i.e., in which the user controls an effector. This was demonstrated by Chavarriaga and collaborators, who used the EEG responses to errors committed by the active BCI (a.k.a. Error-Related Potentials) to adapt the whole system and increase its performance.[41]

PERSPECTIVES

Passive brain–computer interfaces offer a very promising means to achieve online objective mental state monitoring of users and operators. Therefore, it is of great interest to pursue their development for neuroergonomics applications. Even though the literature on passive BCIs has increased drastically these last few years, most research is still conducted in the laboratory and not in ecological settings. Yet portable recording devices that are quite robust to environmental noise have been released (e.g., dry EEG systems[42]). Therefore, this may be due to the lack of neural features and learning algorithms robust to changes in tasks, settings, and subjects. Research should, therefore, focus on these matters, as well as try and develop systems that actually work online but also adapt both at the signal processing level and at the interface level.

REFERENCES

1. Clerc M, Bougrain L, Lotte F. *Brain-computer interfaces 1: foundations and methods.* ISTE-Wiley; 2016a.
2. Wolpaw J, Loeb G, Allison B, Donchin E, do Nascimento O, Heetderks W, Nijboer F, Shain W, Turner JN. BCI meeting 2005- workshop on signals and recording methods. *IEEE Transactions on Neural Systems and Rehabilitation Engineering* 2006;**14**:138–41.
3. Zander T, Kothe C. Towards passive brain-computer interfaces: applying brain-computer interface technology to human-machine systems in general. *Journal of Neural Engineering* 2011;**8**.
4. Clerc M, Bougrain L, Lotte F. *Brain-computer interfaces 2: technology and applications.* ISTE-Wiley; 2016b.
5. Lotte F, Congedo M, Lecuyer A, Lamarche F, Arnaldi B. A review of classification algorithms for EEG-based brain-computer interfaces. *Journal of Neural Engineering* 2007;**4**:R1–13.
6. Bashashati A, Fatourechi M, Ward RK, Birch GE. A survey of signal processing algorithms in brain-computer interfaces based on electrical brain signals. *Journal of Neural Engineering* 2007;**4**:R35–57.
7. Makeig S, Kothe C, Mullen T, Bigdely-Shamlo N, Zhang Z, Kreutz- Delgado K. Evolving signal processing for brain-computer interfaces. *Proceedings of the IEEE* 2012;**100**:1567–84.

8. Fatourechi M, Bashashati A, Ward R, Birch G. EMG and EOG artifacts in brain computer interface systems: a survey. *Clinical Neurophysiology* 2007;**118**:480–94.

9. Pregenzer M, Pfurtscheller G. Frequency component selection for an EEG-based brain to computer interface. *IEEE Transactions on Rehabilitation Engineering* 1999;**7**:413–9.

10. Blankertz B, Tomioka R, Lemm S, Kawanabe M, Müller K-R. Optimizing spatial filters for robust EEG single-trial analysis. *IEEE Signal Processing Magazine* 2008;**25**:41–56.

11. Rivet B, Souloumiac A, Attina V, Gibert G. xDAWN algorithm to enhance evoked potentials: application to brain computer interface. *IEEE Transactions on Biomedical Engineering* 2009;**56**:2035–43.

12. Roy RN, Bonnet S, Charbonnier S, Jallon P, Campagne A. A comparison of ERP spatial filtering methods for optimal mental workload estimation. In: *Proc IEEE Conf Eng Med Biol.* IEEE; 2015. p. 7254–7.

13. Dähne S, Meinecke FC, Haufe S, Höhne J, Tangermann M, Müller K-R, Nikulin VV. SPoC: a novel framework for relating the amplitude of neuronal oscillations to behaviorally relevant parameters. *NeuroImage* 2014;**86**:111–22.

14. Samek W, Kawanabe M, Muller K. Divergence-based framework for common spatial patterns algorithms. *IEEE Reviews in Biomedical Engineering* 2014;**7**:50–72.

15. Lotte F. Signal processing approaches to minimize or suppress calibration time in oscillatory activity-based brain-computer interfaces. *Proceedings of the IEEE* 2015;**103**(6):871–90.

16. Urigüen JA, Garcia-Zapirain B. EEG artifact removal: state-of- the-art and guidelines. *Journal of Neural Engineering* 2015;**12**:031001.

17. Schlögl A, Keinrath C, Zimmermann D, Scherer R, Leeb R, Pfurtscheller G. A fully automated correction method of EOG artifacts in EEG recordings. *Clinical Neurophysiology* 2007;**118**:98–104.

18. Daly I, Scherer R, Billinger M, Müller-Putz G. FORCe: fully Online and automated artifact Removal for brain-Computer interfacing. *IEEE Transactions on Neural Systems and Rehabilitation Engineering* 2015;**23**:725–36.

19. Blankertz B, Lemm S, Treder M, Haufe S, Müller K-R. Single-trial analysis and classification of ERP components: a tutorial. *NeuroImage* 2010;**56**:814–25.

20. Lotte F. A tutorial on EEG signal-processing techniques for mentalstate recognition in brain-computer interfaces. In: *Guide to brain-computer music interfacing.* Springer; 2014. p. 133–61.

21. Yger F, Berar M, Lotte F. Riemannian approaches in brain-computer interfaces: a review. *IEEE Trans Neur Syst Rehab Eng* 2017;**25**(10):1753–1762.

22. Barachant A, Bonnet S, Congedo M, Jutten C. Multiclass brain-computer interface classification by riemannian geometry. *IEEE Transactions on Biomedical Engineering* 2012;**59**:920–8.

23. Shenoy P, Krauledat M, Blankertz B, Rao R, Müller K-R. Towards adaptive classification for BCI. *Journal of Neural Engineering* 2006;**3**:R13.

24. Mladenovic J, Mattout J, Lotte F. A Generic Framework for Adaptive EEG-based BCI Training and Operation. In: Nam CS, Nijholt A, Lotte F, editors. Brain-Computer Interfaces Handbook: Technological and Theoretical Advances. Taylor & Francis; 2018.

25. Fairclough SH. Fundamentals of physiological computing. *Interacting with Computers* 2009;**21**:133–45.

26. Roy RN, Frey J. Neurophysiological markers for passive brain- computer interfaces. *Brain–Computer Interfaces 1: Foundations and Methods* 2016:85–100.

27. Roy RN, Bonnet S, Charbonnier S, Campagne A. Mental fatigue and working memory load estimation: interaction and implications for EEG-based passive BCI. In: *Proc IEEE Conf Eng Med Biol.* IEEE; 2013. p. 6607–10.

28. Mühl C, Allison B, Nijholt A, Chanel G. A survey of affective brain computer interfaces: principles, state-of-the-art, and challenges. *Brain–Computer Interfaces* 2014;**1**:66–84.

29. Frey J, Appriou A, Lotte F, Hachet M. Classifying EEG signals during stereoscopic visualization to estimate visual comfort. *Computational Intelligence and Neuroscience* 2016;**2016**:2758103.

30. Allison BZ, Polich J. Workload assessment of computer gaming using a single-stimulus event-related potential paradigm. *Biological Psychiatry* 2008;**77**:277–83.

31. Roy RN, Bonnet S, Charbonnier S, Campagne A. Efficient workload classification based on ignored auditory probes: a proof of concept. *Frontiers in Human Neuroscience* 2016;**10**.

32. Dehais F, Roy RN, Gateau T, Scannella S. Auditory alarm misperception in the cockpit: an EEG study of inattentional deafness. In: *Int Conf Augm Cog.* Springer International Publishing; 2016. p. 177–87.

33. Zander TO, Shetty K, Lorenz R, Leff DR, Krol LR, Darzi AW, Gramann K, Yang G-Z. Automated task load detection with electroencephalography: towards passive brain-computer interfacing in robotic surgery. *Journal of Medical Robotics Research* 2016:1–10.

34. Yeo MV, Li X, Shen K, Wilder-Smith EP. Can {SVM} be used for automatic {EEG} detection of drowsiness during car driving? *Safety Science* 2009;**47**:115–24.

35. Gateau T, Durantin G, Lancelot F, Scannella S, Dehais F. Real-time state estimation in a flight simulator using fNIRS. *PLoS One* 2015;**10**:e0121279.

36. Heger D, Putze F, Schultz T. Online workload recognition from EEG data during cognitive tests and human-machine interaction. *Advances in Artificial Intelligence* 2010:410–7.

37. Spüuler M, Walter C, Rosenstiel W, Gerjets P, Moeller K, Klein E. EEG-based prediction of cognitive workload induced by arithmetic: a step towards online adaptation in numerical learning. *ZDM* 2016;**48**:267–78.

38. Prinzel LJ, Freeman FG, Scerbo MW, Mikulka PJ, Pope AT. A closed-loop system for examining psychophysiological measures for adaptive task allocation. *The International Journal of Aviation Psychology* 2000;**10**:393–410.

39. Ewing KC, Fairclough SH, Gilleade K. Evaluation of an adaptive game that uses EEG measures validated during the design process as inputs to a biocybernetic loop. *Frontiers in Human Neuroscience* 2016;**10**.

40. Aricò P, Borghini G, Di Flumeri G, Colosimo A, Bonelli S, Golfetti A, Pozzi S, Imbert J-P, Granger G, Benhacene R, et al. Adaptive automation triggered by EEG-based mental workload index: a passive brain-computer interface application in realistic air traffic control environment. *Frontiers in Human Neuroscience* 2016;**10**.

41. Chavarriaga R, Jdel Millán R. Learning from EEG error-related potentials in noninvasive brain-computer interfaces. *IEEE Transactions on Neural Systems and Rehabilitation Engineering* 2010;**18**:381–8.

42. Nijboer F, van de Laar B, Gerritsen S, Nijholt A, Poel M. Usability of three electroencephalogram headsets for brain-computer interfaces: a within subject comparison. *Interacting with Computers* 2015;**27**:500–11.

Chapter 8

Neuroergonomics of Simulators and Behavioral Research Methods

Carryl L. Baldwin
George Mason University Fairfax, VA, United States

OVERVIEW

Neuroergonomics research generally involves some metric of brain functioning, either directly through brain imaging or indirectly through physiological metrics such as eye tracking and heart rate monitoring. However, neuroergonomics also involves designing systems based on models of brain functioning and use of behavioral measures to infer brain functioning.[1,2] In this chapter, I will discuss some methodological issues to consider when designing experiments using behavioral measures and simulators.

SIMULATIONS AND NEUROERGONOMICS

Most neuroergonomics research is aimed at addressing real-world issues outside the laboratory. Although this is the ultimate goal, task simulations and simulators of various levels of fidelity still play a key role in neuroergonomics research. Some key advantages and limitations to keep in mind are discussed.

Advantages

One key advantage of using simulators is safety. Often in neuroergonomics research, we may wish to examine performance and brain function in complex, high-risk situations such as piloting a plane through challenging airways or driving a car in dense traffic and bad weather. It may simply be too dangerous to expose people to these conditions for experimental purposes. For example, if a new invehicle display has the potential to cognitively overload the driver or be too distracting under certain driving conditions, it is certainly better to examine its impact in a driving simulator rather than risking causing a crash on the road.

Simulations also allow a high degree of experimental control. The complexity of naturalistic conditions can be highly variable. One cannot predict with certainty how much traffic will be on the road at any given time or what the weather (e.g., temperature and visibility) will be like on any given day. These types of variables can be programmed in simulations to be identical for each participant. They can further be programmed to systematically change at precise times so that a researcher can examine the impact of varying levels of the factor of interest. This level of control is virtually impossible to achieve outside the laboratory.

Simulations are often much less expensive than more naturalistic conditions. Getting access to planes and flight hours can be costly. Flight simulators offer the advantage of greatly reduced cost. Thus, we see that simulations have several advantages. However, results obtained in simulations must be examined to determine the extent to which they generalize to settings that are more natural.

Challenges

The primary disadvantage of using simulations is that results obtained in laboratory simulations may have little relevance (fail to generalize) to naturalistic settings. This is sometimes referred to as validity[3] and will be discussed in more depth later in this chapter. Using driving simulations as an example again, numerous investigations aim at determining the type of collision warning that most assists drivers in avoiding crashes. The dangerous nature of the potential crash situation makes examination in a driving simulator warranted. However, all too often research designs are implemented that expose drivers

Neuroergonomics. https://doi.org/10.1016/B978-0-12-811926-6.00008-7
Copyright © 2019 Elsevier Inc. All rights reserved.

to multiple repeated hazard situations over the course of a relatively short time. This allows the researcher to compare various warning types. The trouble is, once the participant has been exposed to the first potential collision situation, he or she is likely to expect that it may happen again and will be likely to be overly cautious during subsequent trials. In our laboratory, we have found that response time to hazardous events often drops significantly after the first exposure regardless of what type of warning is presented on subsequent trials.[4,5] Comparing responses across multiple events of the same type may have less variability than responses between the first and all subsequent events of the same type. This "learning" after the first exposure is important to keep in mind when designing simulator experiments.[6]

Another disadvantage of using simulations is that they can rarely represent the multitude of factors that may be at play in naturalistic settings. They can rarely simulate all atmospheric or weather conditions that may impact performance, the unpredictability of other agents (i.e., traffic, animals, communications), or the realistic dangers involved with poor performance. In fact, the complexity of naturalistic conditions may include variables that impact behavior but that researchers are not aware of (e.g., presence of glare, shadows, and background noises). It is difficult to understand the potential impact of such a rich array of factors present in naturalistic conditions.

Further, regardless of how realistic a simulator may be, the operator knows that if they crash the simulated plane or car they are not going to die or even be seriously injured. No real bodily harm will come from even a major performance error. This inherent knowledge likely impacts performance on some level.

IMPORTANT ISSUES

Transfer of Training

A major issue of importance in using simulators is the degree to which performance in the simulator transfer to more naturalistic conditions. For training paradigms, this is termed transfer of training. In research that is more general, we use the term generalizability. Determining which aspects need to be simulated to facilitate transfer and generalizability is essential. To establish this, performance for key tasks may be tested in the simulator and naturalistic conditions to examine the degree of association between the two. For example, scientists at Embry Riddle examined the extent to which training on a low-cost desktop simulator for upset–recovery transferred to skill development in actual aircraft.[7] Compared to a control group that did not receive the training, the simulator-trained pilots exhibited improved upset–recovery maneuvering. However, for the most important skill—that of minimizing altitude loss—simulator training was not nearly as effective as training in an actual airplane.

Fidelity

Some skills may be trained and tested with low-fidelity simulators, whereas others may require higher-fidelity simulations. Determining the level of fidelity necessary for the research question of interest is important. Higher-fidelity simulations are generally much more expensive and less portable than low-fidelity simulations.

Validity

Simulation validity is a multidimensional construct. It refers not only to the physical characteristics of the simulation and its relation to naturalistic driving, but also to the perceived subjective experience and objective performance metrics (Allen et al., 2010). Allen and colleagues illustrated the multitude of components that contribute to simulation validity (see Fig. 8.1). Important aspects include the cueing of the operator's sensory and perceptual systems (e.g., visual, auditory, proprioceptive, etc.) and the synchronization of these cueing systems. To be effective, there needs to be a close coupling between the cueing systems and the operator's control inputs. In other words, the simulation must be responsive to control inputs within naturalistic time expectations, e.g., the plane should pitch up at roughly the same time as a real plane after receiving throttle input from the pilot.

Allen and colleagues (2010) also suggest that new psychological testing standards most important for validity to the simulation environment include several distinct types of validity, including content, response processes, internal structure, and consequences as discussed in a review article by Goodwin and Leech.[8]

Generalizability

The extent to which conclusions obtained by examining performance in simulations can be used to make statements about performance in naturalistic conditions is referred to as generalizability. Because the ultimate goal of most, if not all, simulation research is to inform knowledge of performance in real-world situations, generalization is a critical aspect.

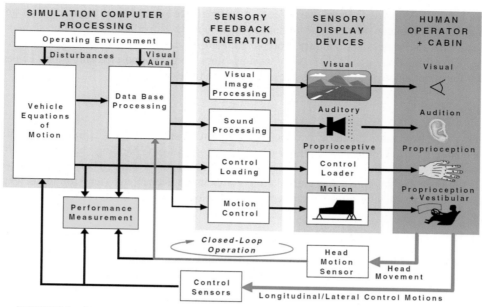

FIGURE 8.1 Components of a driving simulation. *(Borrowed from Allen et al. (2010) with permission.)*

Part Task

Some simulations may involve only part of the overall task; these part-task simulators may assist in testing or training specific skills. One example of the use of part-task simulator training can be found in the work of training older drivers to improve their Useful Field of View or UFOV.[9] Roenker and colleagues had older drivers complete either a UFOV training or traditional training using a driving simulator. They then compared the performance of the different training groups on an onroad evaluation. The UFOV training group improved their UFOV performance and had fewer risky driving maneuvers on the onroad evaluation, relative to the driving-simulator training group. Thus, even though the UFOV simulation training only involves part-task simulation, it has been found to generalize to realistic behaviors that matter in naturalistic driving conditions.

Depending on the psychological construct of interest, part-task simulations can often provide informative cues to performance in real-world conditions. Another example of the efficacy of part-task simulations can be found in the work of Rizzo and colleagues studying patients with neurological damage.[10] They have found that a number of part-task driving simulations can be used to both predict and train safe-driving behaviors in older adults.

Whole Task

More commonly, simulation experiments utilize tasks that are designed to contain all primary aspects of task performance. High-fidelity flight and driving simulations are examples of this. Operators are asked to behave as much as possible as they would in more naturalistic conditions and the simulation is designed to contain all stages of operation (e.g., take-off, level flight, and landing while responding to air traffic control communications, weather disturbances, etc.). Whole-task simulations have the highest face validity and more easily generalize to naturalistic operations. However, they are generally more expensive and time-consuming and therefore may not always be required. Deciding between using part-task or whole-task simulations largely depends on the construct under investigation.

Simulator Sickness

Simulator sickness, or as it is sometimes referred to—simulator adaptation syndrome—is a challenge for researchers. A certain percentage of research participants will experience mild to severe discomfort from operating the simulator. It is generally thought that simulator sickness is largely a result of a mismatch between visual and vestibular cues. The visual display may be indicating a great degree of movement, but the operator's vestibular system is providing incongruent cues. The operator may not be experiencing any motion at all, or the motion experienced may not be aligned with the extent of the visual movement perceived. Different populations are more susceptible to simulator sickness with females and older adults being the most vulnerable. People with extensive video game experience generally perceive the least amount of simulator sickness. A number of simulator sickness questionnaires have been developed and undergoing some screening prior to experimentation can rule out those that are highly prone to simulator sickness. One commonly used screening tool is the Simulator Sickness Questionnaire, or SSQ.[11] It can be used as either a pre- or a postscreening tool. Experimenters should remain vigilant for the appearance of simulator sickness throughout the experiment.

Adaptation Period

Because simulators differ from participants' experience in more naturalistic settings, it is important to provide an adaptation period. An adaptation period allows the operator to get accustomed to the handing characteristics of the simulation. Adaptation times may vary depending on the complexity of the simulation, the mismatch between simulator and actual controls, and for different maneuvers within a given simulation. For example, Ronen and Yair[12] found that experienced drivers needed different adaptation periods for different types of roadways depending on the complexity of the roadway curvature.

BEHAVIORAL RESEARCH METHODS

Most of the research design methods necessary for research in any environment are also important to maintain in simulation research. Some key factors are discussed that are important for research in general, and not limited to simulation research.

Baseline Performance

Establishing baselines of performance and physiological metrics is an important practice of experimental design. If the simulation is long, baselines should be taken at several points throughout the experiment. In the case of long simulations, baselines obtained only at the beginning of an experimental protocol may not be representative of task performance because operators may still be getting accustomed to or learning the simulation tasks (e.g., vehicle handing and control characteristics) or may be nervous in anticipation of the upcoming experiment. Baselines obtained at the end of an experimental protocol may be influenced by fatigue. Ideal baseline conditions are sampled periodically throughout the protocol and averaged together or compared to ensure that neither learning nor fatigue effects contaminate baselines.

Establishing effective baseline metrics enable comparisons that are more effective across heterogeneous participant populations. Baseline metrics can be subtracted from metrics obtained during the experimental sessions to provide a delta or change score that accounts for underlying variability between participants.

Control Conditions

Baseline metrics can sometimes serve as control conditions. However, some research questions will need to compare a treatment or condition to a control condition. For example, if one is interested in the impact of a new display or countermeasure, it is essential to be able to compare the condition including the new display to a control condition without the display. For example, Dehais, Causse, and Trembaly[13] found that a countermeasure involving information removal from an unmanned ground-vehicle display reduced attentional tunneling and both improved performance and reduced heart rate relative to a control condition with the standard display. Without the use of the control condition, the researchers would have been unable to determine the added benefit of the countermeasure.

CONCLUSION

Simulators offer the advantage of being able to safely and relatively inexpensively examine complex behaviors and neuroergonomic metrics. Further, they afford much more control over experimental variables than can be obtained in conditions that are more natural. At the same time, it is important to ensure there is adequate validity of the simulated conditions to enable generalization of experimental results to more-naturalistic conditions. Including baseline physiological and performance measures and appropriate control conditions are important experimental design factors for all neuroergonomics research.

REFERENCES

1. Parasuraman R. Neuroergonomics: research and practice. *Theoretical Issues in Ergonomics Science* 2003;**4**:5–20.
2. Parasuraman R. Neuroergonomics: brain, cognition, and performance at work. *Current Directions in Psychological Science* 2011;**20**(3):181–6. https://doi.org/10.1177/0963721411409176.
3. Allen R, Park G, Cook M. Simulator fidelity and validity in a transfer-of-training context. *Transportation Research Record: Journal of the Transportation Research Board* 2010;**2185**:40–7. https://doi.org/10.3141/2185-06.
4. Baldwin CL. Verbal collision avoidance messages during simulated driving: perceived urgency, alerting effectiveness and annoyance. *Ergonomics* 2011;**54**(4):328–37. https://doi.org/10.1080/00140139.2011.558634.
5. Baldwin CL, May JF. Loudness interacts with semantics in auditory warnings to impact rear-end collisions. *Transportation Research Part F-traffic Psychology and Behaviour* 2011;**14**(1):36–42. https://doi.org/10.1016/j.trf.2010.09.004.
6. Engström J, Aust ML, Viström M. Effects of working memory load and repeated scenario exposure on emergency braking performance. *Human Factors* 2010;**52**(5):551.
7. Rogers RO, Boquet A, Howell C, DeJohn C. *An experiment to evaluate transfer of low-cost simulator-based upset-recovery training.* Washington, DC: Washington, DC: Federal Aviation Administration, Office of Aviation Medicine; 2009.
8. Goodwin LD, Leech NL. The meaning of validity in the new "standards for educational and psychological testing": implications for measurement courses. *Measurement and Evaluation in Counseling and Development* 2003;**36**(3):181–91.
9. Roenker DL, Cissell GM, Ball KK, Wadley VG, Edwards JD. Speed-of-processing and driving simulator training result in improved driving performance. *Human Factors* 2003;**45**(2):218–33.
10. Lees MN, Cosman JD, Lee JD, Fricke N, Rizzo M. Translating cognitive neuroscience to the driver's operational environment: a neuroergonomic approach. *The American Journal of Psychology* 2010;**123**(4):391.
11. Kennedy RS, Lane NE, Berbaum KS, Lilienthal MG. A simulator sickness questionnaire (SSQ): a new method for quantifying simulator sickness. *International Journal of Aviation Psychology* 1993;**3**(3):203–20.
12. Ronen A, Yair N. The adaptation period to a driving simulator. *Transportation Research Part F: Psychology and Behaviour* 2013. https://doi.org/10.1016/j.trf.2012.12.007.
13. Dehais F, Causse M, Tremblay S. Mitigation of conflicts with automation. *Human Factors: The Journal of Human Factors and Ergonomics Society* 2011;**53**(5):448–60. https://doi.org/10.1177/0018720811418635.

Chapter 9

Neuroergonomics for Aviation

Daniel E. Callan[1,2], Frédéric Dehais[2]

[1]*Center for Information and Neural Networks (CiNet), National Institute of Information and Communications Technology (NICT), Osaka University, Osaka, Japan;* [2]*ISAE-SUPAERO, Université de Toulouse, Toulouse, France*

INTRODUCTION

Aerospace cerebral experimental science (ACES) is the neuroergonomic investigation of how the brain works in natural complex real-world environments involving aviation and space operations. Neuroergonomics is defined as the study of brain structure and function in relation to human cognition and behavior in real-world settings.[1] Although the primary focus of ACES is related to the neuroergonomic investigation of processes underlying aviation and space operations, the paradigm has far-reaching implications relevant to all neuroscience and may constitute a shift in the way research is conducted and discoveries are made about global neural processing. Two primary goals of neuroergonomics are the following: (1) To determine the interactive neural processes underlying perception, motor control, cognition, and emotion occurring in the context of robust real-world situations. (2) To develop neural-based technology that can be implemented in real-world situations to improve human performance. It is maintained that this neuroergonomic approach to neuroscience will afford insight into underlying processes and provide a basis for developing technology that is not possible using standard reductionist methodology.

CHALLENGES

The standard reductionist experimental method has been instrumental in discovering many aspects of how the brain caries out specific processes. The strength in this approach lies in its ability to manipulate independent variables to better control potential confounds. Most experimental research based on this approach break stimuli and tasks down into very basic elements to more easily understand the underlying neural processes and provide better control over confounds. It is assumed that more complex perceptual, motor, and cognitive processes, even in real-world situations, can be understood by combining these basic elements. This is an assumption yet to be verified and is, perhaps, untrue.

Although there are considerable advantages for the use of the standard reductionist approach with regard to experimental manipulation and control, there are also considerable disadvantages, especially in the context of development of neuroergonomic technology. The tasks and stimuli used in standard experimental research are far removed from real-life experience. They are not ecologically valid. The only time a person is likely to experience the conditions presented in these experiments is in the laboratory. Additionally, another disadvantage of the standard approach is that the experimental tasks are usually not engaging, causing degradation of data due to fatigue. Even subjects with the best intentions and motivation have difficulty truly engaging in the typical experimental tasks employed. Differential engagement in experimental tasks can lead to differential activity related to arousal instead of the condition under investigation. As one subject remarked "Why are your experiments so boring it is like torture. All you are recording in your experiment is brain activity involved with being bored to death versus sleeping."

Would it not be nice if we could investigate neural processes of complex engaging real-world tasks and have good experimental control? There are several advantages afforded by such a neuroergonomic paradigm. Experiments that are more engaging result in a greater degree of motivation in subjects. Therefore, there is less degradation of data due to fatigue. Because the experiments are engaging and the subjects are motivated, longer experiments can be conducted allowing for larger data collection. Additionally, the results can be directly applied to real-world conditions.

In stark contrast to the standard experimental approach, the neuroergonomic approach maintains that to investigate complex real-world behavior, it is necessary to understand the processes within the context of the underlying interacting brain networks rather than under reduced isolated conditions that only occur in the laboratory.[2] Multiple brain-imaging and stimulation methods (electroencephalography [EEG], magnetoencephalography [MEG], functional near infrared spectroscopy [fNIRS], functional magnetic resonance imaging [fMRI], transcranial magnetic stimulation [TMS], and transcranial direct-current stimulation [tDCS]) are utilized. The objective of neuroergonomics research, in general, is to determine the neural correlates of perceptual, motor, and cognitive processing as well as mental states (including alertness, fatigue, workload, and anxiety) that are difficult to quantify behaviorally. By understanding the underlying neural processes in the context of complex real-world tasks, brain–computer

Neuroergonomics. https://doi.org/10.1016/B978-0-12-811926-6.00009-9
Copyright © 2019 Elsevier Inc. All rights reserved.

interfaces (BCIs) can be used to control adaptive automation and give feedback to modulate brain activity and behavior to facilitate learning, situational awareness, and decision-making to promote performance, safety, efficiency, and well-being.

Human operations in aviation and space constitute an ideal paradigm to implement this neuroergonomic approach. Unlike many human activities that involve running, jumping, throwing, etc., in which there is considerable movement of the body, piloting an aircraft/spacecraft involves control of multiple degrees of freedom with relatively little movement of the hands and feet. This is critical when using high-resolution brain-imaging techniques such as MEG and fMRI that are highly susceptible to movement artifacts with some body motions not being possible during scanning at all. It is important to be able to use these high-resolution brain-imaging techniques under simulated real-world situations to determine the relevant underlying brain processes involved with the task. This information can serve to guide and constrain analysis of brain processes made with highly mobile brain-recording devices such as EEG and fNIRS. Flight simulation programs allow for control and output of hundreds of parameters in real time that can be used for experimental manipulation and analysis. This level of control allows one to design experiments with the ability to address potential task-relevant and extraneous variables. The addition of motion platform-based flight simulation and force-feedback controls adds an additional level of similarity with real-world conditions such that brain–computer interface machine-learning procedures can be developed that may transfer successfully into operation in real aircraft. Because of the diversity of tasks involved in aviation and space operations, diverse topics can be investigated ranging from motor control, attention, learning, alertness, fatigue, workload, decision-making, situational awareness, anxiety, etc. This diversity will allow for better understanding of how the brain works in real-word situations and will provide for a large number of potential neuroergonomic applications to be developed. One final reason why aerospace is an ideal paradigm to investigate brain processes under real-world conditions is that Aviation/Aerospace is one of the biggest industries in the world with the consequences of failure being quite severe. With technological advancement in aircraft/spacecraft, it is necessary for the application of neuroergonomic technology to improve the synthesis of human and machine to enhance performance and safety.

It is important for neuroergonomic approaches to utilize constraints based on neuroscience, instead of purely applying an engineering-based solution. In engineering, a data-driven hypothesis-free approach is often employed to make task-related predictive models based on the brain recordings (e.g., EEG, fNIRS data). However, without utilizing proper constraints based on neuroscience, there is a greater chance of contamination by artifacts instead of true brain processes. This could inherently reduce the ability of the BCI to generalize to novel situations. Without utilizing constraints from neuroscience research, crucial features may not be integrated into the analysis to make successful predictions that will generalize from the laboratory to real-world situations.

ELECTRO-ENCEPHALOGRAPHY

Several experiments have been conducted using EEG in simulated and real aviation and space operations.[3–10,10a] One area of extensive research has been to determine neural correlates that assess and predict mental workload. An extensive review of the literature related to workload in aircraft pilots is given in Borghini etal.,.[8] Their review indicated that high mental workload is generally characterized by an increase in EEG power in the theta band (4–8 Hz) and a decrease in alpha-band power (8–15 Hz). Furthermore, a transition between high mental workload and mental fatigue is characterized by increased EEG power in theta as well as delta (<4 Hz) and alpha bands.[8] Extensive research using aviation and space-related tasks with high-resolution brain imaging (fMRI and MEG) is necessary to better determine the brain regions and underlying neural processes involved with mental workload and transition to mental fatigue. Research into the relationship between attention, working memory, and affective states such as arousal and drive will bring considerable insight into an understanding and functional specification of mental workload that goes beyond simple "capacity" definitions.

For EEG to be effectively utilized in aviation and space operations it must be highly portable, easy to wear, comfortable, wireless, and use dry sensors.[7,10a,11] It has been demonstrated that dry-wireless EEG can be utilized in motion-platform simulated and real inflight aviation-related situations despite the considerable inherent vibration and noise. Utilizing artifact cleaning (Automatic Subspace Reconstruction)[12]; and removal (Independent Component Analysis)[13]; techniques, it was possible to train a classifier to detect the presence or absence of an audio stimulus with around 79.2% predictive performance even in an open cockpit biplane inflight with considerable vibration, wind, acoustic noise, and physiological artifacts.[11] Additionally, the ability to detect the occurrence of pilot-induced oscillations with 79% accuracy inflight with dry-wireless EEG has also been successful.[14]

FUNCTIONAL NEAR INFRA RED SPECTROSCOPY

Another approach to consider for neuroergonomics aviation research is the use of fNIRS. fNIRS is a noninvasive and easy-to-use optical brain-imaging device that is suitable to monitor cortical activity under highly ecological settings. Since the work of Takeuchi,[15] this technique has gained momentum to measure pilots' cognitive performance in flight simulator[16–18,18a]; or actual flight conditions[19] (Kobayashi, Tong and Kikukawa, 2002).[20] It can provide objective measurements to assess pilot's training[21] or system design,[22] but as for EEG, most of the efforts are put into the investigation of the neural correlates of mental workload. Several studies have pointed out that changes in oxygenated-hemoglobin concentration in the prefrontal cortex are relevant markers of mental workload variation.[16–18] Interestingly enough, the spatial resolution of fNIRS allows the measure of specific brain areas such as the dorsolateral prefrontal cortex in which disengagement predicts drop in performance.[23–25] Eventually, the implementation of fNIRS-based BCI in ecological settings remains challenging, but the design of adapted-filtering techniques[26] allow discrimination of different levels of working-memory loads with up to 75% accuracy in a motion flight simulator. However, one perspective is the combination of EEG and fNIRS to offer a unique insight on the neurovascular coupling[27] to better understand and predict the pilot's performance.

Brain Computer Interface and Neuro-Adaptive Technology

Interfaces and automation (artificial intelligence) will continue to grow in augmenting human information processing and communication both among individuals and between individuals and machines. For seamlessly efficient interaction, it is necessary to develop intelligent interfaces that optimally deliver relevant information and control automation based on environmental demands and decoded brain/mental states of the user. Neuroadaptive automation based on integrating decoded operator neural states (using BCI technology) in relation to situation assessment is used to enhance overall system performance, safety, and efficiency. The ability to utilize BCI technology practically in neuroergonomic applications requires that it does not interfere with the normal operation of the task, does not increase workload, and improves overall system performance/efficiency in some manner. Active BCIs (often demanding some type of mental imagery) require extensive concentration and increase workload without performance benefits in task operation carried out by normal means.[7,28] For these reasons it is maintained that passive BCI that utilizes spontaneous neural activity related to the task is more appropriate for neuroergonomic applications for aviation and space operations (see Coffey etal.).[7]

One example of neuroadaptive automation that can improve human performance using spontaneous brain activity on a piloting task is given in Callan etal.[29] The objective of this research was to design neuroadaptive automation that can decode motor intention in response to an unexpected perturbation in flight attitude while ignoring ongoing motor activity related to piloting the airplane. The goal was not to take control away from the pilot but rather to facilitate the response speed of the pilot through the use of automation. The BCI was trained on a simple task and was found to be able to generalize to more complex tasks with the ability to differentiate between motor intention to an unexpected perturbation from that used during normal maneuvering. The neuroadaptive automation was able to enhance the response speed (to superhuman levels in some cases) to recover from the perturbation without additional workload utilizing only brain activity naturally occurring during the perceptual motor-piloting task.[29] Although this experiment was conducted offline, it demonstrates the feasibility of utilizing spontaneous brain activity naturally occurring during a task for neuroergonomic applications.

NEUROSTIMULATION

One of the primary components of neuroergonomic research is to determine brain structures and functions underlying perceptual, motor, and cognitive processes that occur during real-world type situations. There have been several experiments carried out using high-resolution brain imaging concerning aviation tasks to learn about underlying neural processes involved with the task and/or skill under investigation.[2,30–33,33a,b] Insight into the neural processes underlying transcranial direct-current stimulation (tDCS; shown to enhance human abilities in a neuroergonomic context)[34] to modulation of resting-state brain activity was investigated using simultaneous fMRI and tDCS on an aviation-related visual search task.[35] It was found that the degree of functional connectivity from the site of stimulation in the precuneus to the substantia nigra predicts future enhancement in visual performance induced by tDCS.[35] The substantia nigra is part of the dopaminergic system and is involved with value-dependent learning. This study gives insight to the possible neural mechanisms by which tDCS enhances human performance.

CONCLUSION

This chapter gave a brief description of neuroergonomics for aviation. Because of the limited space, a focus was given to our own research. There are considerable contributions to this field of neuroergonomics that are moving closer to the goal of establishing a paradigm shift in the way neuroscience research is conducted. That is, an understanding of the how the brain functions in natural settings will lead to development of technology to improve human performance, efficiency, safety, and well-being.

REFERENCES

1. Parasuraman R, Rizzo M. *Neuroergonomics: the brain at work*. New York (NY): Oxford University Press; 2008.
2. Callan D, Gamez M, Cassel D, Terzibas C, Callan A, Kawato M, Sato M. Dynamic visuomotor transformation involved with remote flying of a plane utilizes the 'Mirror Neuron' system. *PLoS One* 2012;**7**(4):1–14.
3. Sem-Jacobsen C, Nilseng O, Patten C, Eriksen O. Electroencephalographic recording in simulated combat flight in a jet fighter plane. *Electroencephalography and Cinical Neurophysiology* 1959;**11**:154–5.
4. Maulsby R. Electroencephalogram during orbital flight. *Aerospace Medicine* October 1966:1022–6.
5. Sterman M, Mann C. Concepts and applications of EEG analysis in aviation performance evaluation. *Biological Psychology* 1995;**40**:115–30.
6. Wilson G. An analysis of mental workload in pilots during flight using multiple psychophysiological measures. *The International Journal of Aviation Psychology* 2002;**12**(1):3–18.

7. Coffey E, Brouwer A, Wilschut E, Erp J. Brain-machine interfaces in space: using spontaneous rather than intentionally generated brain signals. *Acta Astronautica* 2010;**67**:1–11.

8. Borghini G, Astolfi L, Vecchiato G, Mattia D, Babiloni F. Measuring neurophysiological signals in aircraft pilots and car drivers for the assessment of mental workload, fatigue, and drowsiness. *Neuroscience and Biobehavioral Reviews* 2012;**44**:58–75. https://doi.org/10.1016/j.neurobiorev.2012.10.003.

9. Marusic U, Meeusen R, Pisot R, Kavcic V. The brain in micro- and hypergravity : the effects of changing gravity on the brain electrocortical activity. *European Journal of Sport Science* 2014;**14**(8):813–22.

10. Dehais F, Roy, Durantin, Gateau, Callan. EEG-engagement index and auditory alarm misperception: an inattentional deafness study in actual flight condition. In: *AHFE conference, procedia manufacturing*. Elsevier; 2017.

10a. Callan DE, Gateau T, Durantin G, Gonthier N, Dehais F. Disruption in neural phase synchrony is related to identification of inattentional deafness in real-world setting. *Human brain mapping* 2018;**39**(6):2596–2608.

11. Callan D, Durantin G, Terzibas C. Classification of single-trial auditory events using dry-wireless EEG during real and motion simulated flight. *Frontiers in Systems Neuroscience* 2015;**9**(11):1–12. https://doi.org/10.3389/fnsys.2015.00011.

12. Mullen T, Kothe C, Chi YM, Ojeda A, Kerth T, Makeig S, Cauwenberghs G, Jung TP. Real-time modeling and 3D visualization of source dynamics and connectivity using wearable EEG. In: *EMBC, 25th annual international conference of the IEEE*. 2013. p. 2184–7.

13. Delorme A, Makeig S. EEGLAB: an open source toolbox for analysis of single-trial EEG dynamics. *Journal of Neuroscience Methods* 2004;**134**:9–21.

14. Scholl C, Chi Y, Elconin M, Gray W, Chevillet M, Pohlmeyer E. Classification of pilot-induced oscillations during in-flight piloting exerceises using dry EEG sensor recordings. In: *2016 38th annual international conference of the IEEE engineering in medicine and biology society (EMBC), Orlando, FL, 2016*. 2016. p. 4467–70. https://doi.org/10.1109/EMBC.2016.7591719.

15. Takeuchi Y. Change in blood volume in the brain during a simulated aircraft landing task. *Journal of Occupational Health* 2000;**42**(2):60–5.

16. Çakır MP, Vural M, Koç SÖ, Toktaş A. Real-time monitoring of cognitive workload of airline pilots in a flight simulator with fNIR optical brain imaging technology. In: *International conference on augmented cognition*. Springer International Publishing; July 2016. p. 147–58.

17. Gateau T, Durantin G, Lancelot F, Scannella S, Dehais F. Real-time state estimation in a flight simulator using fNIRS. *PLoS One* 2015;**10**(3):e0121279. https://doi.org/10.1371/journal.pone.0121279.

18. Ayaz H, Shewokis PA, Bunce S, Izzetoglu K, Willems B, Onaral B. Optical brain monitoring for operator training and mental workload assessment. *Neuroimage* 2012;**59**(1):36–47.

18a. Verdière KJ, Roy RN, Dehais F. Detecting Pilot's Engagement Using fNIRS Connectivity Features in an Automated vs. Manual Landing Scenario. *Frontiers in human neuroscience* 2018;**12**:6.

19. Gateau T, Ayaz H, Dehais F. In silico versus over the clouds: On-the-fly mental state estimation of aircraft pilots, using a functional near infrared spectroscopy based passive-BCI. *Frontiers in human neuroscience* 2018;**12**:87.

20. Kobayashi A, Tong A, Kikukawa A. Pilot cerebral oxygen status during air-to-air combat maneuvering. *Aviation, Space, and Environmental Medicine* 2002;**73**(9):919–24.

21. Choe J, Coffman BA, Bergstedt DT, Ziegler MD, Phillips ME. Transcranial direct current stimulation modulates neuronal activity and learning in pilot training. *Frontiers in Human Neuroscience* 2016;**10**.

22. Andéol G, Suied C, Scannella S, Dehais F. The spatial release of cognitive load in cocktail party is determined by the relative levels of the talkers. *Journal of the Association for Research in Otolaryngology* 2017. https://doi.org/10.1007/s10162-016-0611-7.

23. Durantin G, Dehais F, Delorme A. Characterization of mind wandering using fNIRS. *Frontiers in Systems Neuroscience* 2015;**9**.

24. Durantin G, Gagnon JF, Tremblay S, Dehais F. Using near infrared spectroscopy and heart rate variability to detect mental overload. *Behavioural Brain Research* 2014;**259**:16–23.

25. Harrivel AR, Weissman DH, Noll DC, Peltier SJ. Monitoring attentional state with fNIRS. *Frontiers in Human Neuroscience* 2013;**7**:861.

26. Durantin G, Scannella S, Gateau T, Delorme A, Dehais F. Processing functional near infrared spectroscopy signal with a Kalman filter to assess working memory during simulated flight. *Frontiers in Human Neuroscience* 2015;**9**.

27. Mandrick K, Chua Z, Causse M, Perrey S, Dehais F. Why a comprehensive understanding of mental workload through the measurement of neuro-vascular coupling is a key issue for neuroergonomics? *Frontiers in Human Neuroscience* 2016;**10**.

28. Zander T, Kothe C. Toward passive brain-computer interfaces: applying brain-computer interface technology to human-machine systems in general. *Journal of Neural Engineering* 2011;**8**(2). https://doi.org/10.1088/1741-2560/2/025005.

29. Callan D, Terzibas C, Cassel D, Sato M, Parasuraman R. The brain is faster than the hand in split-second intentions to respond to an impending hazard: a simulation of neuroadpative automation to speed recovery to perturbation in flight attitude. *Frontiers in Human Neuroscience* 2016b;**10**(187):1–21. https://doi.org/10.3389/fnhum.2016.00187.

30. Callan D, Terzibas C, Cassel D, Callan A, Kawato M, Sato M. Differential activation of brain regions involved with error-feedback and imitation based motor simulation when observing self and an expert's actions in pilots and non-pilots on a complex glider landing task. *Neuroimage* 2013;**72**:55–68.

31. Causse M, Péran P, Dehais F, Caravasso CF, Zeffiro T, Sabatini U, Pastor J. Affective decision making under uncertainty during a plausible aviation task: an fMRI study. *Neuroimage* 2013;**71**:19–29.

32. Ahamed T, Kawanabe M, Ishii S, Callan D. Structural differences in gray matter between glider pilots and non-pilots. A voxel based morphometry study. *Frontiers in Neurology* 2014;**5**(248):1–5.

33. Adamson M, Taylor JL, Heraldez D, Khorasani A, Noda A, Hernandez B, Yesavage JA. Higher landing accuracy in expert pilots is associated with lower activity in the caudate nucleus. *PLoS One* 2014;**9**(11):e112607. https://doi.org/10.1371/journal.pone.0112607.

33a. Durantin G, Dehais F, Gonthier N, Terzibas C, Callan D. Neural signature of inattentional deafness. *Human Brain Mapping* 2017;**38**(11):5440–5455.

33b. Falcone B, Wada A, Parasuraman R, CallanD. Individual differences in learning correlate with modulation of brain activity induced by transcranial direct current stimulation. *PLoS One* 2018;**13**(5):e0197192. https://doi.org/10.1371/journal.pone.0197192.

34. Parasuraman R, McKinley R. Using noninvasive brain stimulation to accelerate learning and enhance human performance. *Human Factors* 2014;**56**(5):816–24.

35. Callan D, Falcone B, Wada A, Parasuraman R. Simultaneous tDCS-fMRI identifies resting state networks correlated with visual search enhancement. *Frontiers in Human Neuroscience* 2016a;**10**(72):1–12. https://doi.org/10.3389/fnhum.2016.00072.

Chapter 10

MoBI—Mobile Brain/Body Imaging

Evelyn Jungnickel[1], Lukas Gehrke[1], Marius Klug[1], Klaus Gramann[1,2]

[1]Biological Psychology and Neuroergonomics, Technische Universitaet Berlin, Berlin, Germany; [2]Center for Advanced Neurological Engineering, University of California San Diego, San Diego, CA, United States

INTRODUCTION

Mobile brain/body imaging (MoBI) is a method to record and analyze brain dynamics and motor behavior under naturalistic conditions. It is thus suitable to investigate a wide range of scientific problems, including analyses of human brain dynamics with the aid of information derived from movement and studies with an interest in motor behavior using brain imaging as an additional source of information. An integrative approach to brain and movement dynamics distinguishes MoBI from mobile electroencephalography (EEG) studies that do not consider body movements as an informative input allowing understanding of the relationship of movement, cognition, and brain dynamics. Mobile EEG often uses low-spatial-density recordings to improve mobility, and analyzes approaches based on the sensor level.[1,2]

Investigation of the brain as the organ that organizes complex behaviors to optimize the consequences of our actions effectively and efficiently in a multiscale and ever-changing environment[3–5] is based on the concept of embodied cognition. The embodied cognition paradigm proposes that cognitive processes are mainly rooted in and aimed at the body's interactions with the world.[6] This view is supported by the fact that brain dynamics change when insects, mammals, and humans actively move in the real world or in virtual reality (VR) environments as compared to rest or stationary settings.[7–10] It is thus reasonable to assume that movements in the workplace or other naturalistic conditions are accompanied by changes in brain dynamics that are as yet unknown. To understand and reliably estimate the brain's dynamic changes during active movement, MoBI combines brain imaging with recordings of motor behavior and ongoing environmental changes. In this regard, movement should be as unrestricted and natural as possible to facilitate investigation of the relation between the musculoskeletal system and the brain as well as the underlying principles of motor action, control, and planning.

To enable inferences about the relationship of behavior, cognition, and brain activity, synchronized recordings of brain and movement dynamics are fundamental.[4] MoBI accounts for the shortcomings of stationary experimental setups and utilizes mobile hardware solutions to measure brain activity and body movements. A dedicated software architecture allows the recording of multiple datastreams wirelessly, and synchronizes them online for near real-time feature extraction or recording of all datastreams for later analysis.[11]

Importantly, allowing active behavior introduces nonbrain activity like eye movements and muscle activity as well as mechanical artifacts that mix with the signals generated in the brain. Thus to analyze brain dynamics or muscle and eye movement dynamics as well as their contributions to the sensor level, the different signal sources have to be dissociated. This renders pure sensor-based analyses approaches infeasible.[12–14] Instead, data-driven analysis approaches like independent component analysis (ICA)[4a,15] are necessary to derive meaningful information from the volume-conducted sensor data. Once the sources that contribute to the sensor data are decomposed, analyses can focus on sources of interest to investigate brain and nonbrain dynamics in the time and frequency domain,[16] to back-project sources of interest to the sensor level,[9] or to use other approaches such as network analysis (e.g., Hassan et al., 2014)[16a].

Summing up, MoBI can be defined as a multimethod approach to imaging human brain dynamics during movement in and interaction with the environment. This requires adequate hardware and software solutions to record brain dynamics, motor behavior, and other datastreams simultaneously and new processing methods to provide sensor and source-level analyses.

PHYSIOLOGICAL PRINCIPLES

The strong connection between the human brain and body as predicted by the embodiment paradigm is supported by strong anatomical links between both systems. There is a direct connection from each motor unit of the musculoskeletal system through spinal neurons to the brainstem and the motor cortex, which generates all movement patterns.[17] Most motor commands are sent

Neuroergonomics. https://doi.org/10.1016/B978-0-12-811926-6.00010-5
Copyright © 2019 Elsevier Inc. All rights reserved.

59

through the primary motor cortex, with more complex movements being generated in the premotor cortex and supplementary motor areas.[18] The motor cortex has direct connections to the cerebellum, which supports balance, coordination, and fine-tuning of movements, and the basal ganglia, which are responsible for activating and suppressing movement plans based on intentions managed in the prefrontal cortex.[19,20] Several other brain regions play a role in human movement and locomotion.

Sensory input filtered by the thalamus is forwarded from the primary sensory cortices to the posterior parietal cortex, where spatial aspects are processed for movement planning.[21] Movement plans, and thus very likely also movement execution, impacts perception, such that sensory input is selected that best supports action goals.[22] This action-dependent modulation of perception is reflected in an anatomical overlapping of action-related and perception-related brain areas Rizzolatti, 2002[22a]. In addition to a modulation of brain dynamics accompanying changes in perception based on action plans, movement execution is associated with movement-related sensory feedback that is used to control and adapt movement parameters. Kinematics allow an accurate quantitative description of movement parameters, including displacements, velocities, and accelerations of body parts and the body's center of mass.[23] Kinesthetic information, including proprioception and feedback from the vestibular system, is processed and represented in different brain regions including the frontal and parietal cortices (e.g., Andersen et al., 1997)[23a]. As a consequence, movement does not only impact perception and the accompanying brain dynamics via top-down regulation, but also produces sensory feedback that is processed by the very brain structures that control movement. Investigating these cortical modulations requires brain-imaging methods that allow movement of the participant.

INSTRUMENTATION

Unrestrained movement presupposes lightweight mobile sensors and wireless data transfer. In brain imaging, only EEG (see chapter 2) and functional near-infrared spectroscopy (see chapter 3) have sensors suitable for measuring brain activity during active movement. To facilitate adequate analysis of brain and body dynamics, a variety of input and output datastreams must be collected and interpreted. Examples are visual, auditory, and tactile stimulation as inputs, button presses, physical motion capture, force measures, or scene and gaze tracking as outputs, and EEG or fNIRS as mobile brain-imaging data. Recently an increasing interest in head-mounted VR displays has evolved based on the possibility of controlling visual input during movements while being more naturalistic than classic two-dimensional displays. Because room-scale VR uses tracking systems intrinsically, they carry motion-capture data which can be collected easily. Accessory trackers extend the available head and hand tracking, and allow multiple rigid body parts to be tracked simultaneously with six degrees of freedom (position and orientation). Additionally, equipping VR glasses with an eye-tracking system is possible (e.g., Tobii). VR environments for experiments can be created with different game engines (e.g., Unity, WorldViz Vizard, Unreal) which allow implementation of datastreaming options in their routines. Taken together, VR facilitates a rich and comprehensive MoBI environment with visual, auditory, and tactile stimulus modalities as well as motion capture, scene capture, and possible eye-tracking data collection options.

For meaningful interpretation of brain activity, movement, and performance data, it is necessary to preserve the context of the recorded modalities, i.e., the temporal and potentially spatial placing of each acquired sample. This challenge can be met by time-stamping each sample to add the time of the measurement to each acquired data point. However, different machines for creating and storing the datastreams have different clock offsets. In addition, hardware constraints and changing CPU load can cause temporal jitter, missed sample points, and, as a consequence, irregular time-stamps. To validate the measured data it is necessary to measure and store for each datastream the clock offset of this stream's time-stamps, the age of sample (i.e., the delay between the physical occurrence of, for example, an electrode potential to the time-stamp of its digitized sample), and the amount of jitter in this age of sample. If the underlying time series are presumed to be regularly sampled (e.g., EEG data with 1000 Hz sampling rate), temporal jitter can be met with equal spacing of time-stamps.

The Lab Streaming Layer (https://github.com/sccn/labstreaminglayer) and OpenViBE[24] software platforms provide solutions for the unified collection of several data types, from different human interface devices over consumer and research-grade EEG systems to various motion-capture systems, eye-tracking technology, and simple text string markers (libraries for MATLAB, C++, C#, Python, and Java exist). Storage and exchange of multimodal time series data are facilitated by general-purpose data containers, e.g., the Extensible Data Format (https://github.com/sccn/xdf) and the European Data Format,[25] with potential for extensive and flexible metainformation specification of each recorded time series.

Signal Processing and Analysis Approaches

The aim of MoBI data analyses is to understand how the brain optimizes the outcome of behavior on all time scales in a natural and stimulus-rich environment.[5] To understand and model the relationship between the observed brain activity and behavior, i.e., what the brain controls, modality-specific analyses as well as early and late data-fusion methods are used.

Movement analyses. Analyses of measured motor behavior progress from registering changes in position and orientation of the measured body parts and their temporal derivatives, i.e., velocity, acceleration, and jerk, to more abstract representations of movement. Standard preprocessing procedures include the interpolation of missing sample points and appropriate filtering of the time series to address jitter and systematic noise.[23,26] Subsequently movement parameters, e.g., peak velocity, peak acceleration, and time to peak acceleration, can be extracted and investigated.[27] However, it is reasonable to assume that the brain does not represent position and orientation of an object in a mathematical three-dimensional space, and does not see movements in a Cartesian coordinate or quaternion representation.[28] Thus the transformation of measured positions and orientations to meaningful representations, i.e., estimating the inverse kinematics based on a human biomechanical model,[23,29] extraction of gait cycles and other abstract movements, and modeling the physical and muscular forces of human movements,[30,31] is central to the MoBI analysis pipeline. Existing software[32–34] facilitates the analysis. Processing other measurements of conscious and autonomic functions of the body, including eye tracking, electrocardiography (ECG), and galvanic skin response (GSR) among others, is discussed elsewhere.[35,36]

EEG analyses. The recorded EEG data can be analyzed on the sensor level (the respective electrode data channels) or the source level, for which spatial filters can be computed that decompose the recorded data into EEG-effective brain, muscle, eye, noise, and other sources.[37] Very often, different algorithms for independent component analysis (ICA) are used to find components that are maximally statistically independent from each other.[15] However, other possibilities like spatio–spectral decomposition[38] and joint decorrelation[39] exist, which maximize the signal-to-noise ratio. In conjunction with accurately measured locations of electrodes and forward models with realistic conductivity factors of different tissues and skull, the spatial filters for each component can also be interpreted with respect to their presumed location in the brain. If desired, the decomposition can help in selecting and rejecting clearly artifactual components of the data, i.e., characteristic patterns of EOG, ECG, EMG, and channel noise, and projecting the remaining components of the data back to the sensor level. This linear mixing model is different for each participant and EEG recording session, however, and the quality of the results may vary greatly with different tasks due to significant noise from cable pull and sway, severe motion artifacts, and other interferences polluting the data. Interpreting the components thus has to be done with care, and helpful guidelines exist.[40,41] The signal decomposition can be improved by appropriate preprocessing pipelines.[13] Unfortunately, noisy data segments frequently appear at points in the time series which are of interest to the experimental paradigm (e.g., arm-reaching movements or steps). Thus rejecting artifacts in a classical fashion may lead to significantly less data to analyze and might distort the results.

Fusion of movement and EEG data. Identifying links between the multimodal information measured in MoBI experiments can be approached from several angles. First, classifying the measurements of ongoing motor behavior allows to make inferences about cognitive states, e.g., arm moving (engaged) versus hanging by the side (disengaged). Subsequently, brain-imaging data can be interpreted and differentiated with respect to cognitive states[9] and decoded online.[42] Second, extracting discrete time points from the preprocessed continuous body movement data, e.g., peak velocity or acceleration, enables the use of event-related mass univariate analysis procedures, as is common throughout the EEG and fMRI communities. However, MoBI research is ultimately interested in finding models that represent the continuous interaction, i.e., coupling, of observed brain source activity and ongoing motor behavior. Describing multimodal interactions of interest may prove difficult due to potential noninstantaneous and nonlinear coupling. Considerations when choosing an applicable analysis method include the data preprocessing prerequisites in each individual data domain with its domain-specific intricacies, assumptions made about the data, and proper understanding and modeling of the data interaction (coupling) of the measured phenomenon under investigation. Recent works on applying domain-independent multimodal early data-fusion methods to neuroimaging data have focused on the simultaneous recording of EEG and fMRI data.[43–45] Due to the generality of these approaches, applications to MoBI data are a promising endeavor for future research.

APPLICATIONS

As MoBI is motivated from different perspectives it covers a wide range of possible applications. Ongoing developments in EEG hardware now allow for fully untethered capture of brain electrical potentials, and make it lightweight and energy efficient enough to support data collection "in the wild."[46,47] Simultaneously, dedicated motion-capture equipment has become very affordable. As noted above, consumer-grade head-mounted VR systems and full body suits equipped with inertial measurement units (IMUs) at each joint can be used to collect full body motion capture. Here we describe two EEG studies demonstrating why imaging human brain dynamics during natural cognition with unrestricted movement and interaction with the environment is beneficial above and beyond classical neuroscientific methods.

The first study, by Jungnickel and Gramann,[9] used high-density EEG recordings synchronized to motion capture to investigate the brain dynamics in participants physically interacting with a dynamically changing system. Even though physical interaction with a system is a common task in many workplaces (e.g., assembly lines), there is no knowledge of how physical activity influences cognition (and vice versa) and the accompanying brain dynamics. To gain a better understanding, participants had to intercept a sphere that was moving on a large screen in front of them whenever it changed its color to a target color (but not a standard color). The results demonstrate that even during rapid volatile movements the contribution of brain activity, eye movements, and muscle activity can be quantified and the event-related P3 can be analyzed, presupposing extensive data-driven preprocessing. This P3 is increased for pointing as compared to standard button-press responses, indicating different brain dynamics for different behavioral states. The results open up the space for investigating brain dynamics at workplaces that require physical interactions with or adaption to an ever-changing environment. Importantly, they point out the necessity of recording high-density data and movement-context information, and dissociating brain activity from nonbrain activity. Fortunately, this dissociation allows extension of the analysis to examine muscle activity and eye and body movement in relation to particular brain-signal sources.

The second study, by Wagner et al.,[48] combines MoBI with VR in rehabilitation settings by investigating participants walking with a robotic gait orthosis. A 61-channel EEG system and three EOG electrodes were used to record brain activity and eye movement while mechanical foot switches at the foot sole measured the heel strikes during five different and randomized visual feedback conditions. Sources of brain activity were calculated from the preprocessed data using ICA and a clustering algorithm, revealing three clusters in sensorimotor areas. Movement-related interactive feedback provided through VR from a first- or third-person perspective led to different brain dynamics than those provided by no or unrelated visual input or mirroring of the walking participants. These results indicate an enhancement of motor planning, as reflected in decreased spectral power in parietal clusters, when interactive feedback is provided. This emphasizes the relevance of different levels of interaction with the environment and the immersiveness of VR for cognitive processing. To analyze brain dynamics in relation to variable and adaptive gait, the single-trial spectrograms have to be time-warped with a linear interpolation function. This method is one important approach to discover the close linking between body and brain predicted by the embodiment paradigm.

REFERENCES

1. Debener S, Emkes R, De Vos M, Bleichner M. Unobtrusive ambulatory EEG using a smartphone and flexible printed electrodes around the ear. *Scientific Reports* 2015;**5**:16743.
2. Wascher E, Heppner H, Hoffmann S. Towards the measurement of event-related EEG activity in real-life working environments. *International Journal of Psychophysiology* 2014;**91**(1):3–9.
3. Gramann K, Gwin JT, Ferris DP, Oie K, Jung T-P, Lin C-T, et al. Cognition in action: imaging brain/body dynamics in mobile humans. *Reviews in the Neurosciences* 2011;**22**(6). https://doi.org/10.1515/rns.2011.047.
4. Gramann K, Ferris DP, Gwin J, Makeig S. Imaging natural cognition in action. *International Journal of Psychophysiology* 2014;**91**(1):22–9.
4a. Makeig S, Bell AJ, Jung TP, Sejnowski TJ. Independent component analysis of electroencephalographic data. *Adv. Neural Inf. Process. Syst* 1996;**8**:145–51.
5. Makeig S, Gramann K, Jung TP, Sejnowski TJ, Poizner H. Linking brain, mind and behavior. *International Journal of Psychophysiology* 2009;**73**(2):95–100.
6. Wilson M. Six views of embodied cognition. *Psychonomic Bulletin and Review* 2002;**9**(4):625–36. https://doi.org/10.3758/BF03196322.
7. Arenz A, Drews MS, Richter FG, Ammer G, Borst A. The temporal tuning of the Drosophila motion detectors is determined by the dynamics of their input elements. *Current Biology* 2017;**27**(7):929–44.
8. Bohbot VD, Copara MS, Gotman J, Ekstrom AD. Low-frequency theta oscillations in the human hippocampus during real-world and virtual navigation. *Nature Communications* 2017;**8**.
9. Jungnickel E, Gramann K. Mobile brain/body imaging (MoBI) of physical interaction with dynamically moving objects. *Frontiers in Human Neuroscience* June 2016;**10**(306). https://doi.org/10.3389/fnhum.2016.00306.
10. Niell CM, Stryker MP. Modulation of visual responses by behavioral state in mouse visual cortex. *Neuron* 2010;**65**(4):472–9.
11. Kothe C. *Lab streaming layer (LSL)*. Available online at: 2014. https://github.com/sccn/labstreaminglayer.
12. Gramann K, Gwin JT, Bigdely-Shamlo N, Ferris DP, Makeig S. Visual evoked responses during standing and walking. *Frontiers in Human Neuroscience* 2010;**4**:202.
13. Gwin JT, Gramann K, Makeig S, Ferris DP. Removal of movement artifact from high-density EEG recorded during walking and running. *Journal of Neurophysiology* 2010;**103**(6):3526–34.
14. Wagner J, Makeig S, Gola M, Neuper C, Müller-Putz G. Distinct β band oscillatory networks subserving motor and cognitive control during gait adaptation. *Journal of Neuroscience* 2016;**36**(7):2212–26.
15. Bell AJ, Sejnowski TJ. An information-maximisation approach to blind separation and blind deconvolution. *Neural Computation* 1995;**7**(6):1004. https://doi.org/10.1162/neco.1995.7.6.1129.
16. Delorme A, Makeig S. EEGLAB: an open source toolbox for analysis independent component analysis. *Journal of Neuroscience Methods* 2004;**134**: 9–21. https://doi.org/10.1016/j.jneumeth.2003.10.009.
16a. Hassan M, Dufor O, Merlet I, Berrou C, Wendling F. EEG Source Connectivity Analysis: From Dense Array Recordings to Brain Networks. *PLoS One* 2014;**9**(8):e105041. http://doi.org/10.1371/journal.pone.0105041.

17. Sanes JN, Donoghue JP. Plasticity and primary motor cortex. *Annual Review of Neuroscience* 2000;**23**(1):393–415.
18. Porter R, Lemon R. *Corticospinal function and voluntary movement. Monogr. of the Phys Soc No. 45.* Oxford: Clarendon Press; 1993.
19. Desmurget M, Epstein CM, Turner RS, Prablanc C, Alexander GE, Grafton ST. Role of the posterior parietal cortex in updating reaching movements to a visual target. *Nature Neuroscience* 1999;**2**(6):563–7.
20. Graybiel AM, Aosaki T, Flaherty AW, Kimura M. *The basal ganglia and adaptive motor control.* New York then Washington: Science; 1994. p. 1826.
21. Snyder LH, Batista AP, Andersen RA. Coding of intention in the posterior parietal cortex. *Nature* 1997;**386**(6621):167.
22. Witt JK. Action's effect on perception. *Current Directions in Psychological Science* 2011;**20**(3):201–6.
22a. Rizzolatti G, Fogassi L, Gallese V. Motor and cognitive functions of the ventral premotor cortex. *Current opinion in neurobiology* 2002;**12**(2): 149–54.
23. Winter DA. Biomechanics and motor control of human movement. *Motor Control* 2009;**2**. https://doi.org/10.1002/9780470549148.
23a. Andersen RA, Snyder LH, Bradley DC, Xing J. Multimodal representation of space in the posterior parietal cortex and its use in planning movements. *Annual Review of Neuroscience* 1997;**20**:303–30.
24. Renard Y, Lotte F, Gibert G, Congedo M, Maby E, Delannoy V, et al. OpenViBE: an open-source software platform to design, test, and use brain–computer interfaces in real and virtual environments. *Presence* 2010;**19**(1):35–53.
25. Kemp B, Olivan J. European data format "plus" (EDF+), an EDF alike standard format for the exchange of physiological data. *Clinical Neurophysiology* 2003;**114**(9):1755–61. https://doi.org/10.1016/S1388-2457(03)00123-8.
26. Woltring HJ. On optimal smoothing and derivative estimation from noisy displacement data in biomechanics. *Human Movement Science* 1985;**4**(3):229–45. https://doi.org/10.1016/0167-9457(85)90004-1.
27. Butler EE, Ladd AL, LaMont LE, Rose J. Temporal-spatial parameters of the upper limb during a Reach & Grasp Cycle for children. *Gait and Posture* 2010;**32**(3):301–6. https://doi.org/10.1016/j.gaitpost.2010.05.013.
28. Tanaka H. Modeling the motor cortex: optimality, recurrent neural networks, and spatial dynamics. *Neuroscience Research* 2016;**104**:64–71. https://doi.org/10.1016/j.neures.2015.10.012.
29. Enoka RM. *Neuromechanics of human movement.* Human Kinetics; 2008. Retrieved from: http://books.google.com/books?id=2JI04kdV9isC&pgis=1.
30. Shadmehr R. Learning to predict and control the physics of our movements 2017;**37**(7):1–20. https://doi.org/10.1523/JNEUROSCI.1675-16.2016.
31. Shadmehr R, Mussa-Ivaldi S. *Biological learning and control: how the brain builds representations, predicts events, and makes decisions.* Mit Press; 2012.
32. Delp SL, Anderson FC, Arnold AS, Loan P, Habib A, John CT, et al. OpenSim: open-source software to create and analyze dynamic simulations of movement. *IEEE Transactions on Biomedical Engineering* 2007;**54**(11):1940–50. https://doi.org/10.1109/TBME.2007.901024.
33. Ojeda A, Bigdely-Shamlo N, Makeig S. MoBILAB: an open source toolbox for analysis and visualization of mobile brain/body imaging data. *Frontiers in Human Neuroscience* March 2014;**8**:1–9. https://doi.org/10.3389/fnhum.2014.00121.
34. Thompson MR. *Mocap Toolbox: a MATLAB toolbox for analyzing motion captures data.* 2014.
35. Cacioppo J, Tassinary LG, Berntson GG. The handbook of psychophysiology. *Dreaming* 2007;**44**. https://doi.org/10.1017/CBO9780511546396.
36. Holmqvist K, Nyström M, Andersson R, Dewhurst R, Jarodzka H, de Weijer J. *Eye tracking: a comprehensive guide to methods and measures.* OUP Oxford; 2011.
37. Parra LC, Spence CD, Gerson AD, Sajda P. Recipes for the linear analysis of EEG. *NeuroImage* 2005;**28**(2):326–41. https://doi.org/10.1016/j.neuroimage.2005.05.032.
38. Nikulin VV, Nolte G, Curio G. A novel method for reliable and fast extraction of neuronal EEG/MEG oscillations on the basis of spatio-spectral decomposition. *NeuroImage* 2011;**55**(4):1528–35. https://doi.org/10.1016/j.neuroimage.2011.01.057.
39. De Cheveigné A, Parra LC. Joint decorrelation, a versatile tool for multichannel data analysis. *NeuroImage* 2014;**98**:487–505. https://doi.org/10.1016/j.neuroimage.2014.05.068.
40. Chaumon M, Bishop DVM, Busch NA. A practical guide to the selection of independent components of the electroencephalogram for artifact correction. *Journal of Neuroscience Methods* 2015;**250**. https://doi.org/10.1016/j.jneumeth.2015.02.025.
41. Winkler I, Haufe S, Tangermann M. Automatic classification of artifactual ICA-components for artifact removal in EEG signals. *Behavioral and Brain Functions: BBF* 2011;**7**:30. https://doi.org/10.1186/1744-9081-7-30.
42. Zander TO, Krol LR, Birbaumer NP, Gramann K. Neuroadaptive technology enables implicit cursor control based on medial prefrontal cortex activity 2016;**113**(52):1–6. https://doi.org/10.1073/pnas.1605155114.
43. Dähne S, Bießman F, Samek W, Haufe S, Goltz D, Gundlach C, et al. Multivariate machine learning methods for fusing functional multimodal neuroimaging data. *Proceedings of the IEEE* 2015;**103**(9):1–22. https://doi.org/10.1109/JPROC.2015.2425807.
44. Fazli S, Dähne S, Samek W, Bießmann F, Müller K-R, Dahne S, et al. Learning from more than one data source: data fusion techniques for sensorimotor rhythm-based Brain-Computer Interfaces. *Proceedings of the IEEE* 2015;**103**(6):891–906. https://doi.org/10.1109/JPROC.2015.2413993.
45. Uludağ K, Roebroeck A. General overview on the merits of multimodal neuroimaging data fusion. *NeuroImage* 2014;**102**(P1):3–10. https://doi.org/10.1016/j.neuroimage.2014.05.018.
46. Gramann K, Fairclough SH, Zander TO, Ayaz H. *Editorial: Trends in Neuroergonomics* April 2017;**11**:11–4. https://doi.org/10.3389/fnhum.2017.00165.
47. Mehta RK, Parasuraman R. Neuroergonomics: a review of applications to physical and cognitive work. *Frontiers in Human Neuroscience* December 2013;**7**:1–10. https://doi.org/10.3389/fnhum.2013.00889.
48. Wagner J, Solis-Escalante T, Scherer R, Neuper C, Müller-Putz G. It's how you get there: walking down a virtual alley activates premotor and parietal areas. *Frontiers in Human Neuroscience* 2014;**8**.

Chapter 11

Experiments With Participants: Some Ethical Considerations

Catherine Tessier, Vincent Bonnemains

ONERA/DTIS, Université de Toulouse, France

INTRODUCTION

Experiments involving human participants must abide by standards that have been built on the 10 points of the Nuremberg Code (1947) and laid down in the 1964 Declaration of Helsinki and further revisions. However, the way an experiment should be prepared and conducted may bring contradictions and raise ethical questions. This chapter focuses on key steps of an experiment and some related issues. Section 2 focuses on who is involved in the experiment, i.e., the investigator or experimenter on the one hand, and the participants on the other hand. Section 3 highlights some issues linked to the information that is given to the participants. Data issues are dealt with in Section 4. Finally, Section 5 focuses on publication and related ethical and scientific integrity concerns.

WHO IS INVOLVED IN THE EXPERIMENT?

Two kinds of people are involved in an experiment: the investigator and the participants.

The Investigator

The principal investigator of an experiment must be a permanent researcher in the lab, which means that this person is accountable for the experiment even if it is actually conducted by a Ph.D. student, for example. The investigator's (or experimenter's) behavior, vocabulary, and nonverbal language during the experiment may not be neutral as far as the participants' behaviors and performance in the experiment are concerned.

The Participants

As participants are needed to conduct the experiment, who can be recruited and how they will be recruited have to be specified. Neither of these is ethically neutral.

- Who can be recruited: *inclusion* and *exclusion* criteria must be clearly stated, e.g., only right-handed participants with no hearing problems are invited to take part. This raised two issues: the criteria may be intrusive, so an aggregation of criteria may be relevant to avoid any emphasis on a particular ability or inability; and how the criteria are assessed—is it a mere statement of the participant, or an affidavit, or are the criteria checked by some means, e.g., a hearing test? In the latter case, who performs the test and the way the results of the test are processed may raise ethical or even legal issues (for instance, illegal practice of medicine).
- The way participants are recruited (online or newspaper advert, notice, flyer, email, etc.) and the targeted population for the advertisement (people in the street, office colleagues, students at the university, etc.) may not be neutral as far as experimental results are concerned because, for example, all the participants may share common features (age, education, knowledge of the research being conducted, etc.), which may introduce a bias in the results.

Neuroergonomics. https://doi.org/10.1016/B978-0-12-811926-6.00011-7
Copyright © 2019 Elsevier Inc. All rights reserved.

INFORMATION GIVEN TO THE PARTICIPANTS

Before the experiment itself begins, each recruited participant must be given relevant information by the investigator about the purpose and procedure of the experiment and about the potential benefits and risks for themselves. Information should clarify several criteria,[1] the most important ones being:

- freedom: the participant enrols willingly, i.e., with no constraint or pressure;
- justice (or equity): all participants are considered in the same way;
- benevolence (or no nuisance): participants are treated with respect and kindness;
- privacy and confidentiality.

Even when the experiment does not pertain to medical research, an *informed consent* is often signed by the participants (or their legal representatives).

Can the Criteria Really Be Satisfied?

Several ethical issues about these criteria may be raised.

Freedom is likely to be impaired when there is an authority relationship between the experimenter and the participants, for example a professor with his/her own students, a doctor/patients, a manager/subordinates. As far as students are concerned, whether they enrol or not as participants in an experiment conducted by a professor at their university should not be linked in any way to grades, recommendations for application files, or more broadly to whatever they are involved in as students.

Real *equity* may be difficult to guarantee; indeed, experiments often require that participants are split up into different groups, e.g., a control group and several other groups with different experimental conditions. Consequently some participants may be involved in simple tasks with no physical or psychological constraints, whereas others may be involved in complex tasks undertaken in stressful, tiring, or questioning conditions.

Some neuroscience experiments rest on the fact that participants are put in some special states, such as stress, tiredness, boredom, physical or psychological discomfort, strong emotions, etc. For such experiments the *benevolence* criterion may be questioned.

Privacy and confidentiality are discussed in Section 4.

To What Extent Is the Participant Informed?

A first issue is whether clear and simple information is actually compatible with an explanation of all scientific challenges of the experiment.

Furthermore, some experiments may rest on the fact that the participant is deceived about the real purpose of the study, e.g., the well-known Milgram experiment, as telling the truth would impair the scientific issue. In such cases the participant gets information which does not disclose all aspects of the experiment (omission) or which states purposes that are not the actual ones (deception). Even if it is recommended that participants are informed about the real purpose of the experiment immediately after they have taken part, the debriefing may be counterproductive in so far as the participants may feel worried about what they have done (Milgram-type experiment), angry about having been deceived, etc.

Potential Benefits and Risks

Information must include the potential benefits of the experiment to the participant and the foreseeable risks.

As far as benefits are concerned, the consent should at least mention that participants will be informed of the global results of the experiment if they wish. The participant may learn new skills while taking part, which has to be mentioned. Compensation in cash or in kind may be contemplated provided that it is derisory and the same for all participants whatever their role in the experiment, e.g., even if they decide to leave the experiment before it is finished, according to the equity criterion. Indeed, the motivation of the participant should not be linked to compensation, as this would contradict the very notion of voluntary participation.

Foreseeable risks should be minor, e.g., light tiredness, allergy to adhesive strips used to put sensors on the skin, etc. Any induced risks of the experiment should be considered, e.g., a complex task with added stress or fatigue is likely to bring about a feeling of failure and lowered self-confidence, especially when the participant is a professional and the experiment is linked to their skills, e.g., a pilot facing a difficult landing task in a flight simulator. A key ethical question is to balance the benefit/risk ratio: to what extent is it worth provoking the participant's inconvenience for the sake of science?

Incidental Findings

Neuroergonomics experiments may involve fNIRS (functional near-infrared spectroscopy), MRI (magnetic resonance imaging), or other means that are likely to show pathologies of which the participant is unaware; but the experimenter is not competent to interpret images or data outside their own field. Consequently information given to the participant should mention that they should not expect any diagnosis from the experiment. Nevertheless a crucial issue is whether the participant's data should undergo a clinical check.[2]

DATA

The first goal of an experiment with participants is to collect data from the participants. Several issues are raised by data collection, data processing, and data storage. Indeed, a good starting point is to apply the who, what, why, and when test to the collection and storage of personal information.[3]

Which Data Is Really Necessary?

This is the why, i.e., why is the data required?[3]

If data is not necessary for the study, it should not be collected. Indeed, data minimization is especially important when personal data is at stake. For example, if the participant's picture is not necessary or can be replaced by an avatar, it should not be collected; and if only the participant's age is required, there is no need to collect their date of birth.

Personal Data

Personal data has been defined, for example by the European Union and the IEEE Meaning of I-E-E-EIEEE, pronounced "Eye-triple-E," stands for the Institute of Electrical and Electronics Engineers. The association is chartered under this name and it is the full legal name.However, as the world's largest technical professional association, IEEE's membership has long been composed of engineers, scientists, and allied professionals. These include computer scientists, software developers, information technology professionals, physicists, medical doctors, and many others in addition to IEEE's electrical and electronics engineering core. For this reason the organization no longer goes by the full name, except on legal business documents, and **is referred to simply as IEEE**.

"Personal data shall mean any information relating to an identified or identifiable natural person; an identifiable person is one who can be identified, directly or indirectly, in particular by reference to an identification number or to one or more factors specific to his physical, physiological, mental, economic, cultural or social identity."[4]

"Personally identifiable information (PII) is defined as any data that can be reasonably linked to an individual based on their unique physical, digital, or virtual identity."[3]

Consequently the participant's pictures, voice, and to a certain extent physiological data are personal data. Indeed, recent findings are that neural signals should be treated as a user's PII.[5]

Anonymization and Pseudonymization

Anonymization breaks the link between data and a given participant so that the participant cannot be identified, directly or indirectly (e.g., through cross-referencing), from their data. Anonymized data is not PII any more. Pseudonymization aims at making data less identifiable but allows data to be tracked back to the participant, e.g., through a correspondence table between pseudonyms and identity data, or a random number replacing identity data. Pseudonymized data is still PII.

As a matter of fact, real anonymization methods hardly exist, although there are some attempts.[6]

Several issues are raised by so-called anonymization.

- Information given to the participants often mentions data anonymization, while most data processes that are used are actually pseudonymization.
- Participants are informed that they may leave the study at any time, which suggests that they may ask that their data be destroyed at any time. This is impossible if data is really anonymized, since a given participant's data could no longer be found to be removed and destroyed. Only pseudonymization allows participants to leave *at any time*.

- Video blurring does not seem to bring sufficient anonymization any more.[7] Consequently all identifying features (face, tattoos, etc.) have to be hidden. If they cannot be hidden and the videos are likely to be watched by people other than the researchers involved in the experiment, a specific permission must be obtained from the participants.
- Online questionnaires which are said to be "anonymous" most of the time are not. Indeed, a unique identifier associated with, e.g., the participant's IP address must be implemented so as to avoid multiple participation and ballot-box stuffing. Yet IP addresses are personal data in some cases.[8] Moreover, cross-referencing a given participant's answers is likely to identify this participant.

Data Storage and Access

Anonymized or pseudonymized collected raw data should be stored separately from the participants' consents and the correspondence table between pseudonyms and identity data. The storage conditions must be secure and placed under the responsibility of the principal investigator. In particular, storing data on a cloud on the internet, even if it is said to be "secure," or on a shared space on the lab intranet must be questioned.

As far as access to data is concerned, the following questions have to be raised.[3]

- Who requires access and for what duration?
- What is the purpose of the access? Data collected for a study the participant has consented to should not be reused for another study, unless each participant is asked to sign a new consent. Nevertheless, extending or generalizing the scope of data usage can be considered when giving initial information to the participant provided that what the participant consents to is clear enough.

How long data will be kept and when, how, and by whom it will be destroyed should be specified. The fact that data should be kept for only a limited duration may conflict with the necessity of providing data supporting the content of a publication or with experiment replication.

TOWARD EXPERIMENT RESULTS PUBLICATION

Approval by an Ethics Committee

Experiment projects involving human participants should be reviewed by an ethics committee prior to the beginning of the experiment: appropriate approval, licensing, or registration should be obtained before the research begins and details should be provided in the report (e.g., institutional review board, research ethics committee approval, national licensing authorities for the use of animals, etc.).[9] This is compulsory for collaborative projects, such as European projects,[10] and above all to enable the publication of results. For example, the publishers Elsevier and Springer mention that when reporting studies which involve human participants, authors should include a statement that the studies have been approved by the appropriate institutional and/or national research ethics committee and have been performed in accordance with the ethical standards as laid down in the 1964 Declaration of Helsinki and its later amendments or comparable ethical standards.

Two issues are worth mentioning as far as ethics committee reviews are concerned. First, the significance of ethics committee reviews might be questioned in so far as they depend on who actually reviews, who attends the committee meetings, etc. Second, ethics committees can hardly check that an experiment will actually be conducted as presented for review. The obligation to submit the actual experiment protocol and the informed consent to journals could be relevant.

Conflicts of Interest

Researchers involved in an experiment are supposed to declare any conflict of interest when they design the protocol. Whether this requirement is sufficient or even relevant is illustrated below by three issues. The analysis is based on a general recognized definition.[11]

One issue concerns honesty: even the most upright researcher might be unwilling to declare a conflict of interest if this is likely to cancel a project that matters to them.

Another issue is the involvement of companies with a high degree of bias. For example, a recent study[12] highlighted dramatic statistics in the field of genetically modified organisms (GMOs). During the last two decades, 40% of 672 published articles presented a conflict of interest. It is worth noting that the results of experiments in cases of conflict of interest are more frequently favorable to the interests of the GMO industry, but the link between conflicts of interest and the frequency of favorable opinions for the industry was not established.

Finally, no matter what the research organization (industrial company, public or private laboratory, etc.), scientific results are required to remain competitive and attractive, and this mainly relies on researchers. Moreover, as an individual, the researcher needs to publish to be recognized by the scientific community. Such pressure can lead to unethical behaviors, such as only considering results that can be further exploited; considering a partial result as a result; and believing that the faster a result is obtained, the better. A broader question is then: can an experiment be totally free of conflict of interest, i.e., avoid both industrial interests and personal interest?

More Scientific Integrity Issues

Experiment results may correspond more or less exactly to what was expected. To embellish results, some questionable research practices are likely to be adopted,[13] e.g., failing to report all of an experiment's conditions, excluding data, stopping collecting data earlier than planned because one found the result that one had been looking for,[14] using inappropriate statistics to support one's hypothesis,[13] etc. Some researchers[14,15] have shown that such scientific misconducts are far from being rare. Moreover the temptation may be great to break up or segment data from a single study and create different manuscripts for publication (salami slicing),[16] which is also unethical.

CONCLUSION

While some issues are often highlighted by the ethics committees that are in charge of reviewing experimental projects, others stem from biases inherent in the very research process. Consequently ethics should not boil down to box ticking, which does not stimulate thought, but rather should be a *process* to be implemented all along research and even inside research itself. As far as an experiment is concerned, its purpose and the way it is conducted must be questioned at every step, and each time new investigation or data-processing techniques appear. Indeed, ethics should not be perceived as a constraint to research and innovation, but rather as way of ensuring high-quality results.[17]

ACKNOWLEDGMENTS

This chapter is based on knowledge acquired by the first author as a member of CERNA (the French Commission for Ethics in Information and Communication Technology Research), COERLE (the Ethics Committee of Inria, France), and CERNI (the Ethics Committee of the University of Toulouse, France).

REFERENCES

1. Faden R, Beauchamp T. *A history and theory of informed consent*. Oxford University Press; 1986.
2. Nelson CA. Incidental findings in magnetic resonance imaging (MRI) brain research. *The Journal of Law, Medicine & Ethics: a Journal of the American Society of Law, Medicine & Ethics* 2008;**36**(2):315–9. https://doi.org/10.1111/j.1748-720X.2008.00275.x.
3. The IEEE Global Initiative for Ethical Considerations in Artificial Intelligence and Autonomous Systems. *Ethically aligned design: a vision for prioritizing wellbeing with artificial intelligence and autonomous systems - V2*. Tech. rep. IEEE; 2018. https://standards.ieee.org/develop/indconn/ec/autonomous_systems.html.
4. *The European Parliament and the Council of the European Union, Directive 95/46/EC, Tech. rep. ORNAC Journal*. 1995. p. 281.
5. Bonaci T. *Brains can be hacked. Why should you care?*. ENIGMA; 2017. https://www.usenix.org/conference/enigma2017/conference- program/presentation/bonaci.
6. Prasser F, Bild R, Eicher J, Spengler H, Kohlmayer F, Kuhn KA. Lightning: utility-driven anonymization of high-dimensional data. *Transactions on Data Privacy* 2016;**9**:161–85.
7. McPherson R, Shokri R, Shmatikov V. *Defeating image obfuscation with deep learning*. 2016. arXiv: 1609.00408v2.
8. Munz M, Hickman T, Goetz M. *Court confirms that IP addresses are personal data in some cases*. 2016. https://www.whitecase.com/publications/alert/court-confirms-ip- addresses-are-personal-data-some-cases.
9. Wager E, Kleinert S. Responsible research publication: international standards for authors. A position statement developed at the 2nd World Conference on Research Integrity, Singapore, 2010. In: Mayer T, Steneck N, editors. *Promoting research integrity in a global environment*. Singapore: Imperial College Press/World Scientific Publishing; 2010. p. 309–16.
10. European Commission. *Horizon H2020-how to complete your ethics self-assessment*. Tech. rep., Europe. 2018. http://ec.europa.eu/research/participants/data/ref/h2020/grants_manual/hi/ethics/h2020_hi_ethics-self-assess_en.pdf.

11. International Committee of Medical Journal Editors. Conflict of interest. *CMAJ: Canadian Medical Association Journal* 1993;**148**(12):2141.

12. Guillemaud T, Lombaert E, Bourguet D. Conflicts of interest in GM bt crop efficacy and durability studies. *Plos One* 2016;**11**(12):e0167777.

13. Custers R. *Research misconduct – the grey area of questionable research practices*. 2013. http://www.vib.be/en/news/Pages/Research-misconduct—The-grey-area-of-Questionable-Research-Practices.aspx.

14. John LK, Loewenstein G, Prelec D. Measuring the prevalence of questionable research practices with incentives for truth telling. *Psychological Science* 2012;**23**(5):524–32.

15. Martinson BC, Anderson MS, de Vries R. Scientists behaving badly. *Nature* 2005;**435**:737–8. https://doi.org/10.1038/435737a.

16. Elsevier. *Salami slicing*. 2015. https://www.publishingcampus.elsevier.com/websites/elsevier_publishingcampus.

17. European Commission. *Responsible Research and Innovation – Europe's ability to respond to societal challenges*. Tech. rep., Europe. 2012. https://doi.org/10.2777/11739.

Neuroadaptive Interfaces and Operator Assessment

Chapter 12

Neural Efficiency and Mental Workload: Locating the Red Line

Stephen Fairclough[1], Kate Ewing[2], Christopher Burns[1], Ute Kreplin[3]

[1]Liverpool John Moores University, Liverpool, United Kingdom; [2]BAE Systems, Wharton, United Kingdom; [3]Massey University, Auckland, New Zealand

INTRODUCTION

Mental workload has been widely used in human factors research since the publication of two key collections,[1,2] which have been reviewed for a historical perspective.[3] The measurement of mental workload is particularly important for the assessment of safety-critical performance where high cognitive demand can lead directly to errors and accidents. Thus research into mental workload tends to focus on a state of overload where selective attention is disrupted[4] and performance quality declines.[5] The point where workload becomes overload has been conceptualized as a "red line"[6] which divides good performance, where the operator has sufficient capacity to meet task demands, and declining performance, when the cognitive requirements of the task exceed the information-processing capacity of the operator (see Fig. 2 in Young et al.[3] for an illustration). Accurate measurement of the workload red line across multiple operational contexts remains a major challenge for human factors research.

The concept of overload was derived from the Yerkes–Dodson law[7] and resource-based models of workload and attention.[8–10] This perspective is based on an assumption that humans have finite limits on information-processing capacity (e.g., number of sensory inputs/motor outputs, complexity of inputs/outputs, time available), which can be overwhelmed by task demands. Others have conceptualized overload within an adaptive framework,[11] highlighting the contribution of strategies and volitional self-regulation to the interaction between operator and task demands. The framework developed by Hockey[12] took this approach to a logical conclusion by including the possibility that an overloaded operator could effectively withdraw from task demands by reducing performance quality as a strategy both to conserve mental effort and to reduce task-related stress.

Motivational intensity theory was originally developed to describe the factors that mediate the interaction between task demand and effort investment.[13,14] This theory is particularly relevant for the self-regulatory concepts of mental overload described in the previous paragraph. According to Brehm's original theory, there is a distinction between the level of effort invested in response to demand (motivational intensity) and the maximum effort the individual is willing to invest to satisfy a goal associated with successful performance (success importance).[15] Thus effort is invested in a proportionate fashion in response to increased task demand until a point is reached where the likelihood of successful performance is assessed to be low or the consequences of success are perceived to be unimportant or inconsequential with respect to other task-related goals (e.g., to earn money, to develop mastery), at which point effort is withdrawn from the task.[16] It is important for measures of mental workload to capture the dynamic relationship between effort investment and demand/performance encapsulated by motivational intensity theory. By representing interaction between user skill and task demand as an adaptive act of self-regulation, we can identify red lines that are predictive of performance breakdown, and distinguish between varieties of mental overload with respect to effort investment or conservation.[12] It is also necessary to develop composite measures of mental workload that reconcile different dimensions of workload assessment to predict performance breakdown, particularly in the context of safety-critical behavior.

NEURAL EFFICIENCY

A number of early neuroimaging research projects[17,18] studied the relationship between intelligence quotient (IQ) and neurophysiological activation. Their findings demonstrated that participants with higher IQ exhibited lower levels of cerebral metabolism when performing cognitive tasks; in other words, higher IQ individuals performed with greater neural efficiency compared to those with lower IQ scores. This neural efficiency hypothesis has been refined over the years to reveal

Neuroergonomics. https://doi.org/10.1016/B978-0-12-811926-6.00012-9
Copyright © 2019 Elsevier Inc. All rights reserved.

a number of significant caveats to the original research.[19] For example, neural efficiency (with respect to a differentiation between higher and lower IQ individuals) was only observed when task difficulty fell in the moderate to high range of cognitive demand (see Fig. 2 in Neubauer and Fink[19]); hence a cognitive task must stimulate a minimum level of challenge/complexity before we can observe the phenomenon of neural efficiency. On a related note, it was argued that moderate to high levels of difficulty allowed participants to develop and utilize efficient cognitive strategies that exhibit neural efficiency as a consequence of skill acquisition.[20] With respect to the latter, Haier and colleagues[21] provided participants with 4–8 h of practice on a computer game; they noted a neural efficiency effect that was associated with both performance improvement (i.e., practice led to improved performance and reduced metabolic activity) and intelligence (i.e., the effects of practice were more pronounced for individuals with higher IQ). A later study[22] reported that the phenomenon of neural efficiency was localized to individuals with higher IQ; in other words, the propensity to develop neural efficiency depended on the capacity to learn, which in turn was related to individual variation in intelligence. As a further caveat, the effects of learning on neural efficiency (i.e., reduced neurophysiological activation with practice) may be specific only to cognitive tasks and not extend to sensory or motor tasks.[23] A 2004 study by Neubauer and colleagues[19] also noted that neurophysiological evidence for neural efficiency was localized to the frontal cortex. This effect has been replicated with respect to reduced frontal activation with increased task automaticity[24] and a study on decisional conflict using neurovascular (functional near-infrared spectroscopy, fNIRS) markers of activity in the inferior frontal gyrus.[25]

The neural efficiency hypothesis represents an interaction between neurophysiological activation and task demand/performance effectiveness; this combination of neuroscience and behavioral data captures a basic tenet of neuroergonomics[26] and can be used as the basis for a brain-based index of mental workload wherein measures of performance are combined with neurophysiological activity. Two studies of neural efficiency are presented in the subsequent sections: the first describes an electroencephalogram (EEG)-based study on working memory load in combination with a financial incentive, and the second details an fNIRS-based investigation using an identical task manipulation.

STUDY ONE

A total of 18 participants (9 male) took part in the experiment. Effort was elicited with a continuous matching verbal working memory task known as the n-back task; this particular version was based on the one described by Gevins et al.[27] This task required participants to indicate if the currently presented stimulus matched an earlier stimulus presentation. Stimuli were single capital letters drawn at random from a group of 12: B, F, G, H, K, M, P, R, S, T, X, and Z. Participants were required to indicate whether the letter matched the previous one (1-back: easy), or the letter that had appeared four letters earlier (4-back: hard), or the letter that had appear seven letters earlier (7-back: impossible). This task was performed in two short blocks of 100 sec for each of the three working memory conditions. Responses were given with a keyboard press of 1 for match and 2 for nonmatch, using the right index and middle fingers. The EEG was recorded from 64 Ag–AgCl pin-type active electrodes mounted in a BioSemi stretch-Lycra head cap. Electrodes were positioned using the 10–20 system. For the purposes of this chapter, we focus on activity in the theta frequency band (4–7 Hz) obtained from the fronto–central site (Fz). Performance by participants was scored for the percentage of correct responses in the 1-back, 4-back, and 7-back tasks. The total power in $\mu V2$ was obtained for the theta frequency band using fast Fourier transform (see Fairclough and Ewing[28] for full details of analysis). Both performance and neurophysiological activation in the form of theta power are plotted in a two-dimensional space in Fig. 12.1 for all three levels of working memory demand.

As expected, neural efficiency is highest during the easy 1-back version of the n-back; note how highly accurate performance coincides with low levels of neurophysiological activation. The cognitive demand of the task increases significantly as participants transition from the 1-back task, where a single letter must be retained and updated in working memory, to the demanding 4-back, which requires memorization and continuous updating of a four-letter sequence. As anticipated, performance accuracy falls from 94% to 76%, and this deterioration is accompanied by a significant increase of theta power in the fronto–central region. This transition represents a decline of neural efficiency (i.e., higher neurophysiological activation is required to sustain a lower level of performance), and this pattern is indicative of participants at the limits of their capacity to engage with the cognitive challenge of the task. The 7-back version of the n-back was designed to represent an "impossible" level of working memory demand. The transition from the 4-back to the 7-back task (Fig. 12.1) illustrates participants passing from a state of high mental workload to overload, which is characterized by both falling performance and reduced neurophysiological activation. This pattern is indicative of participants who are no longer engaged with the cognitive demand of the task or the pursuit of task-related goals, e.g., mastery and skill acquisition.

The data from this study demonstrates how a composite measure of neural efficiency, representing an interaction between performance/demand and neurophysiological activation, allows us both to visualize the trajectory from low workload to overload and to differentiate a number of stages along this continuum. Two distinct phases of neural efficiency can

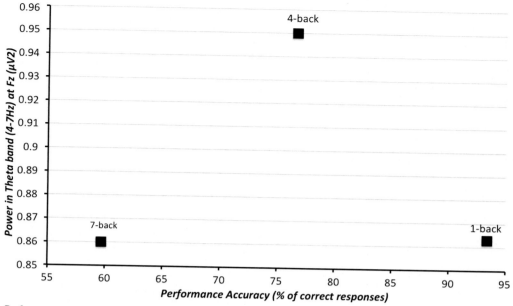

FIGURE 12.1 Performance accuracy and power in fronto–central theta during all three levels of working memory demand (1-back, 4-back, 7-back), N = 18.

be observed in Fig. 12.1: an inverse correlation as neurophysiological activation increased and performance effectiveness declined as the participants reached the limits of their capacity to perform prior to the 4-back task; and a coupling between falling levels of neurophysiological activation and performance quality when the participants were overloaded.

STUDY TWO

This experiment (unpublished at the time of writing) utilized a mixed design wherein working memory load served as a within-participants manipulation. Five versions of the n-back working memory task were used: 0-back (very easy) 1-back (easy), 3-back (hard but success possible), 5-back (very hard and success unlikely), and 7-back (impossible). Thirty people took part in the experiment, recruited from the university population of undergraduates and postgraduate students; the mean age of participants was 26.9 years and the sample included equal numbers of males and females. A verbal version of the n-back was used to create all five conditions. In this task participants are exposed to a sequential presentation of single capital letters, e.g., B, F, R, T, that appear at a rate of approximately one item every 1.5 s. The participant must react to each letter with one of two possible responses: either the letter is the same as the previous letter (a "match") or it is different (a "nonmatch"). Participants are required to perform this task continuously for a period of approximately 2 min.

An fNIR Imager1000 and cognitive optical brain imaging (COBI) data collection suit (Biopac System) were used for data collection. The 16-channel probe is placed on the forehead aligned to Fp1 and Fp2 of the international 10–20 system, and rotated so that Fpz corresponds to the midpoint of the probe. Areas underlying the 16 voxels are right and left superior and inferior frontal gyrii (BA10 and BA46). The current analysis focuses on the right-lateral area of the prefrontal cortex that approximates the right side of BA46. The fNIRS device captures relative changes in oxygenated hemoglobin (Hb0) and deoxygenated hemoglobin (Hbb), and it is assumed that neuronal activation is represented by a process of neurovascular coupling where increased levels of Hb0 are accompanied by decreased Hbb (see Scholkmann et al.[29] for a review and further explanation). The current analysis focuses on decreased Hbb as a marker of neurophysiological activation.

The relationship between neurophysiological activation and task performance across all five levels of working memory load is illustrated in Fig. 12.2. The 0-back condition serves as a control for the motor demands of the task because participants were required simply to press a button when a letter appeared on the screen, hence performance is close to perfect and neurophysiological activation is low. When workload increases from the 0-back to the 1-back task, neurophysiological activation increases sharply but performance remains at a stable and high level. The transition from 1-back to 3-back represents a more substantial increase of task demand. As shown in Fig. 12.2, neurophysiological activation increases slightly but there is a conspicuous decline in performance quality. As workload passes from the realm of challenging demand (3-back) to the rigors of the 5-back, where demand is very high with low likelihood of success, the continued degradation

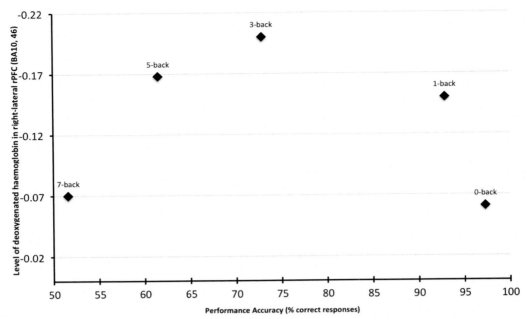

FIGURE 12.2 Performance accuracy and level of deoxygenated hemoglobin in right-lateral area of rostral prefrontal cortex during all five levels of working memory demand (N = 30). Note: decreased levels of deoxygenated hemoglobin are associated with neurophysiological activation.

of performance is accompanied by a fall of neurophysiological activation. Unsurprisingly, this trend is accelerated for the impossible 7-back condition, as neurophysiological activation falls to a similar level as was observed for the 0-back. With respect to identifying regions of mental workload, there are two significant transitions in Fig. 12.2: reduced performance and increased neurophysiological activation from 1-back to 3-back, indicating engagement despite declining performance, and reduced performance in combination with diminished neurophysiological activation from 3-back to 5-back, which is representative of a red line.

SUMMARY

A combination of measures derived from behavior and neuroscience can be used to delineate regions of mental workload, from low demand to overload. These regions are defined by the dynamic relationship between two workload measures, captured here as an index of neural efficiency. By definition, low mental workload is an efficient combination of good performance and low neurophysiological activation. Overload is also characterized by low neurophysiological activation, but in combination with poor performance. It is the identification of critical transitions between these two extremes that represents the value of the current approach. These transitions are defined by the direction of change observed simultaneously in measures of performance and neurophysiological activation. When neurophysiological activity increases or remains stable in the face of declining performance, we can infer that the individual remains engaged with task goals, believes successful performance to be a possibility, and is challenged by the demands of the task. Within this scheme, the workload red line is defined by a triad of high task demands, falling performance quality, and reduced neurophysiological activation.

REFERENCES

1. Hancock PA, Meshkati N. *Human mental workload*. Amsterdam: North-Holland; 1988.
2. Moray N. *Mental workload: its theory and measurement*. New York: Plenum; 1979.
3. Young MS, Brookhuis KA, Wickens CD, Hancock PA. State of science: mental workload in ergonomics. *Ergonomics* 2015;**58**(1):1–17.
4. Lavie N. Perceptual load as a necessary condition for selective attention. *Journal of Experimental Psychology: Human Perception and Performance* 1995;**21**(3):451–68.
5. De Waard D. *The measurement of driver mental workload*. Groningen, The Netherlands: Rijksuniversiteit Groningen; 1996.
6. Wickens CD, Tsang P. Workload. In: Durso F, editor. *Handbook of human-systems integration*. Washington, DC: APA; 2014.
7. Teigen KH. Yerkes-Dodson: a law for all seasons. *Theory and Psychology* 1994;**4**(4):525–47.
8. Kahneman D. *Attention and effort*. Englewood Cliffs, NJ: Prentice-Hall; 1973.
9. Navon D, Gopher D. On the economy of the human-processing system. *Psychological Review* 1979;**86**(3):214–25.

10. Wickens CD. Processing resources and attention. In: Damos DL, editor. *Multiple-task performance.* London: Taylor and Francis; 1991. p. 3–34.
11. Hancock PA, Warm JS. A dynamic model of stress and sustained attention. *Human Factors* 1989;**31**(5):519–37.
12. Hockey GRJ. Compensatory control in the regulation of human performance under stress and high workload: a cognitive-energetical framework. *Biological Psychology* 1997;**45**:73–93.
13. Brehm JW, Self EA. The intensity of motivation. *Annual Review of Psychology* 1989;**40**:109–31.
14. Wright RA. Brehm's theory of motivation as a model of effort and cardiovascular response. In: Gollwitzer PM, Bargh A, editors. *The psychology of action: linking cognition and motivation to behaviour.* New York: Guilford Press; 1996. p. 424–53.
15. Wright RA. Refining the prediction of effort: Brehm's distinction between potential motivation and motivation intensity. *Social and Personality Psychology Compass* 2008;**2**(2):682–701.
16. Richter M, Gendolla GHE, Wright RA. Three decades of research on motivational intensity theory: what we have learned about effort and what we still don't know. *Advances in Motivation Science* 2016;**3**:149–86. https://doi.org/10.1016/bs.adms.2016.02.001.
17. Haier RJ, Siegel BV, Nuechterlein KH, Hazlett E, Wu JC, Paek J, et al. Cortical glucose metabolic rate correlates of abstract reasoning and attention studied with positron emission tomography. *Intelligence* 1988;**12**:199–217.
18. Parks RW, Crockett DJ, Tuokko H, Beattie BL, Ashford JW, Coburn KL, et al. Neuropsychological 'system efficiency' and positron emission topography. *Journal of Neuropsychiatry* 1989;**1**:269–82.
19. Neubauer AC, Fink A. Intelligence and neural efficiency. *Neuroscience and Biobehavioral Reviews* 2009;**33**(7):1004–23. https://doi.org/10.1016/j.neubiorev.2009.04.001.
20. Doppelmayr M, Klimesch W, Sauseng P, Hodlmoser K, Stadler W, Hanslmayr S. Intelligence related differences in EEG-bandpower. *Neuroscience Letters* 2005;**381**:309–13.
21. Haier RJ, Siegel BV, MacLachlan A, Soderling E, Lottenberg S, Buchsbaum MS. Regional glucose metabolic changes after learning a complex visuospatial/motor task: a positron emission topographic study. *Brain Research* 1992;**570**:134–43.
22. Neubauer AC, Grabner RH, Freudenthaler HH, Beckmann JF, Guthke J. Intelligence and individual differences in becoming neurally efficient. *Acta Psychologica* 2004;**116**(1):55–74. https://doi.org/10.1016/j.actpsy.2003.11.005.
23. Kelly AMC, Garavan H. Human functional neuroimaging of brain changes associated with practice. *Cerebral Cortex* 2005;**15**(8):1089–102. https://doi.org/10.1093/cercor/bhi005.
24. Ramsey NF, Jansma JM, Jager G, Van Raalten T, Kahn RS. Neurophysiology factors in human information processing capacity. *Brain* 2004;**127**(3):517–25. https://doi.org/10.1093/brain/awh060.
25. Di Domenico SI, Rodrigo AH, Ayaz H, Fournier MA, Ruocco AC. Decisionmaking conflict and the neural efficiency hypothesis of intelligence: a functional near-infrared spectroscopy investigation. *Neuroimage* 2015;**109**:307–17. https://doi.org/10.1016/j.neuroimage.2015.01.039.
26. Parasuraman R. Neuroergonomics: research and practice. *Theoretical Issues in Ergonomic Science* 2003;**4**(1–2):5–20.
27. Gevins A, Smith ME, Leong H, McEvoy L, Whitfield S, Du R, Rush G. Monitoring working memory load during computer-based tasks with EEG pattern recognition models. *Human Factors* 1998;**40**(1):79–91.
28. Fairclough SH, Ewing K. The effect of task demand and incentive on neurophysiological and cardiovascular markers of effort. *International Journal of Psychophysiology* 2017. https://doi.org/10.1016/j.ijpsycho.2017.01.007.
29. Scholkmann F, Kleiser S, Metz AJ, Zimmermann R, Mata Pavia J, Wolf U, Wolf M. A review on continuous wave functional near-infrared spectroscopy and imaging instrumentation and methodology. *Neuroimage* 2014;**85 Pt 1**:6–27. https://doi.org/10.1016/j.neuroimage. 2013.05.004.

Chapter 13

Drowsiness Detection During a Driving Task Using fNIRS

Rayyan A. Khan[1], Noman Naseer[1], Muhammad J. Khan[2]
[1]*Air University, Islamabad, Pakistan;* [2]*Pusan National University, Busan, Republic of Korea*

INTRODUCTION

A brain–computer interface (BCI), also known as brain–machine interface, is a combination of hardware and software in as communication system that facilitates users interacting with their surroundings using different external devices. A BCI comprises an artificial intelligence system that is trained to recognize a certain set of patterns of brain signals used to generate commands for different applications, such as control of a robotic arm or a wheelchair, detection of state of mind, etc. (Alonso and Gil, 2012).[1a]

Brain signal acquisition can be carried out using invasive or noninvasive modalities. In invasive BCI sensors are placed on the gray matter of the brain to acquire signals, whereas in noninvasive methods detectors are placed on the surface of the head.[1] Invasive BCI has potentially better signal-to-noise ratios and fewer artifacts; however, it is not common in practice as sensors do not cover the whole cortex and are laborious due to the need for surgical invention.[1] Noninvasive brain-signal acquisition modalities include electroencephalography (EEG), functional near-infrared spectroscopy (fNIRS), and functional magnetic resonance imaging (fMRI). Due to its nonportability and high cost, fMRI is not used for most applications, while EEG has been widely used since 1929[2] as a thoroughly researched modality with great temporal resolution, low cost, and portability. But due to vulnerability of its artifacts, EEG is not recommended. fNIRS is a relatively new modality that has greater spatial resolution, portability, and less artifacts. Promising results have been achieved in fNIRS-based brain imaging and BCI.[3–7]

BCI can be subcategorized into three main types: active, passive, and reactive. Active BCI involves direct user control in which signals are intentionally generated, independent of external events—for example mental tasks and motion intentions. Reactive BCI involves reaction to an external event, for example movement of limbs, etc. Passive BCI detects a user's mental state from arbitrary brain activity without any intentional control, for example drowsiness, fatigue detection, etc.[2,4]

BCI is usually performed in five steps: signal acquisition, preprocessing, feature extraction, classification, and application interface. In signal acquisition brain signals are acquired using a suitable modality. In preprocessing, noises such as instrumental, physiological, and experimental errors are removed. The third stage is feature extraction, in which valuable information is extracted. In the fourth stage a suitable classifier is applied to classify the features extracted in the third stage. Finally, these classified signals are used to generate control commands that can be interfaced with different applications.[8] Fig. 13.1 shows a schematic of BCI.

The possibility of falling asleep while driving increases with lack of adequate rest. According to the National Highway Traffic Administration, around 56,000 to 100,000 accidents occur annually due to drowsiness, resulting in more than 1500 fatalities and 71,000 injuries. This produces economic loss of US$230 billion annually, according to the Federal Highway Administration.[9,9a]

Drowsiness can be detected by facial expressions, changes in ocular, sagging eyes, or attention lapses.[10,11] Previous studies have used these expressions to detect drowsiness, but they result in false detection due to the variability of scenarios.[4]

In this chapter we investigate the potential aspects of passive BCI for drowsiness detection. We analyze the drowsy state through the brain's hemodynamic response measured using fNIRS (Hong et al., 2014).[12a] Signals from the dorsolateral prefrontal cortex were used for this purpose. To acquire the maximum classification accuracy, statistical features (signal peak and signal mean),[12] calculated over 0–7 s time windows,[13] were used for a passive BCI.

Neuroergonomics. **https://doi.org/10.1016/B978-0-12-811926-6.00013-0**
Copyright © 2019 Elsevier Inc. All rights reserved.

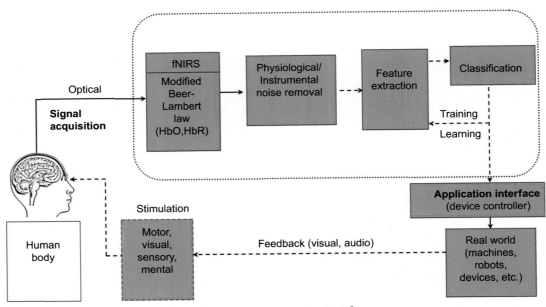

FIGURE 13.1   Schematic of BCI.[8]

LITERATURE REVIEW

In the past decade researchers have made efforts to measure mental workload of multiple cognitive mental tasks for passive fNIRS-BCI studies, and some are discussed in this section. Izzetoglu et al.[14] showed that fNIRS can be used in ecologically valid environments to assess complex cognitive tasks (air traffic control and piloting unmanned air vehicles), indicating that fNIRS measures are sensitive to mental task load and practice level. Menda et al.[15] monitored unmanned air vehicle operators' cognitive workload and situational awareness during simulated missions for fNIRS BCI. Ayaz et al.[16] used mental workload, level of expertise, and task performance to examine the relationship of the hemodynamic response in the dorsolateral prefrontal cortex using fNIRS to monitor level of expertise and mental workload. Harrison et al.[17] examined cognitive workload of air traffic control specialists utilizing a next-generation conflict resolution advisory; their results showed that workload levels continuously increase by increasing the number of aircrafts under control. Causse et al.[18] discussed emotion or stress effects on pilot decision-making: the fMRI modality was used to acquire brain images, and results show that negative emotional consequences cause temporary impairment in decision-making. The problem of detecting auditory warning alarms during flight simulation were discussed by Dehais et al.[19] Their study was conducted with one pilot equipped with a 32-channel EEG. Results showed that the percentage of missed alarms increases with the increase in difficulty level of flight scenarios. Furthermore, it was shown that lack of response to alarms is not intentional but due to cognitive issues. Gateau et al.[20] discussed a negative effect on flight safety due to working memory load. Signals were acquired using a fNIRS modality, while two estimators' Moving Average Convergence Divergence (MACD)-based algorithm to identify the pilot's instantaneous mental state (not on task versus on task) and an SVM-based classifier to discriminate task difficulty (low versus high working memory load) were used to differentiate task difficulties. The obtained accuracies and sensitivities showed that the pilot's estimated mental state matched better than the pilot's real state.

METHODS

Five healthy subjects took part in the current study; one was left-handed, and their mean age was 30.8±2.9. The subjects recruited for this experiment had not previously taken part in any drowsiness experiment and had no psychological, mental, or physical disorders. The participants were sleepless for at least 10h before the start of the experiment; for instance, subjects did not sleep during the night and experiments were performed in the morning. Experiments were performed with a drive simulator in a simulated environment. A presession was conducted for subjects to familiarize them with the virtual driving simulator. Subjects were equipped with an NIRS system (DYNOT, NIRx Medical Technologies, USA) with wavelengths of 760 and 830nm. They were asked to drive the car for 30min while their brain signals were continuously monitored using fNIRS. Sensors were placed on the dorsolateral prefrontal region, with two emitters and six detectors.

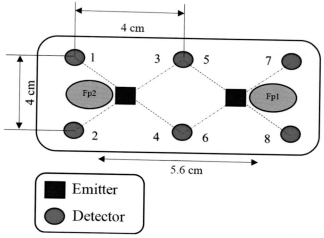

FIGURE 13.2   Schematic of the experimental paradigm: Fp1 and Fp2 are the reference points of the international 10–20 system.[21]

Eight channels were used to acquire brain signals by a combination of emitters and detectors, as shown in Fig. 13.2. A sampling frequency of 1.81 Hz was used to acquire signal data from the DYNOT machine. Concentration changes of oxyhemoglobin and deoxyhemoglobin ($\Delta c_{HbO}(t)$ and $\Delta c_{HbR}(t)$) were obtained using a modified Beer–Lambert law.

$$\begin{bmatrix} \Delta c_{HbO}(t) \\ \Delta c_{HbR}(t) \end{bmatrix} = \frac{1}{l \times d} \begin{bmatrix} \alpha_{HbO}(\lambda_1) & \alpha_{HbR}(\lambda_1) \\ \alpha_{HbO}(\lambda_2) & \alpha_{HbR}(\lambda_2) \end{bmatrix}^{-1} \begin{bmatrix} \Delta A(t, \lambda_1) \\ \Delta A(t, \lambda_2) \end{bmatrix}$$

(13.1)

where $\Delta A(t; \lambda_j)$ $(j = 1, 2)$ is the absorbance (optical density) measured at two values of wavelength, λ_j, $a_{HbX}(\lambda_j)$ is the extinction coefficient of HbX (i.e., HbO and HbR) in $\mu M^{-1} mm^{-1}$, d is the differential path length factor (DPF), and l is the emitter–detector distance (in millimeters).

Signal mean and signal peak of $\Delta c_{HbO}(t)$ signals were calculated during a 7 s time window. These features were normalized between 0 and 1 by

$$x' = \frac{x - \min(x)}{\max(x) - \min(x)}$$

(13.2)

where x' represents the feature values rescaled between 0 and 1, $x \in R^n$ are the original values of the features, and max (x) and min (x) represent the largest and smallest values, respectively. After feature extraction, for simplicity and speed of execution SVM and LDA were used as classifiers.[21,22]

Support Vector Machine

SVM is widely used in fNIRS-based BCI systems[10,23,24] Due to its explicit control of errors and good scalability to large dimensional data, it gives higher accuracies.[6,10,23–26] SVM maximizes the margins between classes by creating hyperplanes. The optimal solution of $r*$ is obtained by minimizing the following cost function between the hyperplane and the nearest training data points.

$$\text{Minimize } \frac{1}{2} \| w \|^2 + C \sum_{i=1}^{n} \xi_i$$

(13.3)

$$\text{Subject to } y_i \left(w^T x_i + b \right)^3 \geq 1 - \xi_i, \ \xi_i \geq 0$$

(13.4)

where w^T, $x_i \in R^2$ and $b \varepsilon R^1$, $\| w \|^2 = w^T w$, C is the tradeoff parameter between margin and error, ξ_i is the measure of training data, and y_i is the class label for the ith sample. SVM can be used for linear as well as nonlinear data classification. To obtain better accuracies we used a nonlinear classifier, as in Naseer et al.,[6] with a third-order kernel function having 10fold cross-validation.

Linear Discriminant Analysis

LDA uses discriminant hyperplane(s) to differentiate two or more classes of data; due to low computation and simplicity, this is preferred for online BCI systems. LDA classifies data by minimizing the interclass variance and maximizing the distance between two classes' means. It assumes equal covariance matrices with normal distribution for both classes.[27] In simple words, it finds a vector in a lower dimensional plane, such that the two classes are well differentiated. This is done by maximizing the Fisher's criterion:

$$J(v) = \frac{v^T S_b v}{v^T S_w v}$$

(13.5)

where S_b and S_w are the between-class and within-class scatter matrices defined as

$$S_b = (m_1 - m_2)(m_1 - m_2)^T,$$

(13.6)

$$S_w = \sum_{x_n \in 1} (x_n - m_1)(x_n - m_2)^T + \sum_{x_n \in 2} (x_n - m_1)(x_n - m_2)^T$$

(13.7)

where m_1 and m_2 represent the group means of classes C1 and C2, respectively, and x_n denotes the samples. A vector v that satisfies (13.5) can be reformulated as a generalized eigenvalue problem:

$$S_w^{-1} S_b v = \lambda v$$

(13.8)

The optimal v is the eigenvector corresponding to largest eigenvalue of $S_w^{-1} S_b$, or can be written as

$$v = S_w^{-1}(m_1 - m_2)$$

(13.9)

provided S_w is nonsingular.

RESULTS

The results have endorse a previous study finding by Naseer[28] that during driving drowsiness causes users to concentrate more as compared to a nondrowsy active state. A significant increase in $\Delta c_{HbO}(t)$ was observed during the drowsy state compared to the normal state, because users try to concentrate more during a drowsy state to compensate for the lack of focus. This results in an increase of brain activity, which causes a change in $\Delta c_{HbO}(t)$. When classifying drowsy and active states, the accuracies were found to be $74.3 \pm 2.5\%$ using SVM and $73.0 \pm 2.7\%$ using LDA. Table 13.1 shows classification accuracies of all subjects using signal mean and signal peak as features. Fig. 13.3 shows two-dimensional feature spaces of signal mean and signal slope for all five subjects.

DISCUSSION

The results of classification were encouraging, suggesting that drowsy and active states can be distinguished well using SVM and LDA classifiers. The classification accuracies were well above the chance level (50% in a two-class problem). According to a previous study by Bogler et al.,[29] a significant peak related to task was obtained between 5 and 8 s, hence we used 0–7 s time

TABLE 13.1 Classification Accuracies of All Subjects Using Signal Peak and Signal Mean as Features

No	LDA (%)	SVM (%)
1	70.2	71.6
2	75.0	75.7
3	71.2	73.4
4	72.1	72.8
5	76.7	77.9
Mean	73.0±2.7	74.3±2.5

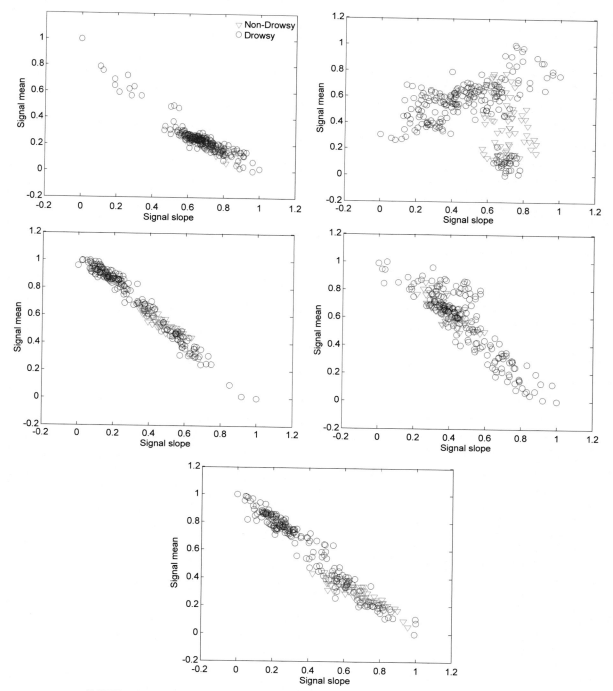

FIGURE 13.3 Classification accuracies of all subjects using signal peak and signal peak as features.

windows to reduce warning generation time.[30] A point to be noted is that none of the subjects went to nonrapid eye movement sleep but instead concentrated on steering the car, which causes the increase in $\Delta c_{HbO}(t)$. The most common features reported in studies by Bhutta et al.[31] and Naseer and Hong[5,8] for fNIRS were signal mean and signal slope. Due to changes in $\Delta c_{HbO}(t)$, even in this study signal mean and peak gave notable accuracies between drowsy and active states.[32]

The classifications obtained for the drowsy state show that the right prefrontal cortex has a more significant result as compared to the left prefrontal cortex. This indicates that the right prefrontal cortex is more active in drowsy-state tasks. Similar results were reported in Aritake et al.[33] and Khan and Hong.[4] It should be noted here that subjects selected for the experiment were not professional drivers, hence variation can be expected in the case of professional drivers.

CONCLUSION

In this study, we took a step forward in detecting drowsy state versus active state during a driving task using functional near-infrared spectroscopy (fNIRS). Drowsiness was detected by monitoring change in concentration of oxyhemoglobin with signal peak and signal mean as features for classification. It was also observed that right dorsolateral prefrontal cortex is more active as compared to left dorsolateral prefrontal cortex for drowsiness detection. These results suggest that it is possible to detect drowsy state of drivers using fNIRS, thereby, increasing the driver safety.

REFERENCES

1. Waldert S. Invasive vs. Non-Invasive neuronal signals for brain-machine interfaces: will one prevail? *Frontiers in Neuroscience* 2016;**10**(June):1–4. https://doi.org/10.3389/fnins.2016.00295.

1a. Nicolas-Alonso LF, Gomez-Gil J. Brain computer interfaces, a review, Sensors 2012;**12**(12):1211–1279, [Electronic Resource]. https://doi.org/10.3390/s120201211.

2. Zander TO, Kothe C. Towards passive brain–computer interfaces: applying brain–computer interface technology to human–machine systems in general. *Journal of Neural Engineering* 2011;**8**(2):25005. https://doi.org/10.1088/1741-2560/8/2/025005.

3. Hong K-S, Naseer N, Kim Y-H. Classification of prefrontal and motor cortex signals for three-class fNIRS–BCI. *Neuroscience Letters* 2015;**587**:87–92. https://doi.org/10.1016/j.neulet.2014.12.029.

4. Khan MJ, Hong K. Passive BCI based on drowsiness detection: an fNIRS study. *Biomedical Optics Express* 2015;**6**(10):4063. https://doi.org/10.1364/BOE.6.004063.

5. Naseer N, Hong K-S. Decoding answers to four-choice questions using functional near infrared spectroscopy. *Journal of Near Infrared Spectroscopy* 2015a;**23**(1):23. https://doi.org/10.1255/jnirs.1145.

6. Naseer N, Hong MJ, Hong K-S. Online binary decision decoding using functional near-infrared spectroscopy for the development of brain–computer interface. *Experimental Brain Research* 2014;**232**(2):555–64. https://doi.org/10.1007/s00221-013-3764-1.

7. Nguyen LH, Hong K-S. Investigation of the hemodynamic response in near infrared spectroscopy data analysis. In: *2010 second International Conference on knowledge and systems engineering.* IEEE; 2010. p. 28–32. https://doi.org/10.1109/KSE.2010.26.

8. Naseer N, Hong K-S. fNIRS-based brain-computer interfaces: a review. *Frontiers in Human Neuroscience* 2015b;**9**(January):1–15. https://doi.org/10.3389/fnhum.2015.00003.

9. Garcés Correa A, Orosco L, Laciar E. Automatic detection of drowsiness in EEG records based on multimodal analysis. *Medical Engineering and Physics* 2014;**36**(2):244–9. https://doi.org/10.1016/j.medengphy.2013.07.011.

9a. NCSDR/NHTSA Expert Panel on Driver Fatigue and Sleepiness "Drowsy Driving and Automobile Crashes". https://one.nhtsa.gov/people/injury/drowsy_driving1/drowsy.html.

10. Hu X-S, Hong K-S, Ge SS. fNIRS-based online deception decoding. *Journal of Neural Engineering* 2012;**9**(2):26012. https://doi.org/10.1088/1741-2560/9/2/026012.

11. Poudel GR, Innes CRH, Jones RD. Cerebral perfusion differences between drowsy and nondrowsy individuals after acute sleep restriction. *Sleep* 2012;**35**(8):1085–96. https://doi.org/10.5665/sleep.1994.

12. Noori FM, Naseer N, Qureshi NK, Nazeer H, Khan RA. Optimal feature selection from fNIRS signals using genetic algorithms for BCI. *Neuroscience Letters* 2017;**647**:61–6. https://doi.org/10.1016/j.neulet.2017.03.013.

12a. Hong KS, Nguyen HD. State-space models of impulse hemodynamic responses over motor, somatosensory, and visual cortices. *Biomedical Optics Express* 2014;**5**(6):1778. https://doi.org/10.1364/BOE.5.001778.

13. Hu X-S, Hong K-S, Ge SS. Reduction of trial-to-trial variability in functional near-infrared spectroscopy signals by accounting for resting-state functional connectivity. *Journal of Biomedical Optics* 2013;**18**(1):17003. https://doi.org/10.1117/1.JBO.18.1.017003.

14. Izzetoglu K, Ayaz H, Menda J, Izzetoglu M, Merzagora A, Shewokis PA, et al. Applications of functional near infrared imaging: case study on UAV ground controller. In: Schmorrow DD, Fidopiastis CM, editors. Schmorrow DD, Fidopiastis CM, editors. *Lecture Notes in computer science (including subseries lecture notes in artificial intelligence and lecture notes in bioinformatics*, vol. 6780. Berlin (Heidelberg): Springer Berlin Heidelberg; 2011. p. 608–17. https://doi.org/10.1007/978-3-642-21852-1_70.

15. Menda J, Hing JT, Ayaz H, Shewokis PA, Izzetoglu K, Onaral B, Oh P. Optical brain imaging to enhance UAV operator training, evaluation, and interface development. *Journal of Intelligent and Robotic Systems* 2011;**61**(1–4):423–43. https://doi.org/10.1007/s10846-010-9507-7.

16. Ayaz H, Shewokis PA, Bunce S, Izzetoglu K, Willems B, Onaral B. Optical brain monitoring for operator training and mental workload assessment. *NeuroImage* 2012;**59**(1):36–47. https://doi.org/10.1016/j.neuroimage.2011.06.023.

17. Harrison J, Izzetoglu K, Ayaz H, Willems B, Hah S, Ahlstrom U, et al. Cognitive workload and learning assessment during the implementation of a next-generation air traffic control technology using functional near-infrared spectroscopy. *IEEE Transactions on Human-Machine Systems* 2014;**44**(4):429–40. https://doi.org/10.1109/THMS.2014.2319822.

18. Causse M, Dehais F, Péran P, Sabatini U, Pastor J. The effects of emotion on pilot decision-making: a neuroergonomic approach to aviation safety. *Transportation Research Part C: Emerging Technologies* 2013;**33**:272–81. https://doi.org/10.1016/j.trc.2012.04.005.

19. Dehais F, Causse M, Regis N, Menant E, Labedan P, Vachon F, Tremblay S. Missing critical auditory alarms in aeronautics: evidence for inattentional deafness? *Proceedings of the Human Factors and Ergonomics Society Annual Meeting* 2012;**56**(1):1639–43. https://doi.org/10.1177/1071181312561328.

20. Gateau T, Durantin G, Lancelot F, Scannella S, Dehais F. Real-Time State estimation in a flight simulator using fNIRS. *PLoS One* 2015;**10**(3):1–19. https://doi.org/10.1371/journal.pone.0121279.
21. Naseer N, Noori FM, Qureshi NK, Hong K. Determining optimal feature-combination for LDA classification of functional near-infrared spectroscopy signals in brain-computer interface application. *Frontiers in Human Neuroscience* 2016;**10**(May):1–10. https://doi.org/10.3389/fnhum.2016.00237.
22. Hong K-S, Santosa H. Decoding four different sound-categories in the auditory cortex using functional near-infrared spectroscopy. *Hearing Research* 2016;**333**:157–66. https://doi.org/10.1016/j.heares.2016.01.009.
23. Abibullaev B, An J. Classification of frontal cortex haemodynamic responses during cognitive tasks using wavelet transforms and machine learning algorithms. *Medical Engineering and Physics* 2012;**34**(10):1394–410. https://doi.org/10.1016/j.medengphy.2012.01.002.
24. Burges CJC. A tutorial on support vector machines for pattern recognition. *Data Mining and Knowledge Discovery* 1998;**2**(2):121–67. https://doi.org/10.1023/A:1009715923555.
25. Sitaram R, Zhang H, Guan C, Thulasidas M, Hoshi Y, Ishikawa A, et al. Temporal classification of multichannel near-infrared spectroscopy signals of motor imagery for developing a brain–computer interface. *NeuroImage* 2007;**34**(4):1416–27. https://doi.org/10.1016/j.neuroimage.2006.11.005.
26. Tai K, Chau T. Single-trial classification of NIRS signals during emotional induction tasks: towards a corporeal machine interface. *Journal of NeuroEngineering and Rehabilitation* 2009;**6**(1):39. https://doi.org/10.1186/1743-0003-6-39.
27. Lotte F, Congedo M, Lécuyer A, Lamarche F, Arnaldi B. A review of classification algorithms for EEG-based brain–computer interfaces. *Journal of Neural Engineering* 2007;**4**(2):R1–13. https://doi.org/10.1088/1741-2560/4/2/R01.
28. Naseer N. Commentary: correlation of prefrontal cortical activation with changing vehicle speeds in actual driving: a vector-based functional near-infrared spectroscopy study. *Frontiers in Human Neuroscience* 2015;**9**(December):895. https://doi.org/10.3389/fnhum.2015.00665.
29. Bogler C, Mehnert J, Steinbrink J, Haynes J-D. Decoding vigilance with NIRS. *PLoS One* 2014;**9**(7):e101729. https://doi.org/10.1371/journal.pone.0101729.
30. Hong K, Naseer N. Reduction of delay in detecting initial dips from functional near-infrared spectroscopy signals using vector-based phase analysis. *International Journal of Neural Systems* 2016;**26**(3):1650012. https://doi.org/10.1142/S012906571650012X.
31. Bhutta MR, Hong MJ, Kim Y-H, Hong K-S. Single-trial lie detection using a combined fNIRS-polygraph system. *Frontiers in Psychology* 2015;**6**(June):709. https://doi.org/10.3389/fpsyg.2015.00709.
32. Hu X-S, Hong K-S, Ge SS. Recognition of stimulus-evoked neuronal optical response by identifying chaos levels of near-infrared spectroscopy time series. *Neuroscience Letters* 2011;**504**(2):115–20. https://doi.org/10.1016/j.neulet.2011.09.011.
33. Aritake S, Higuchi S, Suzuki H, Kuriyama K, Enomoto M, Soshi T, et al. Increased cerebral blood flow in the right frontal lobe area during sleep precedes self-awakening in humans. *BMC Neuroscience* 2012;**13**(1):153. https://doi.org/10.1186/1471-2202-13-153.

FURTHER READING

1. Hong K-S, Nguyen H-D. State-space models of impulse hemodynamic responses over motor, somatosensory, and visual cortices. *Biomedical Optics Express* 2014;**5**(6):1778. https://doi.org/10.1364/BOE.5.001778.
2. Naseer N, Qureshi NK, Noori FM, Hong K. Analysis of different classification techniques for two-class functional near-infrared spectroscopy-based brain-computer interface. *Computational Intelligence and Neuroscience* 2016;**2016**:1–11. https://doi.org/10.1155/2016/5480760.

Chapter 14

Neural Oscillation Dynamics of Emerging Interest in Neuroergonomics

Robert J. Gougelet

Department of Cognitive Science, University of California, San Diego, San Diego, CA, United States; Swartz Center for Computational Neuroscience, University of California, San Diego, La Jolla, CA, United States

NEUROERGONOMISTS SHOULD KNOW ABOUT NEURAL OSCILLATIONS

Neuroergonomists seek to apply knowledge and tools from neuroscience to optimize the tracking and regulation of human factors in everyday and working environments; for example, assessing and tracking mental workload using electroencephalography (EEG).[1] The study of neural oscillations in neuroscience using electrophysiology continues to generate notable interest, knowledge, and tools for neuroergonomists; but problematically, one finds little discussion of the application of what we know about neural oscillations in this nascent field of neuroergonomics. Pioneers such as Gevins and Smith[1] provided a solid foundation that needs building upon, especially since new tools and interpretations have arisen since their work. This chapter introduces neural oscillations and relates neural oscillation dynamics of emerging interest to neuroergonomic states of interest; particularly mental workload,[2] decision making, vigilance,[3] fatigue, situational awareness,[2] skill retention, and user error. The chapter concludes with a practical vision of how neural oscillations might be leveraged in neuroergonomics, suggesting methodology and next steps to interested neuroergonomists.

NEURAL OSCILLATION DYNAMICS OF EMERGING INTEREST

What are neural oscillations? Simply put, when a few to millions of neurons fire together in rhythm, they change their surrounding electromagnetic field. This electromagnetic field change is proportionate to the number of neurons involved and how well coordinated they are as they fire. Electrophysiologists call these rhythmic changes in the electromagnetic field neural oscillations, and use voltmeters or magnetometers to measure them.[4]

Neural oscillations have wide theoretical implications, as they may be a fundamental computational mechanism in the brain.[5,6] They reflect rhythmic fluctuations of underlying neural excitation and inhibition, corresponding to precise and well-regulated time windows of communication. This precisely timed communication supports preferential routing of information in the brain,[7] and better coordinates more fundamental neural processing such as spike synchrony.[8] This coordination can occur across large and distant brain regions, potentially mediating top-down sensorimotor processing more generally.[9] The precise time coordination of neural oscillations could also support bottom-up temporal binding together of features of internal representations.[10]

Electrophysiologists transform time-varying electrophysiology signals into their frequency domain, or spectral, representations to study neural oscillations (see Fig. 14.1A). The Fourier transform does this by breaking down a signal into a spectrum of frequency components, each with a respective amplitude and phase. In practice the spectrum of frequencies is broken down into normatively defined bands: delta (less than 4 Hz), theta (4–8 Hz), alpha (8–12 Hz), beta (12–30 Hz), and gamma (greater than 30 Hz). When a sinusoidal-like neural oscillation is presumed embedded in the time-varying signal, it is verified as a peak in amplitude on the spectrum with its own frequency and phase. Electrophysiologists identify neural oscillations by the band in which their peaks fall, e.g., calling an oscillation or rhythm in the gamma band a "gamma rhythm." Changes in spectral properties of neural oscillations, defined here as amplitude, frequency, and phase modulation, could have broad neuroergonomic consequences.

Amplitude, Frequency, and Phase Modulation of Neural Oscillations

Amplitude modulation (see Fig. 14.1A and B), sometimes known as event-related synchronization or desynchronization,[11] has been well studied and summarized for neuroergonomics[1] and neuroscience more generally.[12] Briefly,

Neuroergonomics. https://doi.org/10.1016/B978-0-12-811926-6.00014-2
Copyright © 2019 Elsevier Inc. All rights reserved.

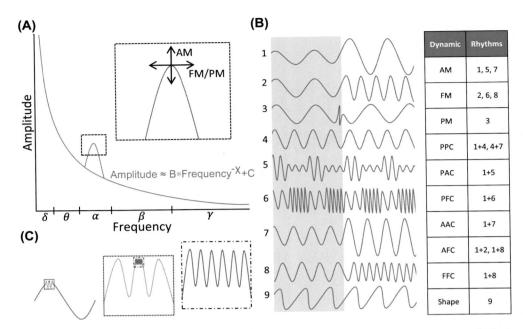

FIGURE 14.1 Spectral representation and dynamics of neural oscillations. *AAC*, amplitude–amplitude coupling; *AFC*, amplitude–frequency coupling; *AM*, amplitude modulation; *FFC*, frequency–frequency coupling; *FM*, frequency modulation; *PAC*, phase–amplitude coupling; *PFC*, phase–frequency coupling; *PPC*, phase–phase coupling; Shape: nonsinusoidal shape of the time-domain waveform. (A) The frequency spectrum representation of an electrophysiological signal. Alpha oscillation embedded as a peak in the alpha frequency band. This oscillation can be subject to amplitude, frequency, and phase modulation. (B) Modulation and coupling dynamics of different oscillations/rhythms in the time domain. Note that coupling cannot occur without modulation. (C) Hierarchical PAC embedding of higher-frequency oscillations in lower- frequency oscillations.

increased amplitude of alpha oscillations is suggested to reflect increased top-down rhythmic inhibitory control of processing.[13] Increased inhibitory control might be used to suppress distracting external or internal information, potentially indexing mental workload, decision making, vigilance, skill retention, and user error. Engel and Fries[14] suggest beta band amplitude increases are associated with motor control, particularly the top-down maintenance of the "status quo" of active sensorimotor processes. Beta band amplitude modulation might therefore index user error, sensorimotor skill acquisition, and decision making. In addition, frontal theta amplitude increases correspond to greater working memory load and attentional demands, as well as to greater long-term learning,[1] and this amplitude is therefore suitable for measuring mental workload and skill retention. Increased alpha and delta amplitude canonically index sleepiness and fatigue, as well.

Frequency modulation refers to the changing of frequency of the peak amplitude of a frequency band (see Fig. 14.1A and B), presumably reflecting an embedded oscillation (see Fig. 14.1A). Frequency modulation research predominantly focuses on peak alpha frequency (PAF).[15,16] Higher trait-level PAF in individuals indicates higher intelligence, reading and verbal ability, and cognitive preparedness, whereas decreasing PAF parallels age-related declines in memory. Frequency modulation as an index of long-term learning and cognitive preparedness might cover skill retention and situational awareness. During task performance, transient increases in PAF reflect accurate working memory retrieval and span, whereas decreases in PAF reflect task difficulty, reaction time, and physical fatigue. Thus transient changes in PAF could index transient changes in mental workload, vigilance, and fatigue. Users who volitionally increase their transient PAF actually enhance their cognitive performance.[17] Few frequency modulation results exist outside the alpha band, but there is a potential relationship between theta peak frequency and working memory load.[18]

Phase modulation of a neural oscillation has been primarily studied as phase resetting (see Fig. 14.1A and B). Klimesch et al.[13] suggest that phase resetting reflects early top-down influences on sensory semantic processing and is actually the evoked phase reset of alpha oscillations, though this issue is contentious.[19] Theta phase resetting has been implicated in user error,[20] trait intelligence,[21] and auditory decision making.[22] Notably, phase and frequency modulation are mathematically related, since the slope of the phase representation of a sinusoidal signal is shallower or steeper if the frequency is slower or faster, respectively, i.e., frequency is the temporal derivative of phase. Thus findings regarding frequency modulation might extend to phase modulation.

Amplitude, Frequency, and Phase Coupling of Neural Oscillations

Cross-frequency coupling is a generic term for when multiple neural oscillations within or between brain regions mutually drive the spectral properties of each other, and may provide profound computational affordances in the brain[23] with broad neuroergonomic consequences. Assuming only two oscillations are involved, coupling can occur in the form of phase–phase, phase–amplitude, phase–frequency, amplitude–amplitude, amplitude–frequency, or frequency–frequency interactions, all discussed below (see Fig. 14.1B). Notably, no forms of coupling can occur without some form of modulation, implying that the underlying physiological mechanisms may be the same.

Phase–phase coupling, a near-zero difference between the phases of two oscillations, is widely studied as phase synchrony, a posited fundamental mechanism of integration in the brain.[24] Phase synchrony potentially mediates attention, multisensory integration, learning, and memory at multiple spatial and temporal scales.[25] It could therefore be relevant to those interested in measuring mental workload, decision making, vigilance, situational awareness, and skill retention. It has also been related to user error.[26]

Phase–amplitude coupling provides a potentially robust, versatile, and multiscale mechanism for a nested multiplexing, or hierarchical structure, of neural oscillations.[27] In this hierarchical structure, higher-frequency oscillations increase in amplitude along the peak phases of lower-frequency oscillations (see Fig. 14.1C). Generally, large-scale network-level activity is indexed by slow and low-frequency oscillations, whereas small-scale local-level activity is indexed by fast and high-frequency oscillations. Phase–amplitude coupling could therefore support the transformation of information across temporal and spatial scales through this nesting of higher-frequency oscillations in lower-frequency oscillations.[28] The phase of low-frequency oscillations could then rhythmically entrain to external and internal stimuli, aligning bursts of increased gamma amplitude along the low-frequency phases and supporting well-timed learning, attention, and memory. Considering the potential role of phase–amplitude coupling in transformation of information at different scales, it could be particularly pertinent to situational awareness and decision making, but could also be useful as an indicator for general information processing, relevant for mental workload, vigilance, fatigue, skill retention, and user error.

Amplitude–amplitude coupling might play an important role in large-scale neuronal interactions subserving sensorimotor decision making and top-down attention.[6] This form of coupling may index mental workload, decision making, vigilance, and situational awareness. Gamma amplitude comodulation could occur as the result of multiple brain regions coupling to the phase of a lower-frequency and more widespread oscillation.[29] Considering that traditional phase–phase coupling, or coherence, measures include an amplitude–amplitude component,[30] many phase–phase coupling findings are worth dissociating from amplitude–amplitude coupling.

Phase–frequency, amplitude–frequency, and frequency–frequency coupling seem to be understudied, perhaps due to the minimal study of frequency modulation of unitary neural oscillations. These exact terms may not emerge in the literature, as many of the other forms of modulation and coupling have acquired different names, e.g., phase synchrony as phase–phase coupling. Thus standardization of terms regarding modulation and property–property coupling is suggested here to emphasize the fundamental spectral properties of oscillations: amplitude, frequency, and phase.

Waveform Shape and Neural Noise

Nonsynchronous or nonoscillatory, properties of the electrophysiological signal, such as waveform shape and wideband shape of the frequency spectrum, might change with potential neuroergonomic consequences but are understudied. Phase–amplitude and phase–phase coupling might actually be conflated with waveform shape,[31] so findings regarding them may actually be a consequence of waveform shape. In addition, the shape of the overall spectrum, parameterized by its fit to a broadband power law relationship between frequency and amplitude, $Amplitude \sim B * Frequency^{-\chi} + C$, could reflect the nonoscillatory "asynchronous," or scale-free, changes in cortical potentials (see Fig. 14.1A). This 1/f "neural noise," in turn, could reflect the balance of excitation and inhibition across the human cortex, interdependently interacting with neural oscillations.[32] Having such a measure of excitation and inhibition through parameterization of the frequency spectrum is important, considering the critical roles excitation and inhibition play in cortical processing.[33] The chi parameter in the power law relationship between frequency and amplitude as an index for global excitation and inhibition might therefore index mental workload, vigilance, fatigue, and situational awareness.

LEVERAGING NEURAL OSCILLATIONS IN NEUROERGONOMICS

Table 14.1 summarizes how neuroergonomists might measure and manipulate neuroergonomic states by measuring and manipulating neural oscillations. Applying the knowledge and tools of neural oscillations could take many forms. Applications in research settings could take the form of interface, workspace, or workflow redesign by measuring

TABLE 14.1 Measuring Ergonomic/Human Factors via Neural Oscillation Dynamics

	AM	FM	PM	PPC	PAC	AAC	PFC	AFC	FFC	1/f	Shape
Mental workload	X	X	?	X	X	X	?	?	?	X	?
Decision making	X	X	X	X	X	X	?	?	?	?	?
Vigilance	X	X	X	X	X	X	?	?	?	X	?
Fatigue	X	X	?	?	X	?	?	?	?	X	?
Situational awareness	X	X	X	X	X	X	?	?	?	X	?
Skill retention	X	X	X	X	X	?	?	?	?	?	?
User error	X	?	X	X	X	?	?	?	?	?	?

1/f: power law shape of Fourier spectrum; *AAC*: amplitude–amplitude coupling; *AFC*: amplitude–frequency coupling; *AM*: amplitude modulation; *FFC*: frequency–frequency coupling; *FM*: frequency modulation; *PAC*: phase–amplitude coupling; *PFC*: phase–frequency coupling; *PPC*: phase–phase coupling; *Shape*: shape of the time-domain waveform. An X indicates that the ergonomic/human factor could be measured by the neural oscillation dynamic, and a query mark (?) indicates no association has been widely reported.

neural oscillation modulation and coupling dynamics during user testing. Fieldwork interventions could be done via measurement and adaptive interface response to, or active manipulation of, neural oscillation modulation and coupling.

Another form of leveraging neural oscillations could be simple neurofeedback.[34] An interface readout of the user's state could be enough to affect task performance significantly. Yet another form could involve enhanced timing of the presentation of important interface and situational information. Indexing optimal brain information-processing windows, e.g., particular amplitudes or phases of ongoing oscillations, could give alerts when adaptive interfaces present information. Lastly, real-time neurostimulation using "brain pacemakers" could affect the spectral modulation and coupling of the user's oscillations as a potential (albeit challenging) means to create optimal neuroergonomic states.[35]

How do we measure neural oscillations in research settings? Cohen gives an excellent introduction to time-series analysis of electrophysiology signals, and how to extract amplitude, frequency, and phase information from unitary oscillations.[36] Phase–phase coupling can be measured using phase-locking statistics.[37] Phase–amplitude coupling has a variety of methods of measurement.[38] Amplitude–amplitude coupling has been measured using amplitude envelope correlation.[30] Unfortunately, methods for measuring amplitude–frequency, phase–frequency, and frequency–frequency coupling have not been thoroughly developed.

How do we measure neural oscillations in the field? EEG is likely best. One suggested method is the use of independent components analysis[39] to decompose EEG data into oscillatory and nonoscillatory brain activity.[40] Next, one discards noisy independent components and identified nonbrain components using well-constrained inverse models.[41] Processing data in this way could overcome many noise issues.

What can neuroergonomists do? Firstly, they can explore the understudied domains of knowledge in neural oscillation research: frequency and phase modulation; phase–frequency, amplitude–frequency, and frequency–frequency coupling; waveform shape; and broadband asynchronous activity. Neuroergonomists can also conduct multimodal studies relating less tractable tools to more tractable tools, e.g., magnetoencephalography to EEG, or functional magnetic resonance imaging to EEG. Because they have the vision to extend the knowledge and tools of neuroscience into the everyday and working environments, neuroergonomists are poised to make much-needed strides in the application of neural oscillation research.

REFERENCES

1. Gevins A, Smith ME. Electroencephalography (EEG) in neuroergonomics. *Neuroergonomics: The Brain at Work* 2006:15–31.
2. Tsang PS, Vidulich MA. Mental workload and situation awareness. In: Salvendy G, editor. *Handbook of human factors and ergonomics*. 3rd ed. 2006. p. 243–68. https://doi.org/10.1002/0470048204.ch9.
3. Warm JS, Parasuraman R, Matthews G. Vigilance requires hard mental work and is stressful. *Human Factors: The Journal of the Human Factors and Ergonomics Society* 2008;**50**(3):433–41.
4. Buzsáki G, Anastassiou CA, Koch C. The origin of extracellular fields and currents—EEG, ECoG, LFP and spikes. *Nature Reviews Neuroscience* 2012;**13**(6):407–20.

5. Buzsáki G, Draguhn A. Neuronal oscillations in cortical networks. *Science* 2004;**304**(5679):1926–9.
6. Siegel M, Donner TH, Engel AK. Spectral fingerprints of large-scale neuronal interactions. *Nature Reviews Neuroscience* 2012;**13**(2):121–34.
7. Fries P. A mechanism for cognitive dynamics: neuronal communication through neuronal coherence. *Trends in Cognitive Sciences* 2005;**9**(10):474–80.
8. Singer W. Neuronal synchrony: a versatile code for the definition of relations? *Neuron* 1999;**24**(1):49–65.
9. Engel AK, Fries P, Singer W. Dynamic predictions: oscillations and synchrony in top–down processing. *Nature Reviews Neuroscience* 2001;**2**(10):704–16.
10. Engel AK, Singer W. Temporal binding and the neural correlates of sensory awareness. *Trends in Cognitive Sciences* 2001;**5**(1):16–25.
11. Pfurtscheller G, Da Silva FL. Event-related EEG/MEG synchronization and desynchronization: basic principles. *Clinical Neurophysiology* 1999;**110**(11):1842–57.
12. Klimesch W. EEG alpha and theta oscillations reflect cognitive and memory performance: a review and analysis. *Brain Research Reviews* 1999;**29**(2):169–95.
13. Klimesch W, Sauseng P, Hanslmayr S. EEG alpha oscillations: the inhibition–timing hypothesis. *Brain Research Reviews* 2007;**53**(1):63–88.
14. Engel AK, Fries P. Beta-band oscillations—signalling the status quo? *Current Opinion in Neurobiology* 2010;**20**(2):156–65.
15. Klimesch W, Schimke H, Pfurtscheller G. Alpha frequency, cognitive load and memory performance. *Brain Topography* 1993;**5**(3):241–51.
16. Angelakis E, Lubar JF, Stathopoulou S, Kounios J. Peak alpha frequency: an electroencephalographic measure of cognitive preparedness. *Clinical Neurophysiology* 2004;**115**(4):887–97.
17. Angelakis E, Stathopoulou S, Frymiare JL, Green DL, Lubar JF, Kounios J. EEG neurofeedback: a brief overview and an example of peak alpha frequency training for cognitive enhancement in the elderly. *The Clinical Neuropsychologist* 2007;**21**(1):110–29.
18. Moran RJ, Campo P, Maestu F, Reilly RB, Dolan RJ, Strange BA. Peak frequency in the theta and alpha bands correlates with human working memory capacity. *Frontiers in Human Neuroscience* 2010;**4**:200.
19. Makeig S, Debener S, Onton J, Delorme A. Mining event-related brain dynamics. *Trends in Cognitive Sciences* 2004;**8**(5):204–10.
20. Yeung N, Bogacz R, Holroyd CB, Nieuwenhuis S, Cohen JD. Theta phase resetting and the error-related negativity. *Psychophysiology* 2007;**44**(1):39–49.
21. Thatcher RW, North DM, Biver CJ. Intelligence and EEG phase reset: a two compartmental model of phase shift and lock. *Neuroimage* 2008;**42**(4):1639–53.
22. Barry RJ. Evoked activity and EEG phase resetting in the genesis of auditory Go/NoGo ERPs. *Biological Psychology* 2009;**80**(3):292–9.
23. Jensen O, Colgin LL. Cross-frequency coupling between neuronal oscillations. *Trends in Cognitive Sciences* 2007;**11**(7):267–9.
24. Varela F, Lachaux JP, Rodriguez E, Martinerie J. The brainweb: phase synchronization and large-scale integration. *Nature Reviews Neuroscience* 2001;**2**(4):229–39.
25. Fell J, Axmacher N. The role of phase synchronization in memory processes. *Nature Reviews Neuroscience* 2011;**12**(2):105–18.
26. Cavanagh JF, Cohen MX, Allen JJ. Prelude to and resolution of an error: EEG phase synchrony reveals cognitive control dynamics during action monitoring. *Journal of Neuroscience* 2009;**29**(1):98–105.
27. Lakatos P, Shah AS, Knuth KH, Ulbert I, Karmos G, Schroeder CE. An oscillatory hierarchy controlling neuronal excitability and stimulus processing in the auditory cortex. *Journal of Neurophysiology* 2005;**94**(3):1904–11.
28. Canolty RT, Knight RT. The functional role of cross-frequency coupling. *Trends in Cognitive Sciences* 2010;**14**(11):506–15.
29. Buzsáki G, Wang XJ. Mechanisms of gamma oscillations. *Annual Review of Neuroscience* 2012;**35**:203–25.
30. Bruns A. Fourier-, Hilbert-and wavelet-based signal analysis: are they really different approaches? *Journal of Neuroscience Methods* 2004;**137**(2):321–32.
31. Cole SR, Voytek B. Brain oscillations and the importance of waveform shape. *Trends in Cognitive Sciences* 2017;**21**.
32. Gao RD, Peterson EJ, Voytek B. Inferring synaptic excitation/inhibition balance from field potentials. *Neuroimage* 2016. bioRxiv, 081125.
33. Yizhar O, Fenno LE, Prigge M, Schneider F, Davidson TJ, O'Shea DJ, et al. Neocortical excitation/inhibition balance in information processing and social dysfunction. *Nature* 2011;**477**(7363):171–8.
34. Heinrich H, Gevensleben H, Strehl U. Annotation: neurofeedback–train your brain to train behaviour. *Journal of Child Psychology and Psychiatry* 2007;**48**(1):3–16.
35. Huang YZ, Edwards MJ, Rounis E, Bhatia KP, Rothwell JC. Theta burst stimulation of the human motor cortex. *Neuron* 2005;**45**(2):201–6.
36. Cohen MX. *Analyzing neural time series data: theory and practice*. Boston, MA: MIT Press; 2014.
37. Lachaux JP, Rodriguez E, Martinerie J, Varela FJ. Measuring phase synchrony in brain signals. *Human Brain Mapping* 1999;**8**(4):194–208.
38. Tort AB, Komorowski R, Eichenbaum H, Kopell N. Measuring phase-amplitude coupling between neuronal oscillations of different frequencies. *Journal of Neurophysiology* 2010;**104**(2):1195–210.
39. Delorme A, Makeig S. EEGLAB: an open source toolbox for analysis of single-trial EEG dynamics including independent component analysis. *Journal of Neuroscience Methods* 2004;**134**(1):9–21.
40. Onton JA, Makeig S. High-frequency broadband modulation of electroencephalographic spectra. *Frontiers in Human Neuroscience* 2009;**3**:61.
41. Baillet S, Mosher JC, Leahy RM. Electromagnetic brain mapping. *IEEE Signal Processing Magazine* 2001;**18**(6):14–30.

Chapter 15

Is Mindfulness Helping the Brain to Drive? Insights From Behavioral Data and Future Directions for Research

Emanuelle Reynaud, Jordan Navarro
Université de Lyon, Bron, France

INTRODUCTION

Mindfulness meditation practices (MMPs) are a particular type of meditation involving full attention to present-moment experiences, both internal and external, and an acceptance of emotional states in a nonjudgmental manner.[1] Although the clinical benefits of MMPs have been studied intensively, in particular through their involvement in reducing anxiety[2] and stress,[3] their impact on cognitive functioning has received less attention. Nonetheless, the cognitive correlates of MMPs have been found on various functions, including attention[4] and executive functions.[5]

The majority of these studies use long-term training for MMPs, wherein subjects are trained over a long period of time (typically several weeks or months). But although this can be achieved under strict experimental control, it is hard to think that the results of these studies can be transposed to real-life situations, as humans are unlikely to engage spontaneously into such long-term training before experiencing any benefits. A few studies have used shorter training periods, over a few days, and have shown that even short-term MMPs can benefit cognitive functioning.[6,7] But, to our knowledge, no study has tried to show the effects of a single MMP session, because it is usually believed that mindfulness has to be learnt over a period of time. Although this assertion is undoubtedly true for gaining long-term benefits of meditational practice, we hypothesize here that a single, one-shot session could have short-term effects on subsequent mood and cognitive activity.

Studies using brief MMP training have revealed benefits for a single, isolated cognitive component,[8] but very few tackle the issue of MMP effects on behavior under complex ecological conditions. One of the most frequent and complex activities we all perform without any trouble, or almost, is driving. It is carried out at high speeds, in a dynamic environment requiring constant anticipation,[9] and uses more than 45 different cognitive functions.[10] Driving is such a common, complex, and potentially lethal activity that improving cognitive processing in the "driving brain" is of primary importance.

Considering this, questions arise. Can a brief, single MMP session improve driving behavior and thus road safety? Can we become better drivers if we engage in MMP just before hitting the road? Following the path initiated by Parasuraman,[11] we also propose here the initial steps for studying the neural correlates of this potential MMP-related improvement in "driving brain" activity.

MMP AND DRIVING

Method

Participants

The experimental group consisted of 38 participants (14 male, age = 21.4 ± 1.9 y) matched to a control group (n = 32, 10 male, age = 22.5 ± 2.1 y) for driving experience (mean = 2.9 y ± 2 versus 3.1 y ± 2.2). All participants declared no neurological condition, no drug consumption, no use of central nervous system medication, no uncorrected visual impairments, and no prior meditation experience, and signed a written informed consent. None experienced simulator sickness. A local ethics committee approved the experiment.

Neuroergonomics. https://doi.org/10.1016/B978-0-12-811926-6.00015-4
Copyright © 2019 Elsevier Inc. All rights reserved.

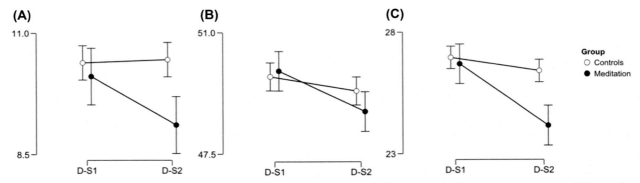

FIGURE 15.1 BMIS scores on the (A) negative–relaxed, (B) pleasant–unpleasant, and (C) arousal–calm dimensions before the two driving sessions.

Driving Simulation Task

A virtual driving environment was built with OpenSD2S software[12] running on a PC Dell Optiplex 380 MT with a Logitech G27 racing kit, projected on a 19" screen located 75 cm from the participants' eyes, giving a visual field of 31 degrees horizontally and 20 degrees vertically.

Participants were asked to drive and follow a vehicle at a safe distance for 3 min. The speed of the lead car varied from 45 to 90 km/h in a sinusoidal fashion with a period of 30 s.

Procedure

After an initial 5 min training, participants performed the driving task (D-S1). Then the experimental group (meditation condition) had a brief MMP practice while the control group performed a task where they had to listen to science popularization podcasts, both for 10 min. Participants were then asked to perform the car-following task again (D-S2).

Emotional states were assessed with the Brief Mood Introspection Scale (BMIS)[13] before each driving session.

Data Analysis

We computed the intervehicle time (IVT), which is the distance between the lead car and the subject's vehicle divided by its longitudinal speed, and the number of steering wheel turns (SWTs), which is the number of directional changes performed on the steering wheel, to assess driving performance.

We also computed BMIS subscores on the arousal–calm, pleasant–unpleasant, and negative–relaxed dimensions to describe effects of MMP on participants' mood.

Repeated-measure ANOVAs were used for statistical analyses.

Results

Mood Changes

Participants' scores on the BMIS negative–relaxed dimension showed that they were more relaxed before D-S2 than before D-S1 in the meditation group but not in the control group ($F(1,66)=4.71$, $P<.05$; see Fig. 15.1A). Scores on the pleasant–unpleasant dimension showed that they were slightly less pleased for the second driving session, but this effect was common to both experimental groups ($F(1,68)=9.84$, $P<.05$; see Fig. 15.1B). Finally, scores on the arousal–calm dimension show that participants were less aroused ($F(1,66)=20.85$, $P<.01$), and this effect was even more pronounced in the meditation condition ($F(1,66)=8.6$, $P<.01$; see Fig. 15.1C).

These results showed that our MMP was efficient over the considered timescale, as MMP seems to relax participants and decrease their arousal level.

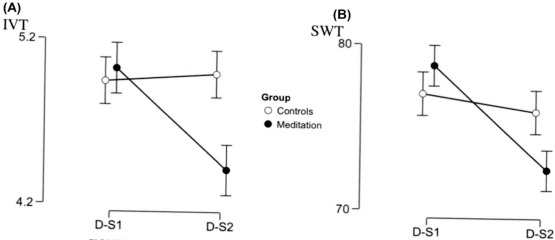

FIGURE 15.2   Driving parameters: (A) IVT and (B) SWT for the two driving sessions.

Driving Behavior

An interaction effect between the driving session and the group was found for the IVT ($F(1,68)=4.86$, $P<.05$; see Fig. 15.2A) as well as for SWTs ($F(1,68)=4.21$, $P<.05$; see Fig. 15.2B).

IVT and SWT decreased significantly after the brief MMP between D-S1 and D-S2, but only for the meditation group.

Discussion

Results showed a reduced IVT and a reduced number of SWTs specific to the meditation group. This can be interpreted as follows: a reduced IVT could indicate a more aware state induced by MMP, as MMP is supposed to focus one's attention on both internal and external states. The driver's safety margin might be decreased by meditation as a result of a more focused attention on the external world and more reactive driving behavior. This is confirmed by the decreased number of SWTs after a brief MMP. Drivers perform fewer adjustments on the steering wheel, suggesting that they anticipate their driving trajectories more and exert a finer control over their vehicle on the road.[14]

It might be argued that the reduced IVT could be the sign of a self-focused state induced by MMP, leading the driver to pay less attention to the traffic and therefore misjudging the appropriate safety margin. If this were the case, SWTs should decrease in D-S2 as a result of the driver being less aware. This contradicts our results, so we do not favor this interpretation.

These results suggest that a very brief, single MMP session could positively affect driving performance, confirming previous results.[15]

MMP AND THE DRIVING BRAIN

Rationale

In an initial attempt to qualify the neural signature of MMP on the brain when driving, we tried to construct a methodology for later transposition to driving contexts once well established on simpler, easier-to-characterize cognitive tasks.

We recorded brain activity through a brain–computer interface (BCI) electroencephalogram (EEG) headset (Emotiv EPOC 14-channel headset) when participants performed two sessions of a motor-imagery-based BCI task, with MMP training in between sessions compared with a control condition.

We used a motor-imagery BCI (MI-BCI) task as a beta version for future driving tasks. This type of task has been well studied in terms of brain activity markers recordable with a simple BCI headset, and it has been shown that motor imagery is related in particular to modulations of sensorimotor EEG rhythm in the alpha band.[16] The hypothesis is that these modulations, related to a power increase or decrease in the alpha band and referred to as event-related (de)synchronization (ERD/ERS), would differ after an MMP training session compared to the control condition (see Pfurtscheller and Lopes[17] for details on this approach).

We hypothesized earlier that the benefits of a one-shot MMP session could be observed on driving performance immediately after practicing, but we were more cautious with the effects of MMP on brain activity. Therefore we replicated the short-term MMP training used in a previous study[18] lasting over 5 days for 20 min every day.

Method

Participants

The experimental group consisted of seven women (age = 24.1 ± 3.2 y.o.). The control group was five women (age = 24.1 ± 3.2 y.o.). All fulfilled the inclusion criteria listed for the driving experiment.

Task

For the MI-BCI two runs of 60 trials were presented in random order. Each trial consisted of a fixation cross for 4 s, a visual cue for 12 s, and then a blank screen for 10 s. Participants were instructed to imagine a movement on the side (left or right) indicated by the visual cue.

Procedure

On Day 1 participants performed the first session of the MI-BCI task; on Days 2–6 participants performed the MMP training (meditation group) or listened to scientific podcasts (control group), during 20 min sessions for 5 days. They were asked on Day 7 to perform the second session of the MI-BCI task.

Data Analysis

We recorded the EEG data with the OpenVibe software,[19] and computed the ERD/ERS markers in the alpha power band (8–13 Hz) under frontal electrodes during the MI-BCI tasks by detecting power increases or decreases in relation to baseline (see Pfurtscheller and Lopes[17] for details on ERD/ERS event detection).

Results

ERD/ERS markers increased for BCI-S2 compared to BCI-S1 ($F(1,10) = 9.16$, $P < .05$; see Fig. 15.3), but this effect was not specific to the meditation group. Interaction between sessions and groups was not statistically significant but only a trend ($P = .09$).

Discussion

The number of ERD/ERS events decreased significantly between the two MI-BCI tasks, regardless of the participants' experimental group. We observed a slight trend toward significance for the interaction between the MI-BCI sessions and the nature of the training, thus we were not able to show a specific impact of MMP on the markers of brain activity. Further research needs to be conducted to eliminate the possibility of this result being due to the small number of participants.

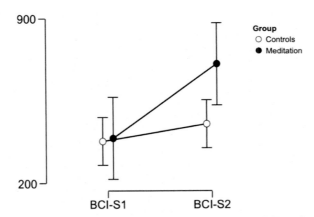

FIGURE 15.3   Number of ERD/ERS events for the two MI-BCI sessions.

GENERAL DISCUSSION AND CONCLUSION

The first study showed that a brief, one-shot MMP session could have a positive effect on driving performance. As a first step toward a neuroergonomics approach to the driving brain, we also investigated the feasibility of a method for observing the effects of MMP on brain activity in a simpler MI-BCI task.

To maximize our chances of observing an effect of MMP training in brain dynamics, we chose to use short-term MMP training taking place over 5 days. The next step will be to reduce MMP training to a single brief session.

Following this, the BCI-based method will be translated to driving tasks similar to that described in the first study. This should allow us to understand the role of MMP in car driving via behavioral and brain activity measures, by using EEG-BCI recordings while participants are driving.

REFERENCES

1. Kabat-Zinn J. Mindfulness-based interventions in context: past, present, and future. *Clinical Psychology: Science and Practice* 2003;**10**(2):144–56.
2. Chiesa a, Serretti a. A systematic review of neurobiological and clinical features of mindfulness meditations. *Psychological Medicine* 2010;**40**:1239–52.
3. Carmody J, Baer RA. Relationships between mindfulness practice and levels of mindfulness, medical and psychological symptoms and well-being in a mindfulness-based stress reduction program. *Journal of Behavioral Medicine* 2008;**31**(1):23–33.
4. Kozasa EH, Sato JR, Lacerda SS, Barreiros MAM, Radvany J, Russell TA, et al. Meditation training increases brain efficiency in an attention task. *Neuroimage* 2012;**59**(1):745–9.
5. Teper R, Inzlicht M. Meditation, mindfulness and executive control: the importance of emotional acceptance and brain-based performance monitoring. *Social Cognitive and Affective Neuroscience [electronic Resource]* 2013;**8**(1):85–92.
6. Prätzlich M, Kossowsky J, Gaab J, Krummenacher P. Impact of short-term meditation and expectation on executive brain functions. *Behavioural Brain Research* 2016;**297**:268–76.
7. Zeidan F, Johnson SK, Diamond BJ, David Z, Goolkasian P. Mindfulness meditation improves cognition: evidence of brief mental training. *Consciousness and Cognition* 2010;**19**(2):597–605.
8. Chiesa A, Calati R, Serretti A. Clinical Psychology Review Does mindfulness training improve cognitive abilities? A systematic review of neuropsychological findings. *Clinical Psychology Review* 2011;**31**(3):449–64.
9. Endsley MR. Toward a theory of situation awareness in dynamic systems. *Human Factors: The Journal of the Human Factors and Ergonomics Society* 1995;**37**(1):32–64.
10. McKnight AJ, Adams BB. *Driver education task analysis. Vol. I: task descriptions.* Washington, DC: Department of Transportation; 1970 (Report DOT HS 800 367).
11. Parasuraman R. Neuroergonomics: research and practice. *Theoretical Issues in Ergonomics Science* 2003;**4**(1–2):5–20.
12. Filliard N, Icart E, Martinez J-L, Gerin S, Merienne F, Kemeny A. Software assembly and open standards for driving simulation. In: *Proceedings of the driving simulation conference Europe 2010.* 2010. p. 99–108. Paris.
13. Mayer JD, Gaschke YN. Brief mood introspection scale (BMIS). *Psychology* 2001;**19**(3):1995.
14. Land MF, Horwood J. Which parts of the road guide steering? *Nature* 1995;**377**:339–40.
15. Kass SJ, VanWormer LA, Mikulas WL, Legan S, Bumgarner D. Effects of mindfulness training on simulated driving: preliminary results. *Mindfulness* 2011;**2**(4):236–41.
16. Jeon Y, Nam CS, Kim YJ, Whang MC. Event-related (De)synchronization (ERD/ERS) during motor imagery tasks: implications for brain-computer interfaces. *International Journal of Industrial Ergonomics* 2011;**41**(5):428–36.
17. Pfurtscheller G, Lopes FH. Event-related EEG/MEG synchronization and desynchronization: basic principles. *Clinical Neurophysiology* 1999;**110**:1842–57.
18. Tang Y-Y, Ma Y, Wang J, Fan Y, Feng S, Lu Q, et al. Short-term meditation training improves attention and self-regulation. *Proceedings of the National Academy of Sciences of the United States of America* 2007;**104**(43):17152–6.
19. Renard Y, Lotte F, Gibert G, Congedo M, Maby E, Delannoy V, Lécuyer A. OpenViBE: an open-source software platform to design, test, and use brain–computer interfaces in real and virtual environments. *Presence: Teleoperators and Virtual Environments* 2010;**19**(1):35–53.

Chapter 16

Tracking Mental Workload by Multimodal Measurements in the Operating Room

Ahmet Omurtag[1], Raphaëlle N. Roy[2], Frédéric Dehais[2], Luc Chatty[3,4], Marc Garbey[3,5]

[1]Nottingham Trent University, Nottingham, United Kingdom; [2]ISAE-SUPAERO, Université de Toulouse, Toulouse, France; [3]Houston Methodist Hospital, Houston, TX, United States; [4]University of Houston, Houston, United States; [5]LaSIE UMR 7356 CNRS, University of La Rochelle, La Rochelle, France

INTRODUCTION

About 15 million operating room (OR) procedures are performed annually in the United States.[1] A "hotspot" for medical errors, inpatient surgery is associated with 0.4%–0.8% rate of death and 3–17% rate of major complications.[2] Studies suggest that about half of surgical complications are avoidable,[3,4] and high-functioning teams have significantly lower numbers of adverse events.[5] New techniques that are being introduced potentially improve patient safety but impose dramatic new demands on surgeons' abilities and workload. More than 1 million laparoscopic surgeries are performed annually in the United States, where a surgeon operates with indirect, narrow visual access and minimal tactile feedback. Such conditions require new skills with different learning curves and new training methods beyond the traditional master–apprentice format.[6] In fact, as healthcare patterns shift toward prevention and quality, previously unexamined aspects of the OR come into sharper focus and surgeons and trainees are scrutinized for their performance.[7–9,9a]

Surgeons use sophisticated instruments for extended periods, often under time pressure, communicate with nurses and anesthesiologists, and interact with the complex interfaces of monitors. They possess technical skills acquired through long training, and also deploy an array of nontechnical skills.[10] These include situation awareness (gathering and understanding information and anticipating future states) and task management (responding to change). A strategic action may be, for example, deciding whether to convert a laparoscopic to an open-incision procedure. If the primary tasks (e.g., suturing) present unusual difficulty, this may impair the detection of an important alarm[11] or undermine proper planning. Even nearly automated mental processes, such as correcting for camera angle[12] or mismatches between an endoscope's optical axis and the instrument's plan on the monitor,[13] may take resources away from the surgeon's overall functions. Changes in mental workload due to training or new instrument design will have far-reaching implications not only for efficiency but also for patient outcomes.

Behavioral and physiological measurements can help improve surgeons' workload monitoring. In developing measures of surgeon workload, hybrid or multimodal approaches are preferable to unimodal ones, since they are able to deliver larger sets of information that illuminate the operator's functioning from multiple perspectives. Distinct measurement methods often have different strengths and shortcomings, and may compensate for each other's artifacts. Furthermore, as hardware becomes increasingly miniaturized and sensor design improves, the cost and effort related to including additional modalities decrease.[14]

Yurko et al.[15] utilized NASA-TLX to analyze the laparoscopic performance of novice trainees and explain the extent of the transfer of their simulator-acquired skills to the OR. They found that the mental and physical demand ratings obtained at the beginning of training predicted part of the subsequent animal OR performance scores (inadvertent injuries and suturing quality). Subjective methods such as NASA-TLX may be disruptive and only provide intermittent information. However, the usefulness of the information highlights the need for unobtrusive, continuous means for tracking surgeons' mental load.

Despite the apparent need, mental workload tracking in the OR using physiological measurements is underexplored. Although some studies have used such measurements to compare standard versus robot-assisted surgery (Hubert et al., 2013) or monitored surgeons using electroencephalography,[16,16a] we are not aware of any study that uses multimodal techniques in this area. We present a system of measurements for the OR whose immediate purpose was to generate a large, multimodal dataset suitable for quantifying mental workload. The acquired datasets were recorded from electroencephalography (EEG), eye tracking, electrocardiography, plethysmography, and instruments with pressure sensors. We computed the pupil diameter[17] and heart rate variability[18] that are commonly used to assess mental effort in response to task demand. We also measured

Neuroergonomics. https://doi.org/10.1016/B978-0-12-811926-6.00016-6
Copyright © 2019 Elsevier Inc. All rights reserved.

auditory evoked potentials, generated by low-probability auditory stimuli, that are useful for investigating the effects of workload on perceptual processing[19] and have proven to be an efficient indirect means to derive mental workload in multitasking scenarios.[20] Expected benefits from this ongoing research effort include improved training programs and certification, more effective development of new technology, real-time safety alerts, and models capable of assisting OR management. Beyond this, the system is intended as a source of data that can be mined to quantify team dynamics and efficiency.

METHODS

We performed experiments with 22 healthy volunteer subjects (4 females) on the fundamentals of laparoscopic surgery (FLS) model of assessment.[21] The subjects varied in level of experience (4 experts or board-certified surgeons; 8 surgical residents from postgraduate years 1–4; 10 nonsurgeons or beginners with no experience with FLS). They performed the standard manual tasks of peg transfer, string pass, and circle cut, listed in order of increasing difficulty.[22] During performance their brain activity was recorded at 500 Hz using a 20-channel dry-sensor EEG at the international 10–20 electrode sites with ear lobes as the reference and ground (Quick-20, Cognionics). Pupil size and point-of-regard data was collected by an infrared-based eye-tracking system at 60 Hz (EyeTribe Tracker). A Hexoskin wearable vest (Carré Technologies, Montreal) was used to monitor heart rate and breathing rate.

In addition, a pressure sensor designed in-house was mounted at both left and right tool handles to monitor the force exerted by the subjects' thumbs while operating the instruments. We formulated several metrics likely to reveal differences of skill and training. Active time segments were defined as those when the pressure was above a fixed threshold. We defined right–left overlap as those time segments when the right and left pressures were both active, and right–left asymmetry as $|R - L| / (R + L)$ where R (and L) was the time average of the amplitude of the right (and left) tool pressure. We verified that the derived measures were not significantly affected by changes in the threshold within a wide range of values.

Fig. 16.1 illustrates a participant using the experimental setup and the types of data that were collected. The figure includes an additional device for monitoring tool trajectories (Smart Trocar[23]) whose data was not used in this chapter.

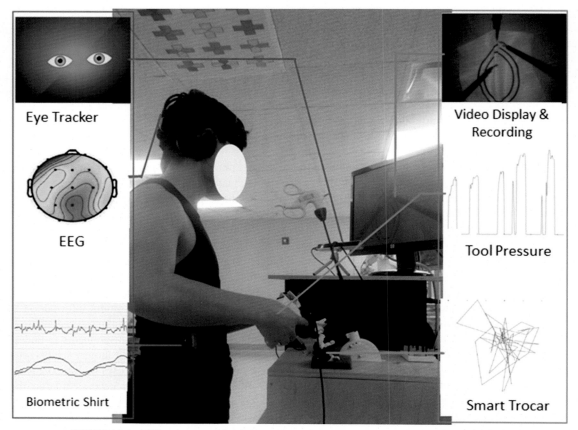

FIGURE 16.1 Different types of data being collected while a participant uses the experimental setup.

Regarding the experimental protocol, sound probes were displayed during the surgical training task (100 ms duration, every 3 s on average, selected randomly from six different frequencies in the range 750–2000 Hz). The subjects were instructed to ignore the sound probes. All modalities were centrally controlled from a graphical user interface (GUI) capable of configuring the type of experiment to be performed, stimulus type (only the sound probe was implemented), and the data modalities to be included in the recording, as well as displaying video feedback from the FLS camera. Implemented in Matlab (MathWorks, Natick, MA, United States), the GUI is part of a software platform that collects real-time synchronized data from all modalities and stores it for offline analysis. To study the effects of expertise and task difficulty, each subject performed the peg transfer, string pass, and circle cut while multimodal data was collected. In a subsequent experiment designed to measure the effects of time on task, groups with different skill levels (expert and surgical resident) performed the peg transfer three times without a break. Each session was preceded by a 1 min resting state recording.

To calculate the heart rate variability (HRV) from the electrocardiogram (ECG) data, a widely used marker of autonomic activity[24] for the time series of the normal-to-normal intervals in the ECG was resampled on a regular time grid with cubic spline interpolation, and the spectral power in its low-frequency (LF) (0.05–0.15 Hz) and high-frequency (HF) (0.15–0.5 Hz) bands was extracted. The HRV was defined as HF/(HF+LF).

As regards the EEG data, although researchers[16] have shown interest in studying frequency measures extracted from the EEG to perform mental workload monitoring during laparoscopic tasks, in this study we investigated whether event-related potentials extracted from ignored auditory probes[19,20] could also be used to monitor training during FLS. Indeed, tasks that require discrimination between classes of stimuli evoke in particular a large positive voltage deflection for about 300–500 ms after stimulus onset, known as the P300 component. The amplitude of the P300 is larger when the subject's attention is more focused on the task, and it is modulated by numerous factors including cognitive ability and mental workload.[25] We therefore chose to focus on P300 in this study, and extracted its amplitude from electrode Cz.

Several statistical comparisons were performed on the extracted markers to determine the impact of expertise, training time, and surgery subtask. The significance of intergroup differences was determined by using the one-way ANOVA test.

RESULTS

Fig. 16.2 shows the results derived from measurements from subjects of varying skill levels during FLS tasks of varying difficulty. The circle cut exercise resulted in subject-averaged pupil size that was significantly different from that during the string pass, an easier task. The average pupil size in the peg transfer, the easiest task, did not significantly differ from the other tasks. HRV varied significantly between operators of different skill levels (expert and nonsurgeon) performing the same set of FLS exercises. The subject-averaged P300 amplitude of response at Cz to the sound probe decreased (regardless of skill level) as subjects spent more time on a repeated peg transfer task.

Fig. 16.3A shows a segment of the tool pressure time series from two participants. The expert time series was shifted up for clarity. As the pegs were picked up one by one by the left tool and passed to the right tool, the expert's left and right pressures indicated a phase-locked wave pattern, absent from those of the resident. Fig. 16.3B shows that pressure-derived metrics (the fraction of active time, right–left overlap, and right–left asymmetry) were significantly different between the expert and resident groups.

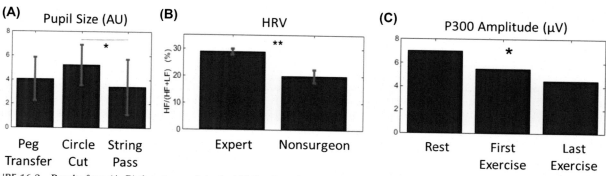

FIGURE 16.2   Results from (A–B) three types of standard FLS tasks and (C) three repetitions of the peg transfer task. (A) Pupil size as a function of type of task. (B) HRV as a function of skill level. (C) Amplitude of the P300 response to sound probes as a function of time on task (*$P < .05$; **$P < .01$).

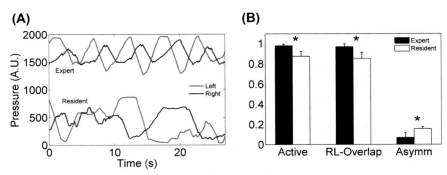

FIGURE 16.3 Results derived from tool pressure measured continuously as subjects (expert and residents) performed the peg transfer task. (A) Pressure time series often differed visibly between experts and residents. (B) The fraction of time during which the pressure was above a fixed threshold (Active), the fraction of time during which right–left pressures were both active (RL-Overlap), and the right–left pressure asymmetry (Asymm) were significantly different between two groups (*$P < .05$).

DISCUSSION

We examined the feasibility of multimodal physiological measures for tracking mental workload during surgery. We developed a platform that can simultaneously collect data from EEG, heart rate, breathing rate, tool handle pressure, and eye trackers from mobile subjects. The FLS assessment model was used as the experimental setting. The results indicate that the FLS task difficulty correlated significantly with pupil size and that HRV was related to operator skill level, indicating that untrained operators experienced a task as being harder.[18]

The P300 response to the ignored sound probes decreased significantly with training and time on task during a repeated FLS task. The effects of time on task and learning are always difficult to disentangle. Fatigue and workload effects often interact,[26] and monitoring systems should be designed to take these phenomena into account. In any case this amplitude decrement could be expected from the mental fatigue monitoring literature.[27] In addition, our setup showed that the applied pressure on the tool handles contained patterns capable of robustly discriminating between experts and residents. Such metrics can be developed further to provide mechanisms for the automated classification of finer gradations of skill, the assessment and certification of surgery trainees, real-time flags and warnings for the OR, and validation of new OR technology.

An advantage of multimodal quantification of operators' activities is that such measurements can reveal varying degrees of effort that may go into similar levels of overt performance. Under some conditions, for example in testing a new instrument, a surgeon may make extra efforts to increase her primary task performance at the expense of additional mental load, which may go undetected. If secondary tasks are introduced they may influence the primary task, or fail to provide accurate estimates because the subject did not reach capacity.[28] Behavioral metrics may also decouple from the mental load when trainees attain a performance plateau, where the *only* effect of additional practice is to decrease the mental load.[29] If trainees stop practicing at this stage, they may be left unprepared for stressful situations that may arise later.[30,31]

CONCLUSION

Despite mounting evidence, the field of surgery has not received sufficient attention from researchers developing physiology-based methods to track operators' mental workloads continuously. Results presented here suggest that quantifying mental workload and other previously unexplored aspects of surgery through multimodal measurements can improve surgery training, and ultimately impact efficiency and safety in the OR.

REFERENCES

1. Weiss AJ, Elixhauser A. Trends in operating room procedures in U.S. Hospitals, 2001–2011: statistical Brief #171. In: *Healthcare cost and Utilization Project (HCUP) statistical Briefs*. Rockville (MD): Agency for Health Care Policy and Research (US); 2006.
2. Haynes AB, Weiser TG, Berry WR, Lipsitz SR, Breizat A-HS, Dellinger EP, et al. A surgical safety Checklist to Reduce Morbidity and Mortality in a Global Population. *New England Journal of Medicine* 2009;**360**(5):491–9.
3. Gawande AA, Thomas EJ, Zinner MJ, Brennan TA. The incidence and nature of surgical adverse events in Colorado and Utah in 1992. *Surgery* 1999;**126**(1):66–75.
4. Kable AK, Gibberd RW, Spigelman AD. Adverse events in surgical patients in Australia. *International Journal for Quality in Health Care* 2002;**14**(4):269–76.

5. Mazzocco K, Petitti DB, Fong KT, Bonacum D, Brookey J, Graham S, et al. Surgical team behaviors and patient outcomes. *The American Journal of Surgery* 2009;**197**(5):678–85.

6. Van Hove PD, Tuijthof GJM, Verdaasdonk EGG, Stassen LPS, Dankelman J. Objective assessment of technical surgical skills. *British Journal of Surgery* 2010;**97**(7):972–87.

7. Kao LS, Thomas EJ. Navigating towards improved surgical safety using aviation-based strategies. *Journal of Surgical Research* 2008;**145**(2):327–35.

8. Kohn L, Corrigan JM, Donaldson MS. *To Err is human: Building a safer Health system.* Washington DC: Committee on Quality of Health Care in America, Institute of Medicine; 2000.

9. Risucci D, Geiss A, Gellman L, Pinard B, Rosser J. Surgeon-specific factors in the acquisition of laparoscopic surgical skills. *The American Journal of Surgery* 2001;**181**(4):289–93.

9a. Pavlidis I, Tsiamyrtzis P, Shastri D, Wesley A, Zhou Y, Lindner P, et al. Fast by nature-how stress patterns define human experience and performance in dexterous tasks. *Sci. Rep* 2012;**2**:305.

10. Yule S, Flin R, Maran N, Rowley D, Youngson G, Paterson-Brown S. Surgeons' non-technical skills in the operating room: reliability testing of the NOTSS behavior rating system. *World Journal of Surgery* 2008;**32**(4):548–56.

11. Dehais F, Causse M, Vachon F, Régis N, Menant E, Tremblay S. Failure to detect critical auditory alerts in the cockpit evidence for inattentional deafness. *Human Factors: The Journal of the Human Factors and Ergonomics Society* 2014;**56**(4):631–44.

12. Klein MI, Riley MA, Warm JS, Matthews G. Perceived mental workload in an endocopic surgery simulator. In: *Proceedings of the Human factors and Ergonomics Society annual Meeting.* Vol. 49. SAGE Publications; 2005. p. 1014–8.

13. Patil PV, Hanna GB, Cuschieri A. Effect of the angle between the optical axis of the endoscope and the instruments' plane on monitor image and surgical performance. *Surgical Endoscopy And Other Interventional Techniques* 2004;**18**(1):111–4.

14. Gramann K, Gwin JT, Ferris DP, Oie K, Jung T-P, Lin C-T, et al. Cognition in action: imaging brain/body dynamics in mobile humans. *Reviews in the Neurosciences* 2011;**22**(6):593–608.

15. Yurko YY, Scerbo MW, Prabhu AS, Acker CE, Stefanidis D. Higher mental workload is associated with poorer laparoscopic performance as measured by the NASA-TLX tool. *Simulation in Healthcare* 2010;**5**(5):267–71.

16. Zander TO, Shetty K, Lorenz R, Leff DR, Krol LR, Darzi AW, et al. Automated task load detection with electroencephalography: towards passive brain–Computer interfacing in robotic surgery. *Journal of Medical Robotics Research* 2016:1–10.

16a. Hubert N, Gilles M, Desbrosses K, Meyer JP, Felblinger J, Hubert J. Ergonomic assessment of the surgeon's physical workload during standard and robotic assisted laparoscopic procedures. *Int. J. Med. Robot* 2013;**9**:142–147.

17. Peysakhovich V, Causse M, Scannella S, Dehais F. Frequency analysis of a task-evoked pupillary response: Luminance-independent measure of mental effort. *International Journal of Psychophysiology* 2015;**97**(1):30–7.

18. Durantin G, Gagnon J-F, Tremblay S, Dehais F. Using near infrared spectroscopy and heart rate variability to detect mental overload. *Behavioural Brain Research* 2014;**259**:16–23.

19. Roy RN, Breust A, Bonnet S, Porcherot J, Charbonnier S, Godin C, Campagne A. Influence of workload on auditory evoked potentials in a single-stimulus paradigm. In: *2nd international confrenrence on physiological computing, PhyCS 2015.* 2015.

20. Roy RN, Frey J. Neurophysiological markers for passive brain–Computer interfaces. In: Bougrain L, Clerc M, editors. *Brain–computer interfaces: methods, Applications, and Perspectives.* Vol. 1. UK: ISTE-Wiley; 2016. p. 85–100.

21. Vassiliou MC, Dunkin BJ, Marks JM, Fried GM. FLS and FES: comprehensive models of training and assessment. *The Surgical Clinics of North America* 2010;**90**(3):535–58.

22. Peters JH, Fried GM, Swanstrom LL, Soper NJ, Sillin LF, Schirmer B, et al. Development and validation of a comprehensive program of education and assessment of the basic fundamentals of laparoscopic surgery. *Surgery* 2004;**135**(1):21–7.

23. Toti G, Garbey M, Sherman V, Bass BL, Dunkin BJ. A Smart Trocar for automatic tool Recognition in laparoscopic surgery. *Surgical Innovation* 2015;**22**(1):77–82.

24. Task Force of the European Society of Cardiology. Heart rate variability standards of measurement, physiological interpretation, and clinical use. *European Heart Journal* 1996;**17**:354–81.

25. Gevins A, Smith ME. Neurophysiological measures of working memory and individual differences in cognitive ability and cognitive style. *Cerebral Cortex* 2000;**10**(9):829–39.

26. Roy RN, Bonnet S, Charbonnier S, Campagne A. Mental fatigue and working memory load estimation: interaction and implications for EEG-based passive BCI. In: *2013 35th annual international Conference of the IEEE Engineering in medicine and Biology Society (EMBC).* IEEE; 2013. p. 6607–10.

27. Dehais F, Duprès A, Di Flumeri G, Verdière K, Borghini G, Babiloni F, Roy RN. Monitoring pilot's cognitive fatigue with engagement features in simulated and actual flight conditions using an hybrid fNIRS-EEG passive BCI. *Forthcoming in Proc. of the 2018 IEEE Int'l Conf. on Systems Man and Cybernetics.* 2018.

28. Byrne A, Tweed N, Halligan C. A pilot study of the mental workload of objective structured clinical examination examiners. *Medical Education* 2014;**48**(3):262–7.

29. Wickens CD, Hollands JG, Banbury S, Parasuraman R. *Engineering psychology & human performance.* Psychology Press; 2015.

30. Carswell CM, Clarke D, Seales WB. Assessing mental workload during laparoscopic surgery. *Surgical Innovation* 2005;**12**(1):80–90.

31. Johnston JH, Cannon-Bowers JA. Training for stress exposure. *Stress and Human Performance* 1996:223–56.

Chapter 17

Toward Brain-Based Interaction Between Humans and Technology: Does Age Matter?

Mathias Vukelić[1,2], Kathrin Pollmann[1,2], Matthias Peissner[1,2]

[1]*University of Stuttgart, Stuttgart, Germany;* [2]*Fraunhofer Institute for Industrial Engineering IAO, Stuttgart, Germany*

INTRODUCTION

In the age of digitalization the influence of technology on our daily lives is increasing, which requires us to learn how to interact with a variety of technical systems in various settings. Thus, there is an increased interest in designing adaptive assistance systems capable of adjusting their behavior to minimize the interaction effort for the user. Due to advances in sensor technology, adaptive systems can use information from the environment to tailor their behavior and interaction mechanisms better to different contexts.[1,2] However, when it comes to gathering and processing information about the user's individual needs and preferences, the potential of adaptive systems has not yet been fully explored.

Brain-based interaction is discussed as a strategy to make assistive technologies more user-oriented. Neuroadaptive assistance systems (NATs) use information about the users' mental and affective state to adjust their behavior and attributes accordingly.[3] While the advantages of NATs are indisputable, so far little empirical evidence exists about neuroelectrical signatures underlying affective user reactions to adaptive system behavior. In addition, the question arises as to whether these signatures are constant across different age groups or change as we grow older. Given the dynamic nature of the brain, it seems likely that a NAT should not follow a "one-size-fits-all approach" but requires age-specific strategies when providing assistance for the user.

The present study investigates whether there are age-related alterations of the cerebral cortex that may influence the ability to detect neuroelectrical signatures underlying affective reactions. We use an experimental paradigm that puts the user in a realistic interaction scenario with an adaptive assistance system (AFFINDU) capable of inducing affective user states.[4]

METHODS

Participants

Thirty-six volunteers (age range 22–71 years) were recruited. They were compensated for their participation and gave their written informed consent which was approved by a review committee (Medical Faculty of the University of Tuebingen, Germany). Participants were divided into two groups based on a median split of their age (median=45): young (Y, N=19, 22–44 years) and old (O, N=17, 48–71 years).

Experimental Procedure and Data Recording

The detailed experimental procedure was described by Pollmann et al.[4]. In short, participants were asked to carry out a number of navigation tasks within the AFFINDU system that generated positive and negative events during the task by either supporting or impeding participants' goal achievement. The experiment lasted 75 min and was divided into three blocks, with the affective events distributed as 20 supportive adaptations in block A, 20 impeding adaptations in block B, and 10 supportive and 10 impeding adaptations in block C. The interaction task with AFFINDU was realized in a controllable, trial-based procedure including different task epochs. The epoch of interest was the time window during which participants observed the adaptation performed by AFFINDU and thus the affect induction took place. This observation

Neuroergonomics. https://doi.org/10.1016/B978-0-12-811926-6.00017-8
Copyright © 2019 Elsevier Inc. All rights reserved.

epoch consisted of two parts: 1 s visualization of the navigation step and 4 s visualization of system adaptation. The onset of each epoch was indicated by a visual cue. After each block a self-assessment manikin (SAM) was used to assess participants' subjective evaluation of the interaction with AFFINDU in terms of valence and arousal. During the experiment, scalp electroencephalography (EEG) potentials were recorded (BrainAmp, Brainproducts, Germany) at 32 positions of the extended international 10–05 system: AFp1, AFp2, AFF5h, AFF6h, F7, F1, F2, F8, FFC3h, FFC4h, FC5, FC3, FC4, FC6, C3, C1, C2, C4, CP3, FCz, CP4, TP10, P7, P3, P1, P2, P4, P8, PPO1h, PPO2h, POO1, and POO2 (actiCAP electrodes, Brainproducts, Germany; the left mastoid was used as common reference, grounded to Cz). All impedances were kept below 20 kΩ. After filtering the EEG signals with a time constant of 10 s the data was recorded with a 1 kHz sampling rate and stored for offline analysis.

EEG Data Analysis

All trials of supportive and impeding system behavior from the three blocks were grouped together. The EEG signals were detrended, zero padded, rereferenced to mathematically linked mastoids[5] and bandpass filtered between 0.5 and 22 Hz. Stimulus-locked epochs ranging from 200 ms before to 2000 ms after the beginning of the visualization of system adaptation were created separately for supportive and impeding system behavior trials. Epochs were rejected when they contained a maximum deviation above 200 μV in any of the frontal EEG channels. For each unrejected epoch we performed an independent component analysis (ICA) using the logistic infomax ICA algorithm implemented in the EEGlab toolbox,[6] and identified ICA components with remaining artifacts by visual inspection.[7] Baseline correction was performed using the mean amplitude during the interval between −200 and 0 ms before the beginning of the system adaptation. Grand averages of event-related potentials (ERPs) were calculated by averaging ERPs across all trials of supportive and impeding system behavior and participants, thus giving an overall view of the temporal neuroelectrical dynamics for each EEG channel.

Discriminative information on ERPs for each EEG channel and time point was computed using the signed r^2 value, as implemented in the BBCI toolbox.[8] This serves as a univariate measure of separability between affective reactions to supportive and impeding adaptations.

RESULTS

Subjective Ratings

The SAM ratings for the valence and arousal dimensions were entered into a series of repeated measures ANOVA (rmANOVA) including blocks (A, B, C) as within-subject factor and age group (Y, O) as between-subject factor. One participant had to be removed from age group Y due to systematic response.

For valence, the rmANOVA revealed a significant effect of measurement block ($F(1.68,55.55)=13.26$; $p<.001$). Post hoc pairwise comparisons showed significant differences between all blocks: block A ($M=6.26$, $SE=.25$) was rated significantly higher than C ($M=5.30$, $SE=.24$) and B ($M=4.56$, $SE=.34$; $p=.006$ and $P<.01$ respectively). Significantly higher ratings were found for block C compared to B ($p=.042$). We also discovered an effect of age group ($F(1,33)=4.37$; $p=.044$), indicating generally higher ratings for group O than for group Y. There was no interaction effect between block and age group ($F(1.68,55.55)=1.07$; $p=.341$).

For arousal, a significant effect of block was found ($F(1.45, 47.75)=5.96$; $p=.01$). Post hoc pairwise comparisons indicated that block A ($M=3.49$, $SE=.34$) was rated significantly lower than block B ($M=4.84$, $SE=.40$; $p=.032$). Neither blocks A nor B differed significantly from block C ($M=4.01$, $SE=.29$; $p=.224$ and $p=.081$ respectively). There was no effect of age group ($F(1, 33)=.82$; $p = 0.371$) nor an interaction effect between measurement block and age group ($F(1.45, 47.75)=.6$; $p=.501$).

Event-Related Potentials

Fig. 17.1 shows the time course of the ERPs as the grand average for groups Y (Fig. 17.1A) and O (Fig. 17.1B) for supportive (Fig. 17.1A and B, left column) and impeding (Fig. 17.1A and B, middle column) system behavior, respectively. Differences for the two types of system behavior are already present for early components of visually induced ERPs, starting after system adaptation onset, such as N100, N200, P300, and the late positive potential (LPP) complex. For the signed r^2 values we found certain time windows to be most discriminative between supportive and impeding adaptations: N100=140–180 ms, N200=185–245 ms, P300=345–375 ms, and LPP=625–660 ms after system adaptation for both age groups (Fig. 17.1A and B, right column).

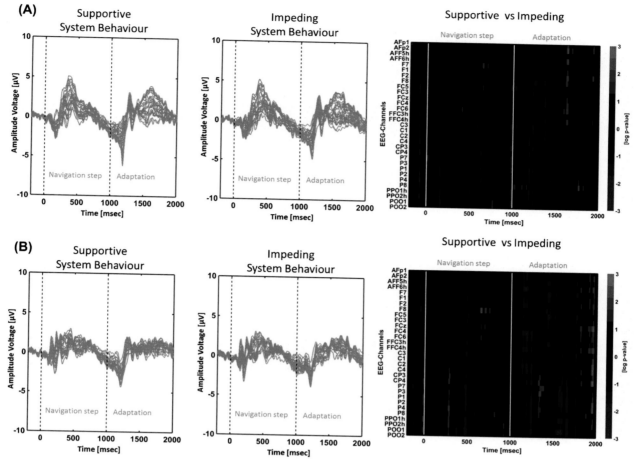

FIGURE 17.1 The plots show the temporal dynamics of event-related potentials (ERPs) for the young (A) and old group (B) Left and middle column in (A) and (B) show the grand-averaged waveforms of ERPs visualized as a butterfly plot for the supportive system behaviour and impeding system behaviour. Every *line* represents single EEG electrodes. The right column in (A) and (B) shows the spatial distribution of strongest discriminability between affective reactions to supportive and impeding system behaviour. The 2D graph represents the grand-averaged signed r^2-values analysed for every time point (abscissa) for all EEG-channels (ordinate). The colours indicate significance level on a log10 scale.

Next we extracted for each participant the maxima of r^2 values for each time window and related this value to their age. We found that age significantly positively correlated with higher discriminability for P300 in electrodes, overlying motor (Fig. 17.2, Spearman's correlation coefficient r=.38, p value=.02) and parieto-occipital regions (Fig. 17.2, Spearman's correlation coefficient r=.37, p value=.02). No correlations were found between age and other ERP time windows.

DISCUSSION

We found that subjective valence ratings were affected by age, while arousal ratings were not. Group O showed a tendency to rate positive events as more positive and negative events as less negative. This finding could be explained by the positivity effect which describes an age-related trend to prefer positive over negative information in cognitive processing.[9]

The analysis of the spatiotemporal dynamics of visually induced ERPs revealed decreased amplitudes for both system behaviors for group O, an alteration that has already been demonstrated for the aging brain.[10,11] Nevertheless, we found affect-related ERPs[12,13] at early latencies (N100 and N200) and later latencies (P300 and LPP) that can be used to dissociate supportive and impeding system behavior as well as identify age-related differences, as they varied to different degrees across electrodes overlying circumscribed cortical regions between the two age groups. P300 is a component that is related to the allocation of attentional resources during affect processing.[14] Our results showed that the discrimination based on P300 between the different types of behaviors of AFFINDU correlated with participants' age. P300 is furthermore known to be sensitive to negative stimuli.[15] The group of younger participants showed higher amplitudes for impeding system behavior while the elderly group showed the opposite effect, a fact that likely reflects the differences in cognitive information processing of affective reactions.[9,16,17]

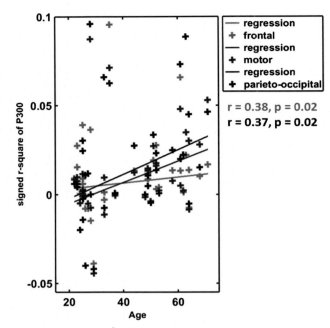

FIGURE 17.2 The scatter plot represents the max signed r^2-value in the P300=345–375 msecs after system adaptation for each participant on the ordinate where the abscissa indicates the age. Colours indicate electrode clusters overlying frontal regions (red): AFp1, AFp2, AFF5h, AFF6h, F1, F2, F7, F8; motor related regions (blue): FC5, FC3, FC6, FC4, FFC3h, FFC4h, C3, C1, C4, C2; and parieto-occipital regions (black): CP4, CP3, P8, P4, P7, P3, PPO2h, PPO1h, POO2, POO1. Lines represent the result of robust regression analysis of signed r^2-values onto age using iteratively reweighted least squares with a bisquare weighting function for motor electrodes (spearman's correlation coefficient r=.38, P-value=.02) and parieto-occipital electrodes (spearman's correlation coefficient r=.37, P-value=.02). No significance for frontal electrodes was found.

To sum up, the ability to process affect during the interaction between humans and assistive technology is unaffected by age-related cortical alterations. However, our results suggest that there are distinct intrinsic neurocognitive strategies for affect processing in the aging brain. Addressing these age-related differences in neural signatures underlying affective reactions may facilitate personalized, user-specific adjustments of assistance technology to the user's current emotional state and preferences, thus promoting the development of more powerful, age-specific NATs.

ACKNOWLEDGMENTS

This research was supported by grants from the German Federal Ministry for Education and Research (BMBF: 16SV7195K) and the European Union's Seventh Framework Programme under FP7 Grant #610510.

REFERENCES

1. Fairclough SH. Fundamentals of physiological computing. *Interacting with Computers* 2009;**21**(1–2):133–45.
2. Picard RW, Vyzas E, Healey J. Toward machine emotional intelligence: analysis of affective physiological state. *IEEE Transactions on Pattern Analysis and Machine Intelligence* 2001;**23**(10):1175–91.
3. Zander TO, Krol LR, Birbaumer NP, Gramann K. Neuroadaptive technology enables implicit cursor control based on medial prefrontal cortex activity. *Proceedings of the National Academy of Sciences* 2016. 201605155.
4. Pollmann K, Ziegler D, Peissner M, Vukelić M. *A New experimental paradigm for affective research in neuro-adaptive technologies*. ACM Press; 2017. p. 1–8.
5. Nunez PL, Srinivasan R. *Electric fields of the brain: the neurophysics of EEG*. 2nd ed. Oxford (New York): Oxford University Press; 2006.
6. Delorme A, Makeig S. EEGLAB: an open source toolbox for analysis of single-trial EEG dynamics including independent component analysis. *Journal of Neuroscience Methods* 2004;**134**(1):9–21.
7. Hipp JF, Siegel M. Dissociating neuronal gamma-band activity from cranial and ocular muscle activity in EEG. *Frontiers in Human Neuroscience* 2013;**7**:338.
8. Blankertz B, Tangermann M, Vidaurre C, Fazli S, Sannelli C, Haufe S, Müller K-R. The Berlin brain–computer interface: non-medical uses of BCI technology. *Frontiers in Neuroscience* 2010;**4**.

9. Mather M, Carstensen LL. Aging and motivated cognition: the positivity effect in attention and memory. *Trends in Cognitive Sciences* 2005;**9**(10):496–502.

10. Kieffaber PD, Okhravi HR, Hershaw JN, Cunningham EC. Evaluation of a clinically practical, ERP-based neurometric battery: application to age-related changes in brain function. *Clinical Neurophysiology* 2016;**127**(5):2192–9.

11. Polich J. EEG and ERP assessment of normal aging. *Electroencephalography and Clinical Neurophysiology* 1997;**104**(3):244–56.

12. Czekóová K, Shaw DJ, Urbánek T, Chládek J, Lamoš M, Roman R, Brázdil M. What's the meaning of this? A behavioral and neurophysiological investigation into the principles behind the classification of visual emotional stimuli: semantic classification in emotion processing. *Psychophysiology* 2016;**53**(8):1203–16.

13. Olofsson JK, Nordin S, Sequeira H, Polich J. Affective picture processing: an integrative review of ERP findings. *Biological Psychology* 2008;**77**(3):247–65.

14. Polich J. Updating P300: an integrative theory of P3a and P3b. *Clinical Neurophysiology* 2007;**118**(10):2128–48.

15. Stewart JL, Silton RL, Sass SM, Fisher JE, Edgar JC, Heller W, Miller GA. Attentional bias to negative emotion as a function of approach and withdrawal anger styles: an ERP investigation. *International Journal of Psychophysiology* 2010;**76**(1):9–18.

16. Addis DR, Leclerc CM, Muscatell KA, Kensinger EA. There are age-related changes in neural connectivity during the encoding of positive, but not negative, information. *Cortex* 2010;**46**(4):425–33.

17. Cassidy BS, Leshikar ED, Shih JY, Aizenman A, Gutchess AH. Valence-based age differences in medial prefrontal activity during impression formation. *Social Neuroscience* 2013;**8**(5):462–73.

Chapter 18

Mobile Neuroergonomics: Action, Interfaces, Cognitive Load, and Selective Attention

Ryan McKendrick

Northrop Grumman, Redondo Beach, CA, United States; George Mason University, Fairfax, VA, United States

ACTION, ENVIRONMENT, AND THE BRAIN

We all experience cognitive physical dual tasking. Going grocery shopping is a simple example. In my personal experience, I often meander up and down the store aisles searching for the obscure items my wife has placed on the shopping list, all the while trying to maintain those last few grocery list items in memory. Others experience these dual-task conditions when the stakes are considerably higher. For example, firefighters and dismounted soldiers must make difficult decisions, navigate difficult terrain, and fight (fires or other humans)—all while laden with heavy gear. Neuroergonomics is a methodology that integrates cognitive neuroscience, cognitive psychology, and human factors to study the brain in relation to performance at work, in everyday settings such as grocery shopping, and in more challenging situations such as firefighting or combat.[1] However, implementing neuroergonomic methods in these settings is problematic mainly due to humans moving around frequently when at work. Thus neuroergonomics requires tools like mobile functional near-infrared spectroscopy (fNIRS) which have a signal that is robust to movement.[2,3]

Movement likely has complex effects on the brain. Such effects are central to the field of embodied cognition. Essentially, the theory suggests that thought, in a basic sense, is influenced by a mind occurring in a body. Strong versions of the doctrine assert that the body determines the thoughts of the mind because the body is in actuality an extension of the mind.[4] Weaker versions of the doctrine assert that thoughts of the mind are heavily constrained by the body, and that certain thoughts cannot occur without a mind interacting with a body. From the embodied cognition perspective, movement is vital to understanding the brain at work, and it further emphasizes the need for neuroimaging tools that do not restrict and are robust to movement.

However, understanding a "mobile brain" means we must also understand what it means for a brain to be situated in an environment. When we think or move, our environment may have profound influences on our thoughts and behavior. A complete approach to neuroergonomics requires us to quantify these effects of environment. We need to understand how our cognitions differ based on what we do, ranging from sitting in a cubicle to walking around city streets or along a forest path.

The integration of neurogonomics and embodied and situated cognition inspired work examining how the brain changes with movement and the environment.[5] Previous work on the interaction between cognitive and physical load has produced complex effects. For instance, it has been observed that when adding a cognitive task to a physical task brain activity increases,[6–8] and conversely brain activity decreases when a physical task is added to a cognitive task.[9,10] The reticular-activating hypofrontality (RAH) hypothesis[11] suggests this is due to a resource distribution hierarchy in the brain. According to the RAH hierarchy, the resource requirement for action take precedence over requirements for resources for executive processes.

To explore the interactions between perceptual load and the cognitive demand of physical tasks on cognitive tasks, three conditions were created.

- In the first condition, participants sat in a chair and performed an auditory memory task. A 1-back memory task was used where participants heard tone triplets and had to maintain and report an internal count of the number of triplets repeated back to back.
- In the second condition, the same memory task was used. However, instead of sitting, participants had to walk through an empty hall.

Neuroergonomics. https://doi.org/10.1016/B978-0-12-811926-6.00018-X
Copyright © 2019 Elsevier Inc. All rights reserved.

- In the final condition, the auditory memory and walking tasks were used, but rather than walking indoors the participants walked outside on a busy college campus.

During each of the three conditions, memory accuracy and brain activity were measured. Measurement of brain activity was made over the prefrontal cortex (PFC) with an fNIR Devices 1100w mobile fNIRS system.

As a result of these manipulations it was observed that cognitive–physical dual tasking had a negative effect on the performance of an easy memory task. When sitting and counting tones, participants recalled the number of matches 90% of the time. When walking, memory accuracy was reduced to about 85%, and walking outdoors reduced accuracy even further to 80%. Walking reduced the total amount of blood normally available to the lateral PFC (LPFC). This finding is consistent with the RAH hypothesis, which proposes walking as a mentally intensive task that requires the brain to redistribute computational resources away from regions not directly involved in walking.

Most surprisingly, the LPFC was less active while cognitive–physical dual tasking outdoors in a more complex environment (Fig. 18.1). One might think the reduction in performance and LPFC activity suggests distraction, or lack of motivation. However, there is a better explanation. Perceptual load theory predicts that under conditions of high perceptual load, such as during outdoor walking, irrelevant stimuli are blocked from attention instead of being resolved by attention.[12] The ventrolateral and especially left ventrolateral PFC has been associated with controlled memory retrieval and interference resolution of irrelevant stimuli.[13] Hence it follows that the ventrolateral PFC should be suppressed when perceptual load is high (i.e., walking outdoors) and distractors are inhibited.

In addition to observations regarding environment and attention, these effects suggest that physical activity and attention take precedence over working memory for computational resources. However, this may only be for cases where the goals of action and attention are linked; the cognitive goals are not causally linked. Future studies, particularly those examining cognition in the contexts of sports, search and rescue, or combat, may observe conflicting results. It could also be observed that this resource distribution will change with expertise.

This work speaks to physical fitness trends during office work. Ample evidence suggests that sitting all day is not conducive to optimal health.[14] The negative effects of sitting have prompted many individuals to work at standing desks. Others have taken even more extreme measures and opted for walking desks. Certainly, if one's goal is to burn calories, walking is superior to standing. However, if one would also like to remain productive and functional (from a cognitive perspective) than a walking desk may not be the best choice. From the above results we can expect a 5%–10% reduction in cognitive performance while walking.

Perhaps employers looking to encourage performance, productivity, and cardiovascular health should encourage their employees to take periodic walking breaks. This could accomplish the goals of a walking desk and provide employees with a designated time for mind wandering (a counterproductive work behavior that occurs more readily during low-demand tasks like walking[15]). Interestingly, both mind wandering and walking have been separately linked to increased creativity.[15,16] Thus designated time for mind wandering and walking would be doubly beneficial in terms of maintaining productivity and potentially enhancing creativity. It is also interesting that Oppezzo and Schwartz observed that walking outdoors relative to other walking conditions produced the greatest number of creative ideas.

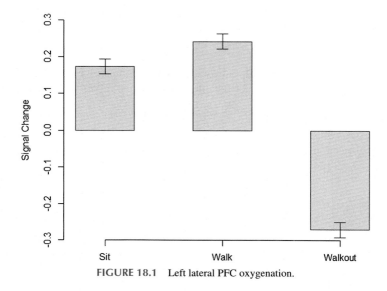

FIGURE 18.1 Left lateral PFC oxygenation.

ACTION, TECHNOLOGY, AND THE BRAIN

The future of technology and computing is leveraging mobility, and advances in "wearables" and augmented reality showcase this trend; examples are tattooed sensors[17] and advanced commercial display systems such as Microsoft Hololens and Meta 2. Neuroergonomics offers a new tool for guiding and advancing technology. Traditionally, ergonomics and human factors psychology have been advocates for the human in the tsunami of technological advancement.[18] Yet in many ways human advocates were "bringing a knife to a gun fight." Neuroergonomics is a key methodological component for leveling the playing field. It provides the needed tools for noninvasive quantification of relevant cognitive concepts and translating these into design recommendations. Two fields of technology that are ripe for neuroergonomic assessment are augmented reality technology and artificial-intelligence-based autonomous systems. Neuroergonomics can contribute to these fields as human factors and ergonomics contributed to the sciences of display design and critiqued the implementation of automation.[19]

In 2014 there was tremendous hype surrounding a new technology product: smart glasses. Effectively, smart glasses were perceived as having the potential to usurp the smartphone as our ubiquitous computational companion. The focus of the hype was Google's Glass. The device was anticipated to inject the smartphone experience into our every waking moment by projecting information directly in our field of view via a head-mounted display. At the time of writing Google Glass, for a myriad reasons, has been deemed a failure. However, the promise of an immersive computing experience when we work and play is still an attractive proposition. To test if Glass was an objective failure, we evaluated it neuroergonomically. We wanted to know which was better for ambulatory navigation in natural settings, an augmented reality wearable display (i.e., Glass) or a smart phone.

We wanted to explore the cognitive capacity required for individuals to navigate with either a smartphone or Google Glass. To measure cognitive capacity objectively, we had to employ functional neuroimaging as well as a dual-task paradigm. Dual tasking is effective because by observing an individual's accuracy on a secondary task, assuming no errors were made on the primary task, we can infer the resources used by the primary task. Combining this with functional neuroimaging enhances observations made with the dual-task method by relating secondary task performance to changes in the strength and locus of brain activity. This linking can lead to deeper explanations of experimental phenomena by connecting disparate literature.

This combination also covers the shortcomings of isolated dual-task or neurophysiological workload assessment. For instance, measuring cognitive capacity without behavioral controls can quickly devolve into an art of reading tea-leaves: because the brain is so complex, it is incredibly difficult to parse out the desired brain signal from many other real brain signals. Further, because there are so many signals, it is very easy to find spurious correlations even when using advanced machine-learning techniques such as support vectors, principal components, or convolutional and recurrent neural nets. And the addition of physiological measurement to dual-task assessment does remediate the criticism that it is dependent on circular reasoning[20] by moderating the relationship between cognitive load and task performance.

Within the dual-task framework, our primary navigation task utilized Google Maps to provide visual directions. Participants navigated across George Mason University's campus, broken into four routes, each about a third of a mile in length. We carefully chose our secondary tasks to emphasize components that are common to everyday ambulatory navigation and the assumed benefits or decrements of a Glass-like system. For these reasons we used an auditory memory task to tax working memory, and a scenery probe task to test situation awareness. The auditory memory task was identical to the one described earlier in the chapter. The scenery probe task was a forced choice, where individuals had to respond "yes" or "no" to a question posed by the experimenter. The questions were focused around whether the participant had or had not seen an object along their route. We posed an equal number of positive and negative questions. Positive questions required a positive response (i.e., "yes") to be correct, and vice versa for negative questions. We alternated between these two secondary tasks during the routes, and the two tasks never cooccurred during navigation.

From the study manipulations it was observed that Google Glass required significantly less mental effort to use during ambulatory navigation. In the Glass condition, memory accuracy was about 80%. The drop in performance when navigating with a phone was quite severe, reducing memory accuracy to about 60%.

The performance differences between the two devices also carried over to the user's brain activity. While there were several interesting neurophysiological effects specific to each device, direct comparisons between the two devices provided the most insight. Specifically, it took considerably less LPFC activity to perform the memory task accurately while using Glass (Fig. 18.2). This difference in oxygenation presents strong evidence that Glass users experienced less mental workload and could perform the navigation task more efficiently than phone users.

Unfortunately, our results for situation awareness could not be as cleanly interpreted as those for cognitive workload. For one, we could not observe a difference in how accurately users could perceive and recall objects between devices. Both groups of users perceived and recalled objects at about 75% efficiency. Given this lack of difference, we looked to our neurophysiological observations to differentiate the devices. When correctly performing the secondary task we did not observe any differentiating effects, but we did observe differentiation when the secondary task was performed incorrectly. Glass users showed an increase in left LPFC activity (Fig. 18.3). We believe this was likely due to irrelevant attention capture, rather than the resolution of perceived irrelevant stimuli such as surrounding conversations. It would be beneficial for future work to validate increases in LPFC oxygenation with experimentally induced cognitive tunneling.

McKendrick et al.[21] focused on the context of google glass use, yet their results are also relevant to smartphone use during ambulatory activity. We know that smartphone use is cognitively taxing during activities like driving,[22] but we demonstrated that it is also highly taxing during walking. There has been a considerable increase in pedestrian deaths in the United

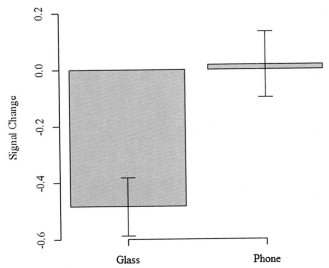

FIGURE 18.2 Auditory memory 1-black oxygenation left LPFC.

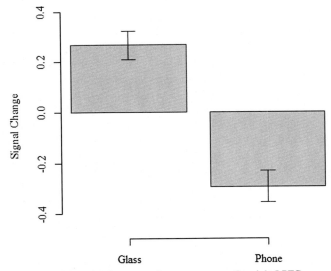

FIGURE 18.3 Scenery probe errors oxygenation right LPFC.

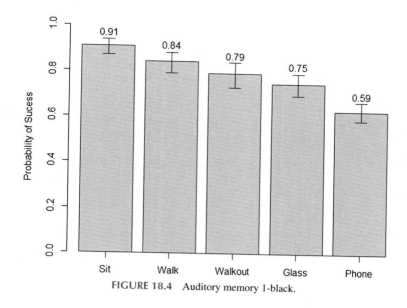

FIGURE 18.4 Auditory memory 1-black.

States recently: from 2015 to 2016 pedestrian deaths were up 11%, and such deaths accounted for 15% of all automotive-related fatalities.[23] Currently this increase is interpreted as reflecting distracted driving, but there is the possibility that the increased fatalities are related to pedestrian distraction. Go to any major US city and one can observe pedestrian smartphone use. It is constant and pervasive, likely more so than smartphone use by drivers. If we combine observations from two studies by McKendrick et al.[5,21] we can see that smartphone use during walking and memorizing takes a heavy toll on cognition (Fig. 18.4). We should therefore increase our exploration of pedestrian smartphone use and automobile-related pedestrian fatalities: while we continue to explore interventions to reduce distracted driving, we can in parallel determine if there is a need to reduce distracted walking.

NEUROERGONOMICS AND SELECTIVE ATTENTION

The interaction between cognition, action, and attention is key to advancing neuroergonomics as a discipline and increasing the benefit of its research products to society. Both studies reviewed above suggest that selective attention plays a key role in our abductive reasoning. During the mobile navigation study, we argue that failure of early selective attention was related to increased brain activity in the LPFC and inadvertent cognitive capture of display elements. In the environmental complexity study we saw less activity in the LPFC, which we argued was related to a suppression of distractor resolution by late selective attention. Both these results reflect the applicability of perceptual load theory to neuroergonomics research. Refining the understanding of what we should expect from attention given the circumstances and an individual's goals will help us better design systems for those consistently involved in cognitive–physical dual tasking. Increased research in this area will particularly benefit dismounted soldiers, who will in the future have to both fight and interact with other intelligent agents. Developing systems and interfaces that enhance their role and provide them with the best safety envelope is critical.

CONCLUSION

Neuroergonomics as a discipline breaks down the traditional boundaries of human factors, cognitive psychology, and cognitive neuroscience. It is beginning to provide understanding and avenues for future research in augmented reality, combat tactics, work–life balance, pedestrian safety, human–machine interaction, and adaptive automation. The field should continue to push forward, enhancing our paradigms for understanding complex behaviors in complex settings with complex technologies to benefit both those who think while mobile and those who interact with technology while mobile.

REFERENCES

1. Parasuraman R, Rizzo M. *Neuroergonomics: the brain at work.* New York (NY): Oxford University Press; 2008.

2. Ayaz H, Onaral B, Izzetoglu K, Shewokis PA, McKendrick R, Parasuraman R. Continuous monitoring of brain dynamics with functional near infrared spectroscopy as a tool for neuroergonomic research: empirical examples and a technological development. *Frontiers in Human Neuroscience* 2013;**7**:871.

3. McKendrick R, Parasuraman R, Ayaz H. Wearable functional near infrared spectroscopy (fNIRS) and transcranial direct current stimulation (tDCS): expanding vistas for neurocognitive augmentation. *Frontiers in Systems Neuroscience* 2015;**9**:1–14.

4. Wilson RA, Foglia L. Embodied cognition. In: Zalta EN, editor. *The stanford encyclopedia of philosophy.* Spring; 2017. https://plato.stanford.edu/archives/spr2017/entries/embodied-cognition/.

5. McKendrick R, Mehta R, Ayaz H, Scheldrup M, Parasuraman R. Prefrontal hemodynamics of physical activity and environmental complexity during cognitive work. *Human Factors* 2017;**59**(1):147–62.

6. Mandrick K, Derosiere G, Dray G, Coulon D, Micallef JP, Perrey S. Prefrontal cortex activity during motor tasks with additional mental load requiring attentional demand: a near-infrared spectroscopy study. *Neuroscience Research* 2013;**76**:156–62.

7. Mehta RK. Stunted PFC activity during neuromuscular control under stress with obesity. *European Journal of Applied Physiology* 2016;**116**:319–26.

8. Mirelman A, Maidan I, Bernad-Elazari H, Nieuwhof F, Reelick M, Giladi N, Hausdorff JM. Increased frontal brain activation during walking while dual tasking: an fNIRS study in healthy young adults. *Journal of Neuroengineering and Rehabilitation* 2014;**11**:1.

9. Mehta RK, Parasuraman R. The effect of mental fatigue on the development of physical fatigue: a neuroergonomic approach. *Human Factors* 2014;**56**:645–56.

10. Mehta RK, Shortz AE. Obesity-related differences in neural correlates of force control. *European Journal of Applied Physiology* 2014;**114**:197–204.

11. Dietrich A, Audiffren M. The reticular-activating hypofrontality (RAH) model of acute exercise. *Neuroscience and Biobehavioral Reviews* 2011;**35**:1305–25.

12. Lavie N, Hirst A, de Fockert JW, Viding E. Load theory of selective attention and cognitive control. *Journal of Experimental Psychology: General* 2004;**133**:339–54.

13. Badre D, Wagner AD. Left ventrolateral prefrontal cortex and the cognitive control of memory. *Neuropsychologia* 2007;**45**:2883–901.

14. de Rezende LFM, Lopes MR, Rey-López JP, Matsudo VKR, do Carmo Luiz O. Sedentary behavior and health outcomes: an overview of systematic reviews. *PLoS One* 2014;**9**(8):e105620.

15. Baird B, Smallwood J, Mrazek MD, Kam JW, Franklin MS, Schooler JW. Inspired by distraction mind wandering facilitates creative incubation. *Psychological Science* 2012. https://doi.org/10.1177/0956797612446024.

16. Oppezzo M, Schwartz DL. Give your ideas some legs: the positive effect of walking on creative thinking. *Journal of Experimental Psychology: Learning, Memory, and Cognition* 2014;**40**(4):1142.

17. Song J-K, Son D, Kim J, Yoo YJ, Lee GJ, Wang L, Choi MK, Yang J, Lee M, Do K, Koo JH, Lu N, Kim JH, Hyeon T, Song YM, Kim D-H. Wearable force touch sensor array using a flexible and transparent electrode. *Advanced Functional Materials* 2017;**27**:1605286.

18. Hancock PA, Hoffman RR. Keeping up with intelligent technology. *IEEE Intelligent Systems* 2015;**30**(1):62–5.

19. Wickens CD, Hollands JG, Banbury S, Parasuraman R. *Engineering psychology and human performance.* Upper Saddle River (NJ): Prentice Hall; 2013.

20. Navon D. Resources: a theoretical soupstone. *Psychological Review* 1984;**91**:216–34.

21. McKendrick R, Parasuraman R, Murtza R, Formwalt A, Baccus W, Paczynski M, Ayaz H. Into the wild: neuroergonomic differentiation of hand-held and augmented reality wearable displays during outdoor navigation with functional near infrared spectroscopy. *Frontiers in Human Neuroscience* 2016;**10**:216.

22. Scott JJ, Gray R. A comparison of tactile, visual, and auditory warnings for rear-end collision prevention in simulated driving. *Human Factors* 2008;**50**(2):264–75.

23. Macek K. *Pedestrian fatalities projected to surge 11% in 2016.* March 30, 2017. Retrieved from http://www.ghsa.org/resources/news-releases/pedestrians-2017.

Chapter 19

Computational Models for Near-Real-Time Performance Predictions Based on Physiological Measures of Workload

Matthias D. Ziegler[1], Bartlett A. Russell[1], Amanda E. Kraft[2], Michael Krein[2], Jon Russo[2], William D. Casebeer[1]

[1]Lockheed Martin Advanced Technology Lab, Arlington, VA, United States; [2]Lockheed Martin Advanced Technology Lab, Cherry Hill, NJ, United States

INTRODUCTION

Over the last decade computational models have gained an increasing presence as techniques to understand human behavior and link it to physiological measures have evolved.[1–3] Studies utilizing computational models have successfully shown links between measures of workload and performance that were not previously apparent due to the large amount of data that current sensors are able to collect.[4,5] While many studies implement such models, they can vary significantly between studies due to the diversity of tasks being tested, number/type of sensors, and analysis techniques. This leads to highly specialized models and corresponding physiological sensors that do not transfer between tasks and individuals. The experiments and models are so specialized that any change to the task or individual being modeled requires complete system retraining, proving impractical in applications beyond controlled experiments. To bring physiologically based computational models outside the lab for practical use in real-world environments it is important to examine how we can minimize the number of physiological features with minimal cost to the predictive power, and reliably process and analyze the features in a manner that is beneficial for understanding both workload levels and performance across individuals and tasks of interest.

Studies have shown that cognitive workload levels can be measured using an increasing number of available sensing techniques. Electroencephalography (EEG) has been one of the most common tools for measuring workload, identifying increased neural activity corresponding to workload levels.[6–8] Eye tracking is also common, as evidence of pupil size and blink rate have been linked to workload levels.[9,10] Electrocardiography (ECG) offers another means to assess workload levels via heart rate variability.[4] By combining sensors some studies have been able to show workload levels consistent across multiple physiological sensors[5,11] and to increase the classification accuracy of any one of these systems alone by accounting for a greater number of physiological systems that respond to changes in workload. In this study we use this combined sensing approach to measure performance in multiple tasks to determine how well general levels of workload are linked to task performance across individuals. While other tools, particularly functional near-infrared spectroscopy, functional magnetic resonance imaging, and biomarkers,[12,13] have also shown to be important measures of workload, we do not address these tools in this study.

Understanding how computational models use the multiple types of physiological information to predict performance is an important way to identify the sensors and features that are most useful. Here we start that process by looking at creating a single subject agnostic generalizable model to predict performance based on the available sensors: EEG, eye tracking, and ECG. We compare accuracy of models trained based on various combinations of three sessions (excluding a practice session) and then tested on a final session. We compare these results with individualized models trained on the first three sessions and tested on the last session for that individual. We then test how sensitive randomized models are to removal of each of the sensors and features within the sensor type. All models were single-layer neural networks built and tested in MATLAB.

Computational models can vary in complexity of programming, amount of data needed for training, time needed to build the model, and number of parameters that need to be adjusted (i.e., layer size, learning rates, etc.). It is important to examine how the accuracy of different modeling approaches and physiological signals affects predictions in an effort to understand the trade-off between accuracy and time needed to set up the testing environment and execute the model.

Neuroergonomics. https://doi.org/10.1016/B978-0-12-811926-6.00019-1
Copyright © 2019 Elsevier Inc. All rights reserved.

Simply running a model with a single training/testing set may cause skewed results, as the training or testing data chosen may not be representative of the overall data. As performance variation between cross-validation runs is an indicator of dataset variability and an estimate for overall method reliability with respect to the data, we present the model results tested over 25 model cross-validation runs and indicate that cross-validation should be standard procedure when testing models.

The results of our study show that a generalized model with no tailoring performs very poorly when applied to a new individual, and some level of model adaptation or personalization is necessary for any level of model prediction to be valid. These findings indicate that a comprehensive study is necessary to understand the trade-offs between generalized model performances versus the costs of adapting models to individuals. Additionally we determined that the predictive power of some sensors (EEG) weighed much heavier in the models than others (ECG). Finally, our sensitivity analysis within the EEG features showed that by analyzing only half the available features the models dropped less than 3% in predictive power, allowing the models to run quicker with minor trade-offs.

METHODS

Participants

A mix of 35 participants were trained and tested on two computer-based video games to measure performance over four nonconsecutive days. EEG systems with 26 electrodes arranged according to the standard 10–20 system were used during testing.[14] Additional BrainVision electrodes placed on the collarbone recorded Electrocardiography (ECG) for Heart rate variability (HRV) analysis. An SMI eye tracker recorded eye movements.

Task Design

We designed two tasks to titrate workload: a simple snake game (see Fig. 19.1 bottom) and Prepar3D flight simulator (Fig. 19.1 top), each of which contained multiple levels of difficulty. Subjects received one 45 min training session to become familiar with the tasks, and returned for 3 days following the training to perform the tasks. The order of the tasks was randomized for each day. Each difficulty level lasted for 5 min in both tasks.

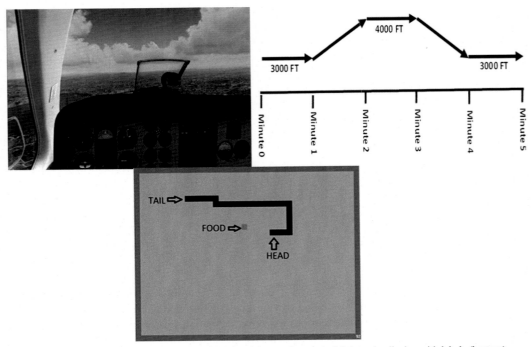

FIGURE 19.1 Prepar3D cockpit (top left) and task instructions (top right); snake display with labels (bottom).

RESULTS

Models were trained on each of the initial sessions over all participants and tested on the final session. For both tasks, the models performed with higher accuracy when trained on a range of sessions rather than a single prior session, e.g., mean areas under the curve (AUCs) of 0.59 and 0.56 when trained only in session 3 versus mean AUCs of 0.64 and 0.61 when trained in sessions 1–3 for the Prepar3D and Snake tasks respectively. Following these results, we built individualized models trained on a participant's first three sessions and tested on the last session for each task. Interestingly, while the generalized models performed better on the Prepar3D task, individualized models performed around chance for Prepar3D (mean AUC=0.52, sd=0.01) and significantly better for Snake (mean AUC=0.68, sd=0.01). Future work may assess how to determine the optimal number of sessions required for accurate predictions of current operator state. Our final models showed the highest predictive power when we tested using a combination of generalized models along with early data from the individual. In these cases the model AUC was 0.80 for Prepar3D and 0.75 for Snake (Fig. 19.2).

We further assessed the individual contributions of each sensor type (Fig. 19.2A). Given the unbalanced number of features for EEG (n=260), ECG (n=5), EOG, electrooculography (n=7), and eye tracking (n=7), EEG data was assessed by band. Regardless of band, EEG features were clearly weighted heavier than the other sensor types in performance prediction. To reduce the computational burden of building and testing models over roughly 300 features, within-feature sensitivity analysis was performed on EEG to assess whether each of the 10 bands initially used was necessary (Fig. 19.2B). Starting with all EEG features, each band was dropped sequentially, starting with the band of lowest contribution. For both tasks, model performance drops by 0.02 when the six lowest bands are dropped, removing 156 features and reducing model run time by roughly 2h.

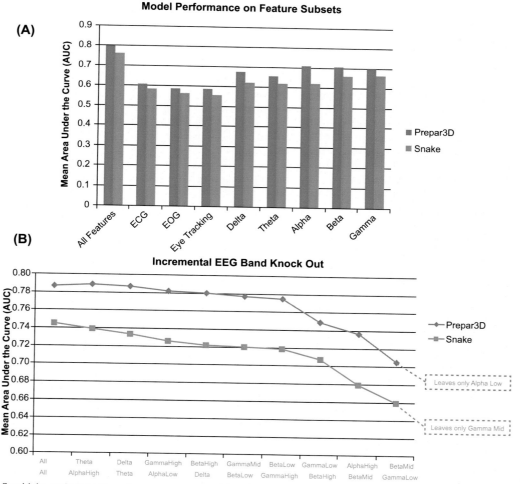

FIGURE 19.2 Sensitivity analysis performed across sensor types (A) and EEG feature bands (B) show differing powers of prediction depending on the task being performed.

DISCUSSION

While the ability to record features from a variety of sensors within the laboratory allows us to create highly accurate predictive models of behavior, it is important to remember that to transition these models to environments outside the research lab we need to minimize the number of sensors a person needs to wear. While it is true that sensors are becoming smaller and cheaper, it is still important to do a cost–benefit analysis on how predictive a sensor may be. In this research we used an automated sensitivity analysis tool that allowed us to determine which sensors were contributing most to our workload-based performance prediction, allowing us to eliminate the sensors with little or no predictive power.

Based on the results from the two simple tasks, we posit that the predictive power of each sensor may differ based on the task being performed and within sensors it may be important to focus on different features. There currently is no one-size-fits-all approach to measuring workload, and each study task should strategically pick sensors and features after performing similar sensitivity analysis to that proposed here.

ACKNOWLEDGMENTS

We would like to acknowledge Bradley Hatfield, Li-Chun Lo, Hyuk Oh, and Kyle Jaquess at the University of Maryland for their assistance in collecting data and technical support. Also we would like to acknowledge Ian Warfield of Lockheed Martin for his work in task development.

REFERENCES

1. Kieras DE, Meyer D. *Computational modeling of human multiple-task performance. No. TR-05/ONR-EPIC-16*. Michigan Univ, Ann Arbor Dept of Electircla Eng. and Comp. Sci.; 2005.
2. Hugo J, Gertman DI. The use of computational human performance modeling as task analysis tool. In: *Proceedings of the Eighth American Nuclear Society International Topical Meeting on nuclear plant instrumentation, control, and human-machine interface technologies, NPIC&HMIT 2012*. 2012. p. 22–6.
3. Meng J, Wu X, Morozov V, Vishwanath V, Kumaran K, Taylor V. SKOPE: a framework for modeling and exploring workload behavior. In: *Proceedings of the 11th ACM Conference on computing frontiers*. ACM; 2014. p. 6.
4. Ke Y, Qi H, He F, Liu S, Zhao X, Zhou P, Ming D. An EEG-based mental workload estimator trained on working memory task can work well under simulated multi-attribute task. *Frontiers in Human Neuroscience* 2014;**8**.
5. Liu Y, Ayaz H, Onaral B, Shewokis PA. Neural adaptation to a working memory task: a concurrent EEG-fNIRS study. In: *Foundations of augmented cognition*. Springer International Publishing; 2015. p. 268–80.
6. Kamzanova AT, Kustubayeva AM, Matthews G. Use of EEG workload indices for diagnostic monitoring of vigilance decrement. *Human Factors: The Journal of the Human Factors and Ergonomics Society* 2014;**56**(6):1136–49.
7. Walter CB. *EEG workload prediction in a closed-loop learning environment* (Doctoral dissertation). Universität Tübingen; 2015.
8. Brouwer AM, Hogervorst MA, Van Erp JB, Heffelaar T, Zimmerman PH, Oostenveld R. Estimating workload using EEG spectral power and ERPs in the n-back task. *Journal of Neural Engineering* 2012;**9**(4):045008.
9. Bodala IP, Kukreja S, Li J, Thakor NV, Al-Nashash H. Eye tracking and EEG synchronization to analyze microsaccades during a workload task. In: *Engineering in Medicine and Biology Society (EMBC), 2015 37th Annual International Conference of the IEEE*. 2015. p. 7994–7.
10. Zheng B, Jiang X, Tien G, Meneghetti A, Panton ONM, Atkins MS. Workload assessment of surgeons: correlation between NASA TLX and blinks. *Surgical Endoscopy* 2012;**26**(10):2746–50.
11. Choe J, Coffman BA, Bergstedt DT, Ziegler MD, Phillips ME. Transcranial direct current stimulation modulates neuronal activity and learning in pilot training. *Frontiers in Human Neuroscience* 2016;**10**.
12. Ayaz H, Shewokis PA, Bunce S, Izzetoglu K, Willems B, Onaral B. Optical brain monitoring for operator training and mental workload assessment. *Neuroimage* 2012;**59**(1):36–47.
13. Just MA, Carpenter PA, Miyake A. Neuroindices of cognitive workload: neuroimaging, pupillometric and event-related potential studies of brain work. *Theoretical Issues in Ergonomics Science* 2003;**4**(1–2):56–88.
14. Jasper HH. Report of the committee on methods of clinical examination in electroencephalography: 1957. *Electroencephalography and Clinical Neurophysiology* 1958;**10**(2):370–5.

Chapter 20

EEG-Based Mental Workload Assessment During Real Driving: A Taxonomic Tool for Neuroergonomics in Highly Automated Environments

Gianluca Di Flumeri[1,2,3,a], Gianluca Borghini[1,2,3,a], Pietro Aricò[1,2,3,a], Nicolina Sciaraffa[1,2,3], Paola Lanzi[4], Simone Pozzi[4], Valeria Vignali[5], Claudio Lantieri[5], Arianna Bichicchi[5], Andrea Simone[5], Fabio Babiloni[1,3, 6]

[1]*BrainSigns srl, Rome, Italy;* [2]*IRCCS Fondazione Santa Lucia, Rome, Italy;* [3]*University of Rome "Sapienza", Rome, Italy;* [4]*DeepBlue srl, Rome, Italy;* [5]*University of Bologna, Bologna, Italy;* [6]*Hangzhou Dianzi University, Hangzhou, China*

INTRODUCTION

The field of neuroergonomics studies the relationship between human behavior and the brain at work.[1] It provides a multidisciplinary translational approach that merges elements of neuroscience, cognitive psychology, human factors, and ergonomics to study brain structure and functions in everyday environments. Applied to the scenario of driving a car, a "neuroergonomic approach" should allow us to investigate the relationship between human mental behavior, performance, and road safety, providing a deeper understanding of human cognition and its role in decision making and possible error commission at the wheel.[2] It has been demonstrated that human error is the main cause of 57% of road accidents and a contributing factor in over 90% of them.[3] Decreased human performance, and consequently error commission, are directly attributable to aberrant mental states, in particular the fact that mental performance degrades in overload situations, which is considered one of the most important human factors constructs in influencing performance.[4,5] The model theorized by De Waard,[6] widely used in automotive research, establishes the relation between task demands and performance depending on driver workload. This model describes driving activity with a hierarchy of tasks on three levels, strategic, tactical, and operational, each divided into different subtasks, indicating that driving is a very complex and often high-demand activity. Thus the cognitive resources required in very complex situations can exceed the available resources, leading to overload and performance impairments.[7,8]

To address this issue and ensure a proper level of user mental workload during his/her operating activities, and thus a higher level of safety, automated systems have been widely investigated and developed to support the user during high-demand activities in complex environments, such as aviation and car driving.[9] In the automotive domain, several systems have been developed and are already standard equipment, such as adaptive cruise control, lane-keeping assistance, and other advanced driver assistance systems, with the final aim being to reach the level of fully automated vehicles.[10] However, the effectiveness of high levels of automation in driving is still debated.[11] Basically, it is argued that automation does not necessarily mitigate difficult situations, and paradoxically that automation can sometimes even make such situations more difficult for the human operator. This occurs when automatic control completely replaces operator actions, while at the same time the user is still required to monitor that the automated tasks are carried out effectively by the system. A recurring notion of Sheridan[9] is that humans have limited ability in monitoring automation because of a decrease in situation awareness or an overreliance on or no confidence in the system, all concepts related to the *out-of-the-loop* phenomenon.

The out-of-the-loop issue in human–automation interaction (e.g., operator performance decreasing when automation fails) has been studied in several domains outside automotives, in particular in aviation.[12] One effective way to mitigate this issue is to develop adaptive automation (AA) solutions: an AA-based system is able to adjust the proper level of automation

---

a. These authors contributed equally to the work.

Neuroergonomics. https://doi.org/10.1016/B978-0-12-811926-6.00020-8
Copyright © 2019 Elsevier Inc. All rights reserved.

continuously, and to assign authority on specific functions to either the human or the automated system depending on the task difficulty and the operator's workload. It has been demonstrated that AA is able to ensure the operator's workload remains within the optimum range, preserve his/her skill level, guarantee continuous task involvement and vigilance, and thus increase his/her performance.[12,13] In this context neurophysiological measures, in particular electroencephalography (EEG), have been demonstrated to be one of the best techniques to infer in real time an objective assessment of the mental workload experienced by the user. Also, recent progress in passive brain–computer interface (pBCI) technology led researchers to consider EEG as the perfect candidate to trigger AA-based systems.[7,13,14]

Despite the large number of studies and evidence about EEG feature variations related to changes in mental workload during driving tasks,[15,16] in scientific literature there is still a lack of a synthetic EEG-based workload index computable online and able to trigger AA-based system.[7]

This preliminary study investigated the possibility of adopting the approach developed by Aricò and colleagues,[13] successfully applied in the air traffic management domain, to evaluate the mental workload experienced by a car driver by means of his/her EEG activity. The rationale was to use the EEG measures, instead of other neurophysiological or subjective measures, because of their specific suitability in objectively assessing human mental states to create a workload index.[14] In addition, eye-tracking technology was employed because of the insights achievable by its offline data analysis, with the aim of verifying the properness of the experimental design, i.e., if two driving conditions, one "easy" and one "hard," were actually perceived differently by the drivers.[4,17] Another added value of this study lies in the investigation of a real driving context. In this regard, all the studies related to workload investigation using EEG have been performed in a simulator or in poor realistic settings. It is important to prove the effectiveness of EEG-based metrics in real contexts, since it has been proven that some experimental tasks are perceived differently, in terms of mental workload, if performed in a simulator or in a real environment.[4]

MATERIAL AND METHODS

Experimental Protocol

Eight male students (24.9 ± 1.7 years old, licensed for 6 ± 1 years) from the University of Bologna were recruited, on a voluntary basis, in this study. They were selected to give a homogeneous experimental group in terms of age and expertise. The experiment was conducted following the principles outlined in the Declaration of Helsinki of 1964, as revised in 2000. Informed written consent was obtained from each subject after explanation of the study.

Two identical cars were used for the experiments: a Fiat 500 L 1.3 Mjt, with diesel engine and manual transmission. The subjects had to drive the car during daylight along a route constituted of urban roads. In particular, the route consisted in a "circuit" about 2500 m long to be covered three consecutive times (i.e. three laps). Brain activity and eye movements of each subject were collected using EEG techniques and an eye-tracking device (Fig. 20.1). Although the subjects received a briefing before the experiment, explaining the whole protocol, the first lap was considered a "lap of adaptation" to the experimental environment and was not taken into account for the data analysis. The circuit was designed to contain two segments

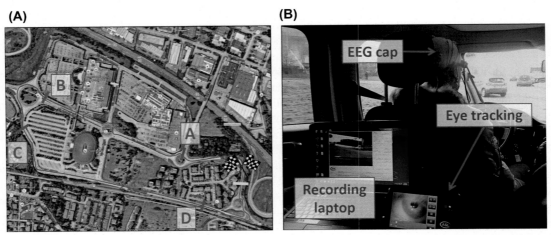

(A) **(B)**

FIGURE 20.1   (A) The circuit designed on urban roads, where the "hard" segment runs from A to B and the "easy" segment from C to D. (B) A picture within the car during the experimental recordings, showing the instrumentation used.

of interest, both about 1000 m long but supposedly different in terms of elicited workload,[7] thus named hereafter "easy" and "hard": "easy" is a secondary road, mainly straight, with an intersection halfway with the right of way, one lane, and low capacity; and "hard" is a main road, mainly straight, with a roundabout halfway, three lanes, and high capacity. Each subject twice performed (the first lap was not considered) both the "easy" driving section and the "hard" one (see Fig. 20.1).

EEG Recording and Processing

The EEG signals were recorded using the digital monitoring BEmicro system (EBNeuro, Italy). Twelve EEG channels (FPz, AF3, AF4, F3, Fz, F4, P3, P7, Pz, P4, P8, and POz), placed according to the international 10–20 system, were collected with a sampling frequency of 256 Hz, all referenced to both earlobes, grounded to the Cz site, and with the impedances kept below 20 kΩ. The acquired EEG signals were digitally bandpass filtered by a fifth-order Butterworth filter [1 - 30 Hz]. The eye-blink artifacts were removed from the EEG using the REBLINCA method.[18] This method is very effective for applications in real contexts, and online, because of its property of removing blink contributions from EEG signals only in correspondence with blinks and not requiring any Electrooculographic (EOG) channel (e.g., FPz was used in the present work).[13] The EEG signal from the remaining 11 electrodes was then segmented in two-seconds-long epochs, with a moving shifting of 0.125 seconds (i.e. for example for 3 seconds of signal you will have 8 epochs, the first from 0 to 2 seconds, from 0.125 to 2.125 s, from 0.250 to 2.250, and so on), both to obtain a high number of observations in comparison with the number of variables and to respect the condition of stationarity of the EEG signal.[19] For other sources of artifacts, specific procedures of the EEGLAB toolbox were applied[20] to remove EEG epochs marked as "artifact." The power spectral density was then estimated using fast Fourier transform in the EEG frequency bands defined for each subject by estimation of the individual alpha frequency (IAF) value.[21] In this regard, before starting the experiment the brain activity of each subject during a minute of rest (closed eyes) was recorded, to calculate the IAF. Thus the theta rhythms $[IAF-6 \div IAF-2]$ over the frontal sites and the alpha rhythms $[IAF-2 \div IAF+2]$ over the parietal sites were investigated, because of their strict relationship with mental workload,[15,22] and used to compute the mental workload index (WL index). As noted earlier, the WL index was calculated by using the machine-learning approach proposed by Aricò and colleagues,[13] the *automatic stop-StepWise Linear Discriminant Analysis* (as-SWLDA) classifier. The algorithm was calibrated by using the "easy" and "hard" conditions of the second lap and testing it on the "easy" and "hard" conditions of the third one, and vice versa. The time resolution of the WL index was fixed at 8 s, since this value was shown to be a good trade-off between the resolution and the accuracy of the measure.[14]

Eye Tracking Recording and Processing

Eye movements of the experimental subjects were recorded through an ASL Mobile Eye-XG device, a system based on lightweight eyeglasses equipped with two digital high-resolution cameras. One camera recorded the scene image and the other the participant's eye. The data was recorded with a sampling rate of 30 Hz (i.e., 33 ms) and a resolution of 0.5–1 degree. ASL software was used to analyze the data, obtaining information about the drivers' fixation points frame by frame (33 ms). A preliminary calibration procedure was carried out for each subject in the car before starting driving, asking them to hold their gaze on 30 fixed visual points spread across the whole scene, to get good accuracy from the eye-movement recorder. The gazes recorded during the driving task were grouped into three different categories: *road infrastructure*, *traffic vehicles*, and *external environment*. For each subject, each lap (second and third), and each condition ("easy" and "hard") the distribution of eye fixations between the three categories was calculated in terms of percentage. In particular, the percentage of fixations over the external environment was investigated, since this indicator has been proven to be inversely correlated with mental workload: the more the workload is experienced, the fewer the number of fixations on the external environment, since the driver gaze will mostly focus on infrastructure and vehicles.[4,23]

Performed Data Analysis

Analysis of subject eye fixations: the fixation percentages over the external environment for the "easy" and "hard" conditions were averaged between the two investigated laps for each subject. A one-tailed paired *t*-test was performed to verify that the two segments of the circuit were correctly designed to reproduce an "easy" and a "hard" condition.

Analysis of the area under curve (AUC) of the receiver operator characteristic curve of the classifier[24]: AUC is a widely used methodology to test performance of a binary classifier, and the classification performance can be considered good with an AUC higher than at least 0.7.[25] The classifier was tested shuffling the testing dataset, to verify that classifier performance on measured data (measured AUC) was significantly higher than that obtained on random data (random AUC). A one-tailed paired *t*-test was performed between measured and random AUCs.

Analysis of the WL index provided by the classifier: the obtained WL indexes were averaged between the two investigated laps for each condition and each subject. A one-tailed paired *t*-test was performed to verify differences between WL indexes, previously normalized by z-score to avoid possible subjective differences related to "easy" and "hard" conditions.

RESULTS

The analysis of the eye fixations, as reported in Fig. 20.2, shows that during the "easy" condition the external environment caught $12 \pm 10.7\%$ of the drivers' fixations, but only $5 \pm 5.5\%$ during the "hard" condition. The paired *t*-test revealed that such a difference was statistically significant ($P = .028$).

The AUC analysis revealed that using this approach it was possible to achieve a mean AUC value of 0.755 ± 0.11, giving a classification accuracy of 75%. In particular, the paired *t*-test demonstrated that measured AUCs were significantly higher ($P = .002$) than random AUCs (see Fig. 20.3A). Also, the paired *t*-test between the WL indexes during the two conditions showed that the WL indexes during the "hard" circuit section were significantly higher ($P = .011$) than those during the "easy" section (see Fig. 20.3B).

FIGURE 20.2 Mean and standard deviation of eye fixations on the external environment, in other words on all these regions of the scene that do not contain information (e.g., infrastructure, vehicles, signage) useful for driving. The fixations on the environment during the "easy" condition (*green bar*) were significant higher than those during the "hard" condition (*red bar*). For each condition two representative frames are reported on the right: on the top the "easy" condition, with the driver's gazed fixed on the external environment; on the bottom the "hard" condition, with the driver's gaze fixed on the precedent vehicles.

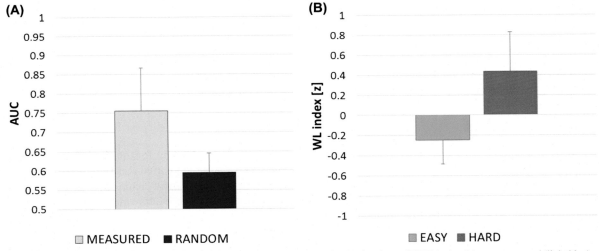

FIGURE 20.3 (A) The AUC values, in terms of mean and standard deviation, obtained using the as-SWLDA algorithm on measured (light blue) and random (dark blue) data. (B) The WL indexes (mean and standard deviation of the z-scores) obtained during the "easy" (green) and "hard" (red) conditions.

DISCUSSION

In the automotive domain, the relationship between human errors and driving performance impairment due to a high mental workload has been widely investigated.[6,7] Although several automated solutions have been introduced to support drivers, there is still a need to develop AA solutions to *adapt* the automation intervention depending on the driver's mental status, i.e., mental workload, to keep him/her always "in the loop."[12]

This study proposes an EEG-based index that is able to assess, even online, driver mental workload and eventually trigger AA-based systems. A machine-learning approach based on the use of as-SWLDA, developed and successfully employed by Aricò and colleagues within the aviation domain,[13] was here used in real traffic conditions along a route designed to elicit different levels of mental workload. The eye-tracking results confirm the hypothesis about the circuit design, highlighting how the assumed conditions of "easy" and "hard" were actually perceived coherently with the assumptions by the drivers (Fig. 20.2), since their eye fixations on the external environment were significantly higher in the "easy" than in the "hard" condition.[4] The results for classifier performance in discriminating the workload conditions demonstrate that the adopted approach achieves considerable performance (AUCs > 0.7 with a time resolution of 8 s), significantly higher than a random classifier ($P < .05$), providing a WL index that correctly assesses the driver's mental workload in the different conditions (Fig. 20.3). The achieved results highlight that it was possible to assess the driver's workload in a real context, and to obtain an objective index of his/her experienced workload each 8 s, thus providing a potential reliable online trigger for AA-based systems.

Despite the small size of the experimental sample, the results appear very promising, since they match a crucial need within the automotive field, where a lot of research had been conducted about neurophysiological features of mental workload but no evidence exists about a systematic and reproducible approach to obtain a workload index usable online and in real context applications.[7] Further analysis is necessary to validate the method with a larger sample and in other driving conditions.

CONCLUSIONS

In this study, the benefits of a multidisciplinary approach typical of neuroergonomics are evident, since it is possible to provide evidence and address issues of specific domains, but also to reemploy solutions in different contexts, such as in our case testing in the automotive field an algorithm developed for the aviation field. In fact, the present study tested a previously validated (in aviation) algorithm for human mental workload assessment with people driving in real traffic conditions. The algorithm performance was remarkable, since it was possible to obtain a correct assessment of the drivers' mental workload. Although further analysis is necessary, the results are very promising and should encourage research on systems based on AA and triggered by a driver's neurophysiological activity.

ACKNOWLEDGMENTS

The authors are grateful to the Unipol Group Spa, and in particular Alfaevolution Technology, for the considerable help given in the research study. The grant provided by the Italian Ministry of University and Education under PRIN 2012, Grant no. 2012WAANZJ scheme to F. Babiloni is also gratefully acknowledged.

REFERENCES

1. Parasuraman R, Rizzo M. *Neuroergonomics: the brain at work.* Oxford University Press; 2008.
2. Lees MN, Cosman JD, Lee JD, Rizzo M, Fricke N. Translating cognitive neuroscience to the driver's operational environment: a neuroergonomics approach. *The American Journal of Psychology* 2010;**123**(4):391–411, ISSN: 0002-9556.
3. Treat JR, Tumbas NS, McDonald ST, Shinar D, Hume RD. *Tri-level study of the causes of traffic accidents. Executive summary.* 1979. Retrieved from: https://trid.trb.org/view.aspx?id=144095.
4. de Winter JCF, Happee R, Martens MH, Stanton NA. Effects of adaptive cruise control and highly automated driving on workload and situation awareness: a review of the empirical evidence. *Transportation Research F: Traffic Psychology and Behaviour* 2014;**27**:196–217. https://doi.org/10.1016/j.trf.2014.06.016.
5. Parasuraman R, Sheridan TB, Wickens CD. Situation awareness, mental workload, and Trust in automation: Viable, Empirically supported cognitive Engineering constructs. *Journal of Cognitive Engineering and Decision Making* 2008;**2**(2):140–60. https://doi.org/10.1518/155534308X284417.
6. De Waard. *The measurement of drivers' mental workload.* Netherlands: University of Groningen, Traffic Research Centre; 1996.
7. Paxion J, Galy E, Berthelon C. Mental workload and driving. *Frontiers in Psychology* 2014;**5**. https://doi.org/10.3389/fpsyg.2014.01344.
8. Robert J, Hockey G. Compensatory control in the regulation of human performance under stress and high workload: a cognitive-energetical framework. *Biological Psychology* 1997;**45**(1–3):73–93. https://doi.org/10.1016/S0301-0511(96)05223-4.
9. Sheridan TB. *Humans and automation: system design and research issues.* Hoboken, N.J: Wiley; 2002. [u.a.].

10. SAE. *Taxonomy and Definitions for terms related to on-road Motor vehicle automated driving systems.* 2014. Retrieved from: http://standards.sae.org/j3016_201401/.

11. Strand N, Nilsson J, Karlsson ICM, Nilsson L. Semi-automated versus highly automated driving in critical situations caused by automation failures. *Transportation Research Part F: Traffic Psychology and Behaviour* 2014;**27**:218–28. https://doi.org/10.1016/j.trf.2014.04.005.

12. Byrne EA, Parasuraman R. Psychophysiology and adaptive automation. *Biological Psychology* 1996;**42**(3):249–68. https://doi.org/10.1016/0301-0511(95)05161-9.

13. Aricò P, Borghini G, Di Flumeri G, Colosimo A, Bonelli S, Golfetti A, Babiloni F. Adaptive automation triggered by EEG-based mental workload index: a passive brain-Computer interface application in realistic Air traffic control environment. *Frontiers in Human Neuroscience* 2016a:539. https://doi.org/10.3389/fnhum.2016.00539.

14. Aricò P, Borghini G, Di Flumeri G, Colosimo A, Pozzi S, & Babiloni F. (2016) A passive brain-computer interface application for the mental workload assessment on professional air traffic controllers during realistic air traffic control tasks. *Progress in Brain Research*, 228, 295–328. https://doi.org/10.1016/bs.pbr.2016.04.021.

15. Borghini G, Astolfi L, Vecchiato G, Mattia D, Babiloni F. Measuring neurophysiological signals in aircraft pilots and car drivers for the assessment of mental workload, fatigue and drowsiness. *Neuroscience and Biobehavioral Reviews* 2014;**44**:58–75. https://doi.org/10.1016/j.neubiorev.2012.10.003.

16. Brookhuis KA, de Waard D. Monitoring drivers' mental workload in driving simulators using physiological measures. *Accident Analysis & Prevention* 2010;**42**(3):898–903. https://doi.org/10.1016/j.aap.2009.06.001.

17. Lantieri C, Lamperti R, Simone A, Costa M, Vignali V, Sangiorgi C, Dondi G. Gateway design assessment in the transition from high to low speed areas. *Transportation Research Part F: Traffic Psychology and Behaviour* 2015;**34**:41–53. https://doi.org/10.1016/j.trf.2015.07.017.

18. Di Flumeri G, Aricò P, Borghini G, Colosimo A, Babiloni F. A new regression-based method for the eye blinks artifacts correction in the EEG signal, without using any EOG channel. In: *Conference Proceedings: Annual International Conference of the IEEE Engineering in Medicine and Biology Society. IEEE Engineering in Medicine and Biology Society. Annual Conference.* 2016.

19. Elul R. Gaussian behavior of the electroencephalogram: changes during performance of mental task. *Science (New York, NY)* 1969;**164**(3877):328–31.

20. Delorme A, Makeig S. EEGLAB: an open source toolbox for analysis of single-trial EEG dynamics including independent component analysis. *Journal of Neuroscience Methods* 2004;**134**(1):9–21. https://doi.org/10.1016/j.jneumeth.2003.10.009.

21. Klimesch W. EEG alpha and theta oscillations reflect cognitive and memory performance: a review and analysis. *Brain Research Brain Research Reviews* 1999;**29**(2–3):169–95.

22. Maglione A, Borghini G, Aricò P, Borgia F, Graziani I, Colosimo A, Babiloni F. Evaluation of the workload and drowsiness during car driving by using high resolution EEG activity and neurophysiologic indices. In: *Conference Proceedings: Annual International Conference of the IEEE Engineering in Medicine and Biology Society. IEEE Engineering in Medicine and Biology Society. Annual Conference, 2014.* 2014. p. 6238–41. https://doi.org/10.1109/EMBC.2014.6945054.

23. Costa M, Simone A, Vignali V, Lantieri C, Bucchi A, Dondi G. Looking behavior for vertical road signs. *Transportation Research Part F: Traffic Psychology and Behaviour* 2014;**23**:147–55. https://doi.org/10.1016/j.trf.2014.01.003.

24. Bamber D. The area above the ordinal dominance graph and the area below the receiver operating characteristic graph. *Journal of Mathematical Psychology* 1975;**12**(4):387–415. https://doi.org/10.1016/0022-2496(75)90001-2.

25. Fawcett T. An Introduction to ROC analysis. *Pattern Recogn Lett* 2006;**27**(8):861–74. https://doi.org/10.1016/j.patrec.2005.10.010.

Chapter 21

Preliminary Validation of an Adaptive Tactical Training Model: Cognitive Alignment With Performance-Targeted Training Intervention Model

Quinn Kennedy[1], Travis Carlson[2], Lee Sciarini[1]
[1]*Naval Postgraduate School, Monterey, CA, United States;* [2]*Office of Naval Research, Arlington, VA, United States*

INTRODUCTION

Military leadership recognizes the importance of agile, adaptive thinkers. The US Army and US Marine Corps have each issued strategic guidance initiatives directing efforts to improve decision-making.[1,2] However, as military experience is hard won—specifically combat experience, where a leader may have only one chance to learn from a decision—the use of adaptive tactical training, in particular for rapid-response decisions, has become the focus of increased study.[3–5]

Effective adaptive tactical training requires assessing whether a trainee is in a cognitive state of exploration or exploitation, and how far their decision performance deviates from optimal decision-making throughout a training event. The Cognitive Alignment with Performance Targeted Training Intervention Model (CAPTTIM) distinguishes between two subject cognitive states: exploration (the subject has not figured out the task and needs to explore the environment) and exploitation (the subject evaluates that s/he has mastered the task and is acting upon acquired knowledge).[4] The model then determines whether cognitive state is aligned or misaligned with observed decision performance (see Fig. 21.1). CAPTTIM utilizes simple behavioral measures to characterize cognitive state and decision performance. It uses variability in latency from decision to decision to determine whether the trainee's cognitive state is exploration (large latency variability) or exploitation (small latency variability). Decision performance is measured by regret: the difference between the trainee's decision and the optimal decision, given perfect knowledge of the task. High regret indicates poor decision performance; low regret indicates near-optimal decision performance (see Ref. 4 for details).

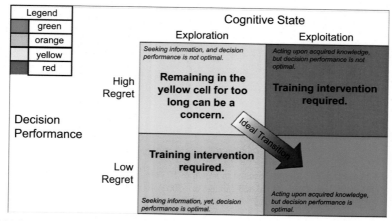

FIGURE 21.1 CAPTTIM categories and corresponding cognitive state and regret information.

Neuroergonomics. https://doi.org/10.1016/B978-0-12-811926-6.00021-X
Copyright © 2019 Elsevier Inc. All rights reserved.

Previous work in which CAPTTIM was applied retrospectively to decision performance data demonstrated that CAPTTIM can accurately classify a subject's cognitive state and decision performance at the trial-by-trial level. This work also showed that CAPTTIM can determine which subjects made the transition to the optimal decision path and which subjects would benefit from individualized feedback.[3,4] The purpose of the current study was to determine if CAPTTIM's categorization of cognitive state and decision performance can be captured on a trial-by-trial basis in real time; whether CAPTTIM-oriented feedback can guide optimal decision-making; and if eye gaze is sensitive to differences between optimal and suboptimal decision-making performance.

METHODS

Participants. The participants were 34 military officers spanning all services (14 US Marine Corps, 8 US Army, 8 US Navy, and 4 US Air Force), who were randomly divided into a control group ($n=17$; 3 females) and feedback group ($n=17$, 3 females). The average age of the control group was 34.71 years ($SD=3.64$), and 32.53 years for the feedback group ($SD=4.08$ years). The control group had slightly more time in service: an average of 13.47 ($SD=4.56$) years versus the feedback group's 10.06 years ($SD=4.13$ years). The median rank was O-3. No significant differences in demographic characteristics were found between the two groups (all $P>0.47$) (see Ref. 6 for more details).

Convoy Task. The convoy task is a militarily relevant version of the Iowa gambling task, which assesses reinforcement learning.[5,7] As depicted in Fig. 21.2, subjects see four identical routes and are instructed to decide on which route to send their convoy over 250 trials. The convoy task has external validity, as a convoy leader in theater views each bend in the road as a new trial: the decision to proceed around the bend or not. Upon making each decision, subjects can incur damage to friendly forces (bad) and/or inflict damage to enemy forces (good). The goal is to achieve the highest total damage score by maximizing enemy damage and minimizing friendly damage over the trials. Through trial and error, subjects should learn that routes 3 and 4 have positive long-term payoffs, whereas routes 1 and 2 have negative long-term payoffs. The outcome of each decision and the current accumulated damage score are provided to the subject after each trial (see Fig. 21.2). The first 50 trials were used to capture each subject's natural (i.e., baseline) latency variability and provide subjects with a period of exploration (see Ref. 6 for more details). Thus the main performance variable is the change in score from trial 51 to trial 250.

FIGURE 21.2 The decision just executed by this subject resulted in a gain of 50 damage points (damage to enemy forces) and a loss of 250 damage points (damage to friendly forces) for a net change to accumulated damage of −200 points.

TABLE 21.1 Messages Provided to Subjects in Feedback Group via Pop-Up Windows

CAPTTIM Category	Message to Subject in Feedback Group
Green (exploit and low regret)	Score is looking good. Stay with your strategy.
Yellow (explore and high regret)	Score could be better; attend to friendly damage.
Orange (explore and low regret)	Score looking good, go ahead and make decisions quickly.
Red (exploit and high regret)	Score could be better, attend to friendly damage and try other routes.

CAPTTIM CLASSIFICATION

Cognitive state. Exploration/exploitation was determined by exponentially weighted moving averages of the variability in latency times across sets of 10 trials. Exploration state was defined as sets of 10 trials in which the standard deviation of that particular set of trials was greater than the 2× baseline latency standard deviation. Exploitation state was defined as sets of 10 trials in which the standard deviation of that particular trial was equal to or less than 2× baseline latency standard deviation.

Decision performance. Regret compares the outcome of the subject's decisions to the outcome generated by playing the optimal policy at each of the *n* trials, and thus provides a quantitative measure of how much someone is deviating from the ideal decision path.[8] Low regret signifies optimal/near-optimal decision-making; high regret indicates suboptimal decision-making. Delineating between high regret and low regret was done through a combination of change point analysis and process means (see Ref. 3 for details).

CAPTTIM-oriented feedback. Starting on trial 60 and every 10 trials thereafter, subjects in the feedback group received one of four feedback messages via pop-up windows depending on their current CAPTTIM category. The pop-up window remained on the screen until the subject clicked on it. Each message corresponds to a different CAPTTIM category (see Table 21.1).

Eye tracker. While completing the convoy task, subjects' eye movements were measured and recorded using the FaceLAB (Seeing Machines, 2009) eye-tracking equipment and Eye Works software.

RESULTS

Can CAPTTIM be categorized in real time? We were able to capture subjects' decision-making data (cognitive state and decision performance) reliably in real time. Fig. 21.3 shows graphical depictions of the trial-by-trial CAPTTIM categorization for two subjects. Control Subject 114 predominantly exploited poor decisions until nearly the end of the game (change in accumulated damage score: −2400). Feedback Subject 211 spent most of the first half of the game exploiting poor decisions but then transitioned to optimal decisions and maintained optimal decision-making for the remainder of the game (change in accumulated damage score: 3950).

Can CAPTTIM-oriented feedback aid decision-making? As expected, the feedback group had a significantly greater mean increase in accumulated damage score than the control group ($M = 1868.8$ ($SD = 1921.0$) versus $M = 165.6$ ($SD = 2197.2$), ($t(30) = 2.33$, $P = .02$). There was no significant difference in the amount of time taken to complete the convoy task (feedback mean (SD) = 265.1 s (86.28) versus control mean (SD) = 299.6 s (126.8). As would be expected, the proportion of green and red trials was significantly associated with total damage score increase ($r = .48$, $P = .006$). Compared to the control group, the feedback group had on average a significantly smaller proportion of red trials ($M = 69.2\%$ ($SD = 26.40\%$) versus $M = 0.85.6\%$ ($SD = 12.00$), $t(20.952) = -2.20$, $P = .039$), with a trend for larger proportions of green trials ($M = 25.80\%$ ($SD = 26.2\%$) versus $M = 12.40\%$ ($SD = 12.20\%$), $t(21.21) = 1.86$, $P = .076$.

Can eye tracking provide additional insights? Of the three pieces of information provided in the game (friendly damage, enemy damage, and total damage), attending to friendly damage is crucial for reaching optimal decision-making, attending to total damage is helpful but not as informative as friendly damage, and attending to enemy damage is detrimental. The feedback group had significantly longer overall fixations, driven by much longer fixations on friendly damage ($M = 0.235$ s ($SD = 0.093$) versus $M = 0.112$ s ($SD = 0.073$), ($t(22) = 3.57$, $P = .002$) and a trend for longer fixations on total

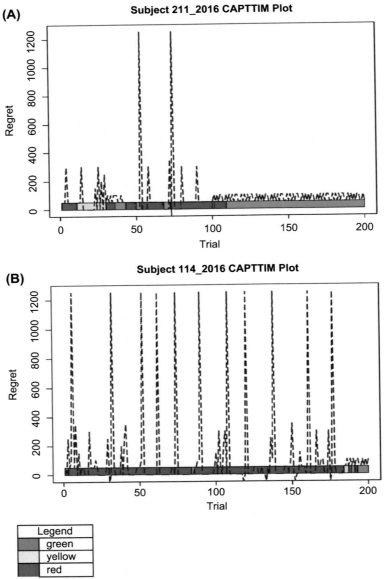

FIGURE 21.3 (A and B) Green, orange, red, and yellow indicate CAPTTIM categorization for a given trial. *Blue vertical spikes* represent trials in which subjects received very heavy friendly damage. Note: neither subject experienced any orange trials.

damage ($M=0.116$ s ($SD=0.108$) versus $M=0.026$ s ($SD=0.009$), $t(5.06)=2.02$, $P=.099$). There also was a trend for the feedback group to look less frequently at enemy damage ($M=4.39$ versus $M=19.2$, $P=.098$). Regardless of group, all high performers spent the majority of their time attending to friendly damage ($M=64.55\%$, $SD=0.082\%$, range 53.1%–71.1%).

DISCUSSION

We have shown that it is possible to capture the cognitive state and decision performance of subjects in real time with simple behavioral measures. By continuing to explore this process, this research moves closer to effective development of continuous, objective measures for long-term tracking of decision-making skills. This experiment consisted of a relatively simple task. However, the concept of categorizing both cognitive state and decision performance in real time can be expanded to existing training simulations that require multiple, complex, chained decisions where each decision can be influenced toward an optimal decision to maximize training value and improve the cognitive skills and effective decision-making of leaders of small units.[9] This intervention is precisely the type of response that was attempted here; when a subject has made one, or a series, of incorrect decisions there is now a mechanism that can alert the subject to the suboptimal performance. More than just pointing out a wrong answer, this research synthesizes a trainee's current cognitive state with their decision performance to understand better *why* suboptimal decision performance occurs and a system's or trainer's ability to act upon that categorization to help the trainee make better decisions. Furthermore, the ability to guide a trainee to understanding the problem at hand and how to act properly within the decision environment, as opposed to merely pointing out the correct answer to a single task or situation, is crucial for learning and developing effective decision-making expertise.[10]

The ability to incorporate objective decision-making measures in any existing simulation and demonstrate the optimal decision path (or the ability to correct deviation from it) may also reduce the time required in the trial-and-error phase of reinforcement learning, resulting in savings of the time and money required to train military decision-makers. Our results suggest that integrating eye tracking into the training model can provide additional insights into why some subjects transition from exploration of the possible outcomes to exploiting optimal decisions and others continue to exploit poor decisions. With further refinement to tailor feedback, this understanding will allow future leaders, instructors, and trainers to leverage the power of this approach to improve the processes and methods used to understand effective decision-making.

ACKNOWLEDGMENTS

This research is supported by Office of Naval Research Grant Number N0001416WX01083. The content is solely the responsibility of the authors and does not necessarily represent the official views of the Office of Naval Research. We thank Major Peter Nesbitt, Dr Ronald Fricker, LTC Jon Alt, Major Cardy Moten, and Captain John Critz for their contributions to the model.

REFERENCES

1. Odierno R, McHugh J. *Army human dimension strategy Draft Version 5.3*. U.S. Army Publication; 2015. Retrieved from: http://usacac.army.mil/sites/default/files/publications/20150524_Human_Dimension_Strategy_vr_Signature_WM_1.pdf.
2. U.S. Marine Corps. *U.S. marine corps S&T strategic plan*. Arlington (VA). 2012.
3. Critz J. *Understanding optimal decision making.*. Monterey (CA): Naval Postgraduate School; 2015.
4. Kennedy Q, Nesbitt P, Alt J, Fricker R. *Cognitive alignment with performance targeted training intervention model*. Los Angeles (CA): Human Factors and Ergonomics Society 2015 International Annual Meeting; 2015.
5. Nesbitt P, Kennedy Q, Alt J, Fricker R. Iowa Gambling task modified for military domain. *Military Psychology* 2015;27(4):252–60.
6. Carlson T. *Can subjects be guided to optimal decisions? The use of a real time training intervention model*. Monterey (CA): Naval Postgraduate School; 2016.
7. Bechara A, Damasio AR, Damasio H, Anderson SW. Insensitivity to future consequences following damage to human prefrontal cortex. *Cognition* 1994;50(1):7–15.
8. Auer P, Ortner R. UCB revisited: improved regret bounds for the stochastic multi-armed bandit problem. *Periodica Mathematica Hungarica* 2010;61(1):55–65.
9. Crichton M, Flin R. Training for Emergency Management: tactical decision games. *Journal of Hazardous Materials* 2001;88:255–66.
10. Archer JC. State of the science in health professional education: effective feedback. *Medical Education* 2010;44:101–8.

FURTHER READING

1. Taylor WA. *Change-point analysis: a powerful new tool for detecting*. 2000. Retrieved April 15, 2015, from: http://www.variation.com/cpa/tech/changepoint.html.

Chapter 22

Neural Efficiency Metrics in Neuroergonomics: Theory and Applications

Adrian Curtin[1,2], Hasan Ayaz[1,3,4]

[1]Drexel University, Philadelphia, PA, United States; [2]Shanghai Jiao Tong University, Shanghai, China; [3]University of Pennsylvania, Philadelphia, PA, United States; [4]Children's Hospital of Philadelphia, Philadelphia, PA, United States

INTRODUCTION

There has long been an implicit understanding that effortful cognition is reflected by changes to brain activity; however, it is only recently with the advent of modern neuroimaging techniques that we have been able to study the relationship between mental workload and its neurological underpinnings. Cognitive psychologists have attempted to harness findings from neuroimaging studies to develop improved training and instructional methods that take advantage of the nature of cognition and its constructs. One particular perspective, cognitive load theory (CLT), proposes that the development of training and instructional methods must take into account the limitations of cognitive capacities, particularly working memory (WM), and that individuals learn most effectively when they allocate an optimal amount of cognitive resources.[1] With the understanding that increased mental effort increases metabolic demands on the brain, neural efficiency (NE) relates neurophysiological measures of brain activity to an individual's active cognitive demands, providing an indispensable link between CLT and accessible measures of neural activity for the measure of cognitive load. Importantly, NE helps capture how performance achieved under a specific cognitive load varies according to the demands of the task as well as the aptitude of the individual.

Reflecting a more global view of NE in cognition and intelligence, Haler et al. introduced the NE Hypothesis,[2] proposing that intelligent individuals have more efficient brain function and, as a result, reduced or more focused neural activity in a given task. Initial descriptions of this hypothesis were quickly amended under observations that measured efficiency was dependent on task difficulty[3] as well as domain knowledge that can be accumulated through practice and experience.[4] The more nuanced interpretation suggests that the intrinsic parameters of a person's cognitive abilities define both that individual's immediate performance on complex cognitive tasks as well as the rate at which they are able to acquire knowledge and develop successful strategies to improve performance. When combined with the viewpoint of CLT, this perspective demonstrates the potential of NE as a defining characteristic of an individual's latent ability as well as their capability to further develop their proficiency by enabling a more sensitive evaluation of individual's cognitive states during the process of learning and task performance. However, the measure of NE requires both contextual behavioral performance along with objective and continuous measures of cognitive load, which requires an understanding of both the cognitive and neurophysiological elements underlying task performance.

According to CLT, the optimization of cognitive resources must adhere to the general architecture and constraints of the WM system. WM, a subcomponent of the Executive system, consists of active memory that is maintained for immediate use during the performance of higher-order cognitive activities. Effective use of WM requires the active maintenance and contextual discrimination of task-relevant from task-irrelevant information to engage in goal-directed behavior.[5] This dynamic system is known to be intimately related with the construct of fluid intelligence, because performance on problem solving/reasoning ability is strongly correlated with WM capacity.[6] Therefore, WM represents a primary consideration in the development of training systems and an important limiting factor on the ability of individuals to transfer knowledge and acquire complex skills.

Although an individual's intelligence may be described in part by WM capacity, CLT suggests that the quantity of the WM resources that are devoted to the task or learning process has a greater effect on how much information is learned and

Neuroergonomics. https://doi.org/10.1016/B978-0-12-811926-6.00022-1
Copyright © 2019 Elsevier Inc. All rights reserved.

retained. Failures of task learning therefore occur when either the demands of the task exceed available capacity, or there is insufficient allocation of mental resources to the task. These conditions of cognitive overload and underload, respectively, are presumed to be liable for task-related performance degradation during both training and execution. Thus the development of training paradigms could employ well-defined measures of cognitive exertion to refine their procedures such that they maintain optimal cognitive allocation and ensure effective task transfer. To do this, techniques for objectively assessing cognitive load must be combined with behavioral performance metrics in a manner that enables the quantification of training efficiency.

Researchers have explored multiple methodologies in the attempt to extract metrics of cognitive load. Rating-scale techniques are by far the most commonly employed techniques due to their ease of application and low-cost nature.[7,8] Despite their utility, these techniques suffer in their reliance on subjective introspection on the part of the participant, as well as the fact that they cannot operate continuously and remain unobtrusive. To address these issues, physiological measures have more recently been studied in the search for continuous and objective measures of cognitive load. Measures such as heart-rate variability and blink rate have primarily been investigated as alternative and complementary measures to rating scales[1] due to their low cost and ease of application, but these techniques lack specificity and, in the search for more specific measures, researchers have turned toward the brain.

Direct measures of brain activity through noninvasive neuroimaging techniques such as functional magnetic resonance imaging (fMRI) and positron emission tomography have revolutionized neuroscience and clinical findings; however, the relatively high cost of these techniques along with restrictions on experimental environment and participant behavior have prevented their direct use in practical applications.[9,10] Fortunately, in recent years, portable noninvasive neuroimaging techniques have dramatically increased their technological capabilities and overall affordability. Measures of brain activity through techniques such as electroencephalography (EEG) and functional near-infrared spectroscopy (fNIRS) may allow a more direct, specific, objective, and continuous assessment of cognitive load required to monitor and adapt training paradigms.[11,12] Both fNIRS and EEG offer distinct advantages due to their ability to measure in both controlled and natural environments, giving promise for roles of noninvasive neuroimaging outside of the laboratory context.[13]

MEASURING NEURAL CORRELATES OF COGNITIVE LOAD: ELECTROENCEPHALOGRAPHY

Electroencephalography relies on the measurement of changes in electric potentials across the scalp, attributed to cortical activity by the underlying neuronal populations.[14] Devices capable of measuring EEG signals can be manufactured at increasingly lower costs, and recent availability of open-hardware platforms have allowed individuals to build their own devices. The major advantages of this technology are that the measurement of interest has a rather high time resolution (~1 ms) and that only a few channels are needed to generate useful cognitive load discrimination. A large number of studies detailing the use of EEG systems to measure mental workload have been published,[11,14–17] and this technique has been at the forefront for development of brain–machine interfaces [11,18] and has demonstrated superiority to peripheral physiological research.[11]

This flexibility of EEG systems to measure cognitive activity in a variety of circumstances has made the technique a popular tool in neuroergonomic and neurocognitive research with a pathway to practical public adoption.[16] EEG measures of workload are conventionally divided into either measurement of the power spectral densities (PSDs) or amplitudes derived from event-related potentials (ERPs). PSD measures are calculated from the power spectrum and typically divided into the alpha (8–13 Hz), beta (13–30 Hz), delta (1–4 Hz), theta (4–8 Hz), and gamma (30–50 Hz) bands. The alpha band has been extensively studied due to its sensitivity to attention and workload.[19,20] Although decreases in alpha activity are positively correlated with increase in task demand and related to attentional processes, their decreases are also often paired with increases in theta-band power,[17] and often this conjunctive relation with alpha power is used as an estimation of workload.[14,21] Furthermore, individual differences in alpha frequency have commonly led investigators to divide the alpha band into two subbands based on the individual alpha frequency,[21] on which the upper alpha band is thought to predominately reflect cognitive ability. Through the measure of temporal changes in band power as reflected by event-related desynchronization and synchronization, increases in workload can be related to both general and individualized alpha measures[20] and may be used to continuously monitor performance and workload. On the other hand, ERP measures concentrate on the average of evoked responses due to stimuli such as the amplitudes of the N100 and P300, which occur at approximately 100 ms and 300 ms, respectively. The P300 amplitude in particular is a demonstrated measure of attention and workload,[14,22] whereas P300 latency instead reflects speed of processing, another important factor in cognition.[23] Although powerful, ERP measures are often more difficult to make in real-world environments due to the requirements of stimuli needed to probe the cognitive state and averaging needed to reduce response variability. In addition to these classical methods, measures such as functional connectivity and others derived from graph theory have opened new area of investigation with the promise of more informed and more sensitive measures.[24–26]

MEASURING NEURAL CORRELATES OF COGNITIVE LOAD: FUNCTIONAL NEAR-INFRARED SPECTROSCOPY

Functional Near-Infrared Spectroscopy (fNIRS) is a more recently developed technique that is capable of measuring the localized changes in oxygenated (HbO) and deoxygenated (Hb) cortical hemoglobin that occur because of cerebral activity.[10,27] These blood-flow changes, termed the hemodynamic response,[28] are the result of coupling between the neurovasculature and metabolic demands of neurons in response to increased activity. fNIRS takes advantage of the relative transparency of tissue to near-infrared light to provide quantification of the changes in relative hemoglobin, thus reflecting the metabolic demands of the brain.[29] Although hemodynamic activity does not occur as quickly as electrophysiological events (4–5 s vs. ~1 ms), fNIRS provides distinct advantages in terms of its ability to detect localized activity, resistance to motion artifacts, and potential to translate shallow cortical biomarkers discovered using fMRI into practical application, thanks to similarity in the measurement used.[30,31] Although fMRI describes activities in terms of the Blood Oxygen Level Dependent (BOLD) response, highly correlated with Hb, measurements using fNIRS can record these neural activities as increases in HbO, decreases in Hb, or as changes between these two parameters such as oxygenation (Oxy) and total hemoglobin (HbT) content. fNIRS systems are capable of measuring surface cortical areas corresponding with numerous functional roles; however, fNIRS systems are particularly well suited for the study of workload due to their ability to conveniently measure the prefrontal cortex (PFC), an area intricately linked with cognitive load and WM.

Through neuroimaging and classic neuroanatomic studies, the PFC has consistently been ascribed roles in WM and the underlying organization of high-level information.[32,33] In one WM model, Repovš and Baddely[34] theorized that WM is constituted from four components including two unimodal storage systems (the phonological loop and visuospatial sketchpad), an episodic buffer for the integration of information, and a central executive responsible for the manipulation of information and the coordination of the unimodal storage systems. The overlap of functionality between this central executive component and the role of the PFC has guided researchers in efforts to map the functional architecture of the PFC. fMRI evidence has suggested a hierarchical relationship between the dorsolateral (dlPFC) and ventrolateral (vlPFC) PFC regions, with the dlPFC responsible for monitoring and identifying task-relevant representations and the vlPFC maintaining those representations.[35] This balance between organization of WM and the maintenance of it was supported by another study examining differences between structured and unstructured sequences, which reported that the consolidation of information to reduce WM load or "chunking," increased vlPFC recruitment.[36] The vlPFC has also been implicated in the retrieval of information,[37] as well as the intention to retrieve.[32] Another area of note, the anterior PFC (encompassing Brodmann's area 10 [BA10]) has been described as the single largest cytoarchitectonic area in the PFC.[38] In humans, BA10 has evolved to be approximately twice relative size in comparison to other primates, leading to intense speculation regarding the roles it plays in human cognition.[39] These roles have ranged from the internal processing of emotions and internal states (mentalizing), to memory retrieval, prospective memory, attentional control, and relational knowledge.[38] Additional metaanalytic work has suggested that the medial–lateral hierarchical organization could be found within BA10 as well, with lateral activity disproportionally associated with memory retrieval and medial activity associated strongly with "multitasking," encompassing executive functions of task-switching, planning, and goal orientation, being associated with medial structures.[40]

The richness of the PFC as an area sensitive to parameters of task complexity, WM load, emotional processing, and planning is a great benefit to fNIRS studies. In particular, due to the natural absence of hair on the anterior PFC, a large area is made more accessible to fNIRS, greatly expediting setup time, reducing system complexity, and often allowing more ecologically valid measurements. Using fNIRS, researchers have validated the relation of PFC activation and WM load under controlled conditions using standardized WM tasks such as the N-Back task.[12,33] The N-Back is a graded demanding memory task that requires the participant to pair stimuli with prior stimuli in a sequence. In fMRI, the N-back has been found to broadly activate the dlPFC, anterior PFC, and vlPFC, with variations in functional specialization according to task-specific modifications.[32] Supporting this, fNIRS has demonstrated sensitivity to various workload levels of the N-Back task with observations of linear increases in HbO,[12] increases in interhemispheric dlPFC connectivity,[41] and, in a visuospatial variant, linear decreases of Hb.[42]

Reliable workload measurements in controlled laboratory conditions using fNIRS have encouraged researchers to generalize these findings to real-world environments and found considerable success under a variety of complex tasks. In one study, Ayaz et al. monitored PFC activity in air traffic controllers who were tasked with supervising an increasing number of virtual aircraft. The results showcased linear changes in left-dlPFC HbO paralleling workload changes observed in the N-Back task.[12] These changes were also significantly correlated with self-reported National Aeronautics and Space Administration–Task Load Index (NASA-TLX) workload, bolstering evidence for the capability of fNIRS to measure workload objectively. Another neuroergonomic study, on the operation of an endoscopic simulator performed by James et al., showed increased lateral PFC activation in expert operators and increased oxygenation in response to more challenging navigation.[43] In a real-life driving task, Yoshino et al. observed that moments that required rapid deceleration were followed by increased cognitive load in BA10 as noted by increased HbO.[44] Similarly, during flight-simulator operations, increased HbO during difficult landings in the initial trial versus reduced HbO observed in the final trial,[45] implied that reduced task demand was observed with practice.[12] Although challenges of continuously measuring workload during prolonged and fatigued states still remain,[46] and the linearity of fNIRS response to workload increases remains an oversimplification,[47] these trends point toward opportunities to develop more-robust statistical methods and imaging technologies as potential solutions.

One such approach might be to combine multiple workload assessment strategies to help identify variabilities and extremes within different operational contexts. In particular, the fact that fNIRS and EEG offer orthogonal perspectives on neural activity has prompted researchers to explore the particular advantages of each technique, as well as the way in which they complement each other. Several studies have explored the use of hybrid neuroimaging strategies reporting enhanced classification accuracy on workload of WM tasks,[48,49] and mental states during the operation of motor vehicles.[50] Although the combination of modalities currently increases complexity both of setup and analysis, future refinements in both technology and methodology may make hybridization a practical and even natural choice, allowing for a more comprehensive assessment of mental workload and operator state.

CALCULATING AND EMPLOYING MEASURES OF NEURAL EFFICIENCY

The goal of NE analysis is to quantify the otherwise hidden relationship between neural activity and performance among individuals over time, under varying task conditions, or alternatively across populations. In this way, NE analysis naturally extends the efficiency view,[51] which states that the relationship between mental effort and performance is subject to many variables of interest such as task parameters, environmental conditions, and individual ability. Under highly efficient conditions, performance is significantly higher than what would otherwise be expected given a specific neural load. On the other hand, inefficient conditions are marked by increased neural load and lower-than-expected behavioral performance. Although a preserved positive effort-to-performance relation is still presumed, efficient systems benefit more in terms of performance per relative measure of neural effort. As a result, efficiency and inefficiency are judged by how far the neurobehavioral measures deviate from the normal effort-to-performance relationship. By measuring the way in which variables of interest impact this effort-to-performance deviation, their impact on overall efficiency can help inform system design and gain insight on the task condition or individual involved.

Behavioral metrics of performance and neural correlates of workload often cannot be directly mapped in a manner that preserves the value and context of their relationship in an individual or system. Therefore, prior to the calculation of efficiency, these individual measures must first be normalized according to the group of interest. Adapting the definition first described by Paas et al.[52] to the context of neural measures, NE is calculated as the projection of normalized (z-score) behavioral performance (P) and brain-derived measures of cognitive effort (CE) onto the identity axis as seen in Fig. 22.1. Converting both behavioral and neural measures into normalized measures allows a clear and comparable interpretation of the immediate measures and their relationship. In this relationship, average neural effort and behavioral performance will by definition be achieved at the origin (0,0). Given the normalized values of each condition, equivalent relative efficiency can be expected to be achieved along the identity line (CE=P) which by definition would have a mean NE of zero. Along this line, performance one standard deviation from the mean would be expected to demand neural effort one standard deviation from that mean. NE can therefore be measured as the distance away from or projection onto the NE identity line as defined in Eq. (22.1). This equation is identical to the definition previously proposed,[52] except with the additional stipulation that measures of mental effort are derived from a linearization of neural measures.

$$ \mathrm{NE} = \frac{z(P) - z(CE)}{\sqrt{2}} $$

(22.1)

To compare the relative efficiency of conditions, as the method was initially described, neural measures and performance (CE, P) should be normalized according to the collected measures of all conditions inclusively. Following this, the new NE metrics can be statistically compared similarly to the original metrics providing they satisfy the assumptions of the statistical test employed.

To explore the way in which NE metrics can be calculated and analyzed, we present results from an earlier study in which 24 healthy participants completed an N-Back WM task at various difficulty levels (0-Back,1-Back,2-Back, 3-Back), while their prefrontal activity was monitored with fNIRS.[53] Throughout the task, we had reported that increases in WM demand resulted in both increased prefrontal oxygenation from baseline in the left dlPFC and decreased behavioral performance. These results are typical of findings from the N-Back WM task and here are presented as an example of neuroefficiency

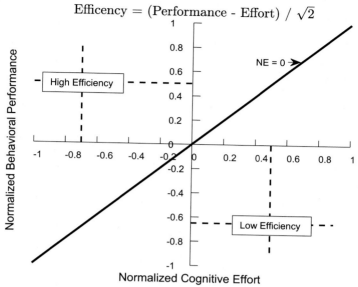

FIGURE 22.1   Neural efficiency in relation to cognitive effort (CE) and behavioral performance (P). High-efficiency quadrant contains relatively low neural demand that results in relatively high behavioral performance, and the low-efficiency quadrant contains the opposite observations.

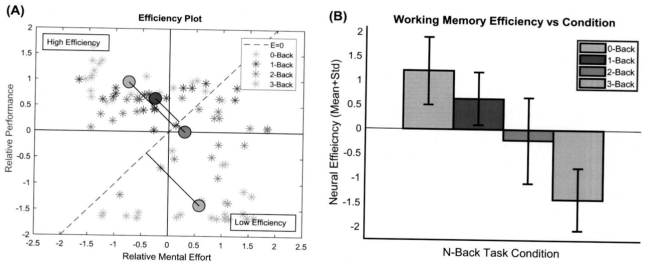

FIGURE 22.2   (A) Normalized oxygenation versus normalized accuracy during the N-Back task. Condition means are indicated by the weighted circles. (B) Mean NE for each N-Back task condition.

analysis as shown in Fig. 22.2. First, N-Back accuracy and prefrontal oxygenation are normalized across all task conditions and participants. Second, NE measures are calculated for each respective trial using Eq. (22.1). Third, newly calculated NE measures are assessed for a main effect of condition using analysis of variance (ANOVA). Fourth, following a significant main effect, planned post hoc tests are performed to evaluate differences between specific conditions.

DISCUSSION: APPLICATION, CHALLENGES, AND LIMITATIONS

In the present sample dataset, we illustrate how NE metrics reflect changing efficiency with increased workload. Clear differences in NE are observed as participants experience increasing WM demand with a trend towards a decrease in efficiency in the presence of increased difficulty. These results show how difficult conditions require a higher investment of effort to maintain performance relative to less demanding conditions. Although the calculation of efficiency metrics is a straightforward mathematical process, interpretation of the results may be more subjective. In this case, the clear changes of both P and CE per Condition strengthen the argument that changes in NE between conditions are statistically significant as well. However, although strength of NE assessment is the increasing sensitivity to conditions that impact the relationship between performance and demand, this may result in significant changes in NE in the absence of significant changes in either P or CE. Whether portrayed as an increased sensitivity to experimental parameters or simply as an additional interpretive perspective, efficiency measures should always be presented and analyzed in the context of the nonnormalized measures that form the basis for the analysis so that the results and implications can be understood in an objective way.

Importantly, NE metrics inherently draw on the reliability and utility of the sensitivity of both the employed CE and P scales. Therefore, any ambiguity inherent in either measure will also present itself within NE analysis. Although many times performance scales for specific tasks are well characterized, neural measures of cognitive workload are an area of ongoing investigation and their behavior across populations and within individuals is often variable. Due to uncertainty of the nature of both scales, NE scores should be compared with some caution even when identical in magnitude and direction. The notion that linear changes in effort result in linear changes in performance rests on the assumption that both scales behave in a linear fashion.[54] With enough knowledge or calibration of the underlying scales, it may be possible to linearize measures of CE and P such that these assumptions are met consistently. This, however, remains a rather substantial challenge, because even the way in which the methodology employed to assess activation may bias workload calculations and therefore the resulting efficiency calculations.[55] Therefore, further improvements in neuroimaging methods and methodologies are needed to not only improve the stable measurement of NE measures, but also enhance their usability as characterizations of cognitive performance.

Although in this example "Condition" refers to the experimental parameter of the task itself, by normalizing scores or neural measures with respect individual subjects across multiple trials it becomes possible to derive NE metrics that may describe differences in individual participant's performance and/or behavior across trials. The ability to use NE metrics as a characterization, not just of a system's behavior, but an individual's capacity within that system, may be a hidden strength of NE analysis. In part, normalization of scales within an individual may remove substantial uncertainty from the assumptions present in NE comparisons introduced by intersubject variability. Individual NE analysis may also play an important part in characterizing an individual's cognition and adaptation of systems or training to their individual strengths.

CONCLUSION

Neuroergonomic design and research benefit substantially from an understanding of the elements that make up individual differences in human cognition and performance. NE, in this case, is a measure that gives meaning to measures of cognitive load by showing how they relate to outcomes of interest. Presently, the use and application of NE analysis is an area of ongoing investigation. However, as neuroergonomic techniques to monitor cognitive workload mature in terms of reliability and usability, an understanding of the relationship that drives translation of CE into tangible performance may well become an integral part of system development. Under these circumstances, a well-designed system could be built in a manner that allows system operators to operate efficiently without exceeding or fatiguing available cognitive resources. On the other hand, well-designed instructional techniques may use NE metrics to enhance learner's information-processing abilities, potentially shortening required training times. Intelligent design using NE characterization offers a substantial and tangible goal for the development of individualized, optimized systems and training that enhance both learning and performance outcomes.

REFERENCES

1. Paas FG, et al. Cognitive load measurement as a means to advance cognitive load theory. *Educational Psychologist* 2003;**38**(1):63–71.
2. Haler RJ, et al. Intelligence and changes in regional cerebral glucose metabolic rate following learning. *Intelligence* 1992;**16**:415–26. https://doi.org/10.1016/0160-2896(92)90018-M.
3. Neubauer AC, Fink A. Fluid intelligence and neural efficiency: effects of task complexity and sex. *Personality and Individual Differences* 2003;**35**(4):811–27. https://doi.org/10.1016/S0191-8869(02)00285-4.
4. Sayala S, Sala JB, Courtney SM. Increased neural efficiency with repeated performance of a working memory task is information-type dependent. *Cerebral Cortex* 2006;**16**(5):609–17. https://doi.org/10.1093/cercor/bhj007.
5. Banich MT. Executive function - the search for an integrated account. *Current Directions in Psychological Science* 2009;**18**(2):89–94.
6. Kyllonen PC, Christal RE. Reasoning ability is (little more than) working-memory capacity? *Intelligence* 1990;**14**(4):389–433. https://doi.org/10.1016/S0160-2896(05)80012-1.
7. Gopher D, Braune R. On the psychophysics of workload: why bother with subjective measures? *Human Factors* 1984;**26**(5):519–32. https://doi.org/10.1177/001872088402600504.
8. Tuovinen JE, Paas F. Exploring multidimensional approaches to the efficiency of instructional conditions. *Instructional Science* 2004;**32**(1):133–52. https://doi.org/10.1023/b:truc.0000021813.24669.62.

9. Gramann K, et al. Editorial: trends in neuroergonomics. *Frontiers in Human Neuroscience* 2017;**11**(April):11–4. https://doi.org/10.3389/fnhum.2017.00165.

10. Ayaz H, et al. Continuous monitoring of brain dynamics with functional near infrared spectroscopy as a tool for neuroergonomic research: empirical examples and a technological development. *Frontiers in Human Neuroscience* 2013;**7**(DEC):871. https://doi.org/10.3389/fnhum.2013.00871.

11. Hogervorst MA, Brouwer AM, van Erp JBF. Combining and comparing EEG, peripheral physiology and eye-related measures for the assessment of mental workload. *Frontiers in Neuroscience* 2014;**8**(OCT):1–14. https://doi.org/10.3389/fnins.2014.00322.

12. Ayaz H, et al. Optical brain monitoring for operator training and mental workload assessment. *Neuroimage* 2012;**59**(1):36–47. https://doi.org/10.1016/j.neuroimage.2011.06.023. Elsevier Inc.

13. McKendrick R, et al. Into the wild: neuroergonomic differentiation of hand-held and augmented reality wearable displays during outdoor navigation with functional near infrared spectroscopy. *Frontiers in Human Neuroscience* 2016;**10**(April):216. https://doi.org/10.3389/fnhum.2016.00216.

14. Brouwer AM, et al. Estimating workload using EEG spectral power and ERPs in the n-back task. *Journal of Neural Engineering* 2012;**9**(4). https://doi.org/10.1088/1741-2560/9/4/045008.

15. Liu Y, et al. Towards a hybrid P300-based BCI using simultaneous fNIR and EEG. In: *Lecture notes in computer science (including subseries lecture notes in artificial intelligence and lecture notes in bioinformatics)*, **vol 8027**. LNAI; 2013. https://doi.org/10.1007/978-3-642-39454-6_35.

16. Berka C, et al. EEG correlates of task engagement and mental workload in vigilance, learning, and memory tasks. *Aviation Space and Environmental Medicine* 2007;**78**(5 Suppl.):B231–44.

17. Gundel A, Wilson GF. Topographical changes in the ongoing EEG related to the difficulty of mental tasks. *Brain Topography* 1992;**5**(1):17–25. https://doi.org/10.1007/BF01129966.

18. Yichuan Liu, et al. Detection of attention shift for asynchronous P300-based BCI. In: *2012 annual international conference of the IEEE engineering in medicine and biology society*. IEEE; 2012. p. 3850–3. https://doi.org/10.1109/EMBC.2012.6346807.

19. Ray W, Cole H. EEG alpha activity reflects attentional demands, and beta activity reflects emotional and cognitive processes. *Science* 1985;**228**(4700):750–2. https://doi.org/10.1126/science.3992243.

20. Fink A, et al. EEG alpha band dissociation with increasing task demands. *Cognitive Brain Research* 2005;**24**(2):252–9. https://doi.org/10.1016/j.cogbrainres.2005.02.002.

21. Klimesch W. EEG alpha and theta oscillations reflect cognitive and memory performance: a review and analysis. *Brain Research Reviews* 1999;**29**(2–3):169–95. https://doi.org/10.1016/S0165-0173(98)00056-3.

22. Polich J, Kok A. Cognitive and biological determinants of P300: an integrative review. *Biological Psychology* 1995;**41**(2):103–46. https://doi.org/10.1016/0301-0511(95)05130-9.

23. Rypma B, et al. Neural correlates of cognitive efficiency. *Neuroimage* 2006;**33**(3):969–79. https://doi.org/10.1016/j.neuroimage.2006.05.065.

24. García-Prieto J, Bajo R, Pereda E. Efficient computation of functional brain networks: toward real-time functional connectivity. *Frontiers in Neuroinformatics* 2017;**11**(February):1–18. https://doi.org/10.3389/fninf.2017.00008.

25. Sun J, Hong X, Tong S. Phase synchronization analysis of EEG signals: an evaluation based on surrogate tests. *IEEE Transactions on Biomedical Engineering* 2012;**59**(8):2254–63. https://doi.org/10.1109/TBME.2012.2199490.

26. Sun Y, et al. Functional cortical connectivity analysis of mental fatigue unmasks hemispheric asymmetry and changes in small-world networks. *Brain and Cognition* 2014;**85**(1):220–30. https://doi.org/10.1016/j.bandc.2013.12.011. Elsevier Inc.

27. Chance B, et al. Cognition-activated low-frequency modulation of light absorption in human brain. *Proceedings of the National Academy of Sciences of the United States of America* 1993/04/15;**90**(8):3770–4.

28. Cauli B. Revisiting the role of neurons in neurovascular coupling. *Frontiers in Neuroenergetics* 2010;**2**(June):1–8. https://doi.org/10.3389/fnene.2010.00009.

29. Ferrari M, Quaresima V. A brief review on the history of human functional near-infrared spectroscopy (fNIRS) development and fields of application. *Neuroimage* 2012;**63**(2):921–35. https://doi.org/10.1016/j.neuroimage.2012.03.049. Elsevier Inc.

30. Heinzel S, et al. Variability of (functional) hemodynamics as measured with simultaneous fNIRS and fMRI during intertemporal choice. *Neuroimage* 2013;**71**:125–34. https://doi.org/10.1016/j.neuroimage.2012.12.074.

31. Steinbrink J, et al. Illuminating the BOLD signal: combined fMRI-fNIRS studies. *Magnetic Resonance Imaging* 2006;**24**(4):495–505. https://doi.org/10.1016/j.mri.2005.12.034.

32. Owen AM, et al. N-back working memory paradigm: a meta-analysis of normative functional neuroimaging studies. *Human Brain Mapping* 2005;**25**(1):46–59. https://doi.org/10.1002/hbm.20131.

33. Peck EM, et al. Using fNIRS to measure mental workload in the real world. In: Fairclough SH, Gilleade K, editors. *Advances in physiological computing*. London: Springer London; 2014. p. 117–39. https://doi.org/10.1007/978-1-4471-6392-3_6.

34. Repovš G, Baddeley A. The multi-component model of working memory: explorations in experimental cognitive psychology. *Neuroscience* 2006;**139**(1):5–21. https://doi.org/10.1016/j.neuroscience.2005.12.061.

35. Wagner AD, et al. Prefrontal contributions to executive control: fMRI evidence for functional distinctions within lateral prefrontal cortex. *Neuroimage* 2001;**14**(6):1337–47. https://doi.org/10.1006/nimg.2001.0936.

36. Bor D, et al. Prefrontal cortical involvement in verbal encoding strategies. *European Journal of Neuroscience* 2004;**19**(12):3365–70. https://doi.org/10.1111/j.1460-9568.2004.03438.x.

37. Wolf RC, Vasic N, Walter H. Differential activation of ventrolateral prefrontal cortex during working memory retrieval. *Neuropsychologia* 2006;**44**(12):2558–63. https://doi.org/10.1016/j.neuropsychologia.2006.05.015.

38. Ramnani N, Owen AM. Anterior prefrontal cortex: insights into function from anatomy and neuroimaging. *Nature Reviews Neuroscience* 2004;**5**(3):184–94. https://doi.org/10.1038/nrn1343.

39. Semendeferi K, et al. Prefrontal cortex in humans and apes: a comparative study of area 10. *American Journal of Physical Anthropology* 2001;**114**(3):224–41. https://doi.org/10.1002/1096-8644(200103)114:3<224::AID-AJPA1022>3.0.CO;2-I.

40. Gilbert SJ, et al. Functional specialization within rostral prefrontal cortex (area 10): a meta-analysis. *Journal of Cognitive Neuroscience* 2006;**18**(6):932–48. https://doi.org/10.1162/jocn.2006.18.6.932.

41. Fishburn FA, et al. Sensitivity of fNIRS to cognitive state and load. *Frontiers in Human Neuroscience* 2014;**8**(February):76. https://doi.org/10.3389/fnhum.2014.00076.

42. Peck EMM, et al. Using fNIRS brain sensing to evaluate information visualization interfaces. In: *Proc. SIGCHI conf. Hum. Factors comput. Syst. - CHI'13*. 2013. p. 473. https://doi.org/10.1145/2470654.2470723.

43. James DRC, et al. The ergonomics of natural orifice translumenal endoscopic surgery (NOTES) navigation in terms of performance, stress, and cognitive behavior. *Surgery* 2011;**149**(4):525–33. https://doi.org/10.1016/j.surg.2010.11.019. Mosby, Inc.

44. Yoshino K, et al. Correlation of prefrontal cortical activation with changing vehicle speeds in actual driving: a vector-based functional near-infrared spectroscopy study. *Frontiers in Human Neuroscience* 2013;**7**(December):1–9. https://doi.org/10.3389/fnhum.2013.00895.

45. Causse M, et al. Mental workload and neural efficiency quantified in the prefrontal cortex using fNIRS. *Scientific Reports* 2017;**7**(1):1–15. https://doi.org/10.1038/s41598-017-05378-x.

46. Boyer M, et al. Investigating mental workload changes in a long duration supervisory control task. *Interacting With Computers* 2015;**27**(5):512–20. https://doi.org/10.1093/iwc/iwv012.

47. McKendrick R, et al. Enhancing dual-task performance with verbal and spatial working memory training: continuous monitoring of cerebral hemodynamics with NIRS. *Neuroimage* 2014;**85**:1014–26. https://doi.org/10.1016/j.neuroimage.2013.05.103. Elsevier Inc.

48. Aghajani H, Garbey M, Omurtag A. Measuring mental workload with EEG+fNIRS. *Frontiers in Human Neuroscience* 2017;**11**(July):1–20. https://doi.org/10.3389/fnhum.2017.00359.

49. Liu Y, Ayaz H, Shewokis PA. Multisubject 'learning' for mental workload classification using concurrent EEG, fNIRS, and physiological measures. *Frontiers in Human Neuroscience* 2017;**11**(July). https://doi.org/10.3389/fnhum.2017.00389.

50. Nguyen T, et al. Utilization of a combined EEG/NIRS system to predict driver drowsiness. *Scientific Reports* 2017;**7**(February):43933. https://doi.org/10.1038/srep43933. Nature Publishing Group.

51. Ahern S, Beatty J. Pupillary responses during information processing vary with scholastic aptitude test. *Science* 1979;**205**(21):1289–92. https://doi.org/10.1126/science.472746.

52. Paas FGWC, Van Merriënboer JJG. The efficiency of instructional conditions: an approach to combine mental effort and performance measures. *Human Factors* 1993;**35**(4):737–43. https://doi.org/10.1177/001872089303500412.

53. Ayaz H, et al. Using brain activity to predict task performance and operator efficiency. In: *Advances in brain inspired cognitive systems*, **vol 7366**. 2012: p. 147–55. https://doi.org/10.1007/978-3-642-31561-9_16.

54. Hoffman B, Schraw G. Conceptions of efficiency: applications in learning and problem solving. *Educational Psychology* 2010;**45**(1):1–14. https://doi.org/10.1080/00461520903213618.

55. Poldrack RA. Is 'efficiency' a useful concept in cognitive neuroscience?. *Developmental Cognitive Neuroscience* 2015;**11**:12–7. https://doi.org/10.1016/j.dcn.2014.06.001. Elsevier Ltd.

Neurostimulation Applications

Chapter 23

Neuromodulatory Effects of Transcranial Direct Current Stimulation Revealed by Functional Magnetic Resonance Imaging

Brian Falcone[1], Daniel E. Callan[2,3]

[1]George Mason University, Fairfax, VA, United States; [2]Center for Information and Neural Networks (CiNet), National Institute of Information and Communications Technology (NICT), Osaka University, Osaka, Japan; [3]ISAE-SUPAERO, Université de Toulouse, Toulouse, France

The field of neuroergonomics[1] seeks to improve performance during real-world tasks at work and in everyday life by applying neuroscience principles. There are various approaches in neuroergonomics, such as the use of neuroimaging methods to assess changes in brain activity during real-world tasks, with the purpose of informing the design of a system or to improve training techniques. Another approach is to affect cognitive function directly through the modulation of neural activity during learning to alter task performance. There are various stimulation methods through which this can be achieved, but this chapter only focuses on transcranial direct current stimulation (tDCS).

TDCS is a noninvasive and low-cost technique that has been shown to enhance many different cognitive components by affecting cortical excitability changes in the neurons in the brain through the depolarization or hyperpolarization of resting membrane potentials. Much of the earlier research involving tDCS was focused on its effects on simple motor learning tasks. The results of these studies showed that tDCS could facilitate motor functions, and these promising results led to the broadening of this research to other areas of human cognition such as memory, attention, planning, language, and mathematical performance in both healthy and unhealthy participants (reviewed elsewhere[2–5]). Generally these studies showed cognitive performance enhancements in humans when a positive current (anodal stimulation) is delivered through an electrode placed on the scalp over specific cortical brain areas, and a performance decrement when a negative current (cathodal stimulation) is delivered.

Recent human and animal studies on the neurophysiological mechanisms of tDCS effects on behavior have provided evidence suggesting that increased excitability in the affected area results in long-term potentiation and synaptic strengthening of task-relevant neural networks.[3,6–8] This theory is supported by the observation that there is often not only an immediate or "active-effect" of tDCS but also an "after-effect" during which cognition will remain enhanced for a short period after the cessation of tDCS. Animal studies have begun to unravel the neural mechanisms of tDCS on the cellular level,[9,10] but how these mechanisms influence larger neural networks in the human brain during task performance is much less clear. To that end, many studies have begun to examine the effects of tDCS on recordings of neural activity using various neuroimaging techniques while participants perform various tasks. Functional magnetic resonance imaging (fMRI) is particularly useful for identifying tDCS-induced modulations of specific neural networks because it provides greater spatial resolution than other neuroimaging techniques.[1]

The most common way to conduct fMRI research with tDCS is to investigate the after-effects of tDCS on brain activity by first applying tDCS for a time, then removing the electrodes and having the subject enter the scanner immediately after stimulation.[11–16] Ellison et al.[16] used this method to show that tDCS could be used to reveal the relationship between the critical nodes within networks during a visual search task. Cathodal stimulation was applied to the right posterior parietal cortex (rPPC) for 15 minutes before a visual search task was performed by participants during an fMRI scan. Results showed that cortical inhibition of the rPPC induced by cathodal stimulation resulted in a decrease in activation in frontal brain regions, specifically the frontal eye fields in the premotor cortex. These results suggest that stimulation of the rPPC can result in activation changes in brain areas throughout the entire attention network involved in visual search rather than just the area being stimulated.

Up until recently, MRI-compatible tDCS devices did not exist. This rendered concurrent tDCS application and fMRI data collection impossible due to tDCS-induced local magnetic field artifacts being introduced into MRI images. With the

Neuroergonomics. https://doi.org/10.1016/B978-0-12-811926-6.00023-3
Copyright © 2019 Elsevier Inc. All rights reserved.

advent of MRI-compatible tDCS devices, researchers can observe the active effects of tDCS by concurrently applying tDCS while collecting fMRI data in the scanner.[17–19] Similar to early behavioral tDCS literature, only basic motor tasks were used in many of these concurrent fMRI studies investigating the active effects of tDCS, but complex tasks are becoming more common. One such study conducted by Holland et al.[20] found that tDCS resulted in decreases in blood oxygen level dependent (BOLD) in Broca's area, which correlated with improved performance on a picture-naming task. They acknowledged that their results conflicted with other studies which found that active tDCS increased BOLD activity on a finger-tapping task and was positively correlated with performance on a complex visual search task.[17,17a] The results of these studies make it clear that how active tDCS affects brain activity will vary depending on the task and whether learning occurs.

TDCS and fMRI have also been paired to investigate the effects of stimulation on resting-state functional connectivity. Functional connectivity analyses use fMRI to look at the temporal synchrony of low-frequency fluctuations in BOLD signal between different brain regions. It is suggested that when these fluctuations are highly temporally correlated between regions, these regions are coupled within the same functional network.[21,22] Using this technique it is possible to explore changes to functional connectivity elicited by anodal tDCS that might contribute to augmented cognition observed in previous behavioral studies. A study by Keeser et al.[23] applied anodal stimulation over the left dorsolateral prefrontal cortex (DLPFC), with resting-state brain connectivity assessed before and after a 20 minutes tDCS session. The left DLPFC is a popular area used in many previous behavioral tDCS studies, as it has been shown to affect working memory, attention, and executive control.[3] This study found that anodal stimulation resulted in the modulation of resting-state functional connectivity in the default-mode network and the left and right frontal–parietal networks. It has been posited that the function of resting-state networks such as these is to maintain a preparatory alertness to react to incoming stimuli more effectively.[24] It is also suggested that increased strength of spontaneous functional connectivity in resting-state networks affects behavior and cognition during task performance.[23]

To investigate the active effects of tDCS modulation of resting-state functional connectivity that is related to complex learning and performance, a study conducted by Callan et al.[25] utilized concurrent tDCS and fMRI paired with training on a high-fidelity visual search task. Resting-state fMRI data was collected over three sessions: a pre-training session which assessed visual search ability, a training session in which anodal tDCS or sham stimulation was delivered along with performance feedback, and a posttraining session which again assessed visual search skill. The training session consisted of visual search trials that provided audio reinforcement and visual target feedback to the participant immediately following their responses to allow for learning and improvement. Anodal stimulation was applied over the rPPC, a region known to be important in attention and visual search. To collect resting-state brain activity, participants passively observed a black screen for a period of 5 minutes of rest immediately following each of the three sessions. This provided resting-state data before, during, and after tDCS.

A seed-driven functional connectivity analysis was used to investigate tDCS-induced changes in functional connectivity with brain regions correlated specifically with the stimulation site (rPPC). Simultaneous fMRI and tDCS allowed for determining areas of modulated spontaneous resting-state activity induced by active tDCS using fractional amplitude of low-frequency (0.01–0.08 Hz) fluctuation (fALFF). The fALFF analysis revealed a region in the rPPC with significant increased resting-state BOLD activation over the sham condition. This analysis was used because it shows that this region of interest is not only theoretically important but is also supported by the data to be significantly modulated by active anodal tDCS during a resting state. From this initial analysis, a voxel was selected from within the rPPC to be used as seed for the functional connectivity analysis. In addition, the baseline performance in the visual search from session 1 was subtracted from the percentage accuracy from session 3 to produce a post-training improvement score for each participant to be used as a covariate in the analysis.

The results revealed that active tDCS during a resting state had a positive correlation between future improved performance and functional connectivity between the site of stimulation and the substantia nigra. This region has been shown to be involved in visuospatial processing,[26] but perhaps more importantly it is part of the dopaminergic subcortical system, which is heavily involved in value-dependent learning.[27–30] In other words, individuals in the stimulation group who displayed increased functional connectivity between the rPPC and this subcortical region associated with attention and learning during active tDCS would on average also have better future visual search performance.

The neuroimaging research discussed in this chapter should encourage the application of combined tDCS and fMRI to investigate the effects of noninvasive brain stimulation in the human brain during complex learning and performance tasks that reflect the demands of the real world. A current limitation of many of these studies is that they have been carried out using very basic cognitive tasks. The conclusions of many of these studies requires the assumption that cognitive improvements elicited by tDCS on basic tasks will also lead to improvements in performance in work or everyday life. However, the translation of this basic research to real-world applications needs to be supported by research that demonstrates enhanced cognition or neurophysiological modulation during more complex cognitive tasks that are representative of naturalistic or work tasks.

REFERENCES

1. Parasuraman R, Rizzo M, editors. *Neuroergonomics: the brain at work.* Oxford University Press; 2008.
2. Brunoni AR, Vanderhasselt MA. Working memory improvement with non-invasive brain stimulation of the dorsolateral prefrontal cortex: a systematic review and meta-analysis. *Brain and Cognition* 2014;**86**:1–9.
3. Coffman BA, Clark VP, Parasuraman R. Battery powered thought: enhancement of attention, learning, and memory in healthy adults using transcranial direct current stimulation. *Neuroimage* 2014;**85**:895–908.
4. Jacobson L, Koslowsky M, Lavidor M. tDCS polarity effects in motor and cognitive domains: a meta-analytical review. *Experimental Brain Research* 2012;**216**(1):1–10.
5. Utz KS, Dimova V, Oppenländer K, Kerkhoff G. Electrified minds: transcranial direct current stimulation (tDCS) and Galvanic Vestibular Stimulation (GVS) as methods of non-invasive brain stimulation in neuropsychology—a review of current data and future implications. *Neuropsychologia* 2010;**48**:2789–810.
6. Coffman BA, Trumbo MC, Clark VP. Enhancement of object detection with transcranial direct current stimulation is associated with increased attention. *BMC Neuroscience* 2012;**13**(1):108.
7. Liebetanz D, Nitsche MA, Tergau F, Paulus W. Pharmacological approach to the mechanisms of transcranial DC-stimulation-induced after-effects of human motor cortex excitability. *Brain* 2002;**125**(10):2238–47.
8. Nitsche MA, Liebetanz D, Antal A, Lang N, Tergau F, Paulus W. Modulation of cortical excitability by weak direct current stimulation–technical, safety and functional aspects. *Supplements to Clinical neurophysiology* 2003;**56**:255–76.
9. Jefferys JGR, Deans J, Bikson M, Fox J. Effects of weak electric fields on the activity of neurons and neuronal networks. *Radiation Protection Dosimetry* 2003;**106**(4):321–3.
10. Radman T, Ramos RL, Brumberg JC, Bikson M. Role of cortical cell type and morphology in subthreshold and suprathreshold uniform electric field stimulation in vitro. *Brain Stimulation* 2009;**2**(4):215–28.
11. Nitsche MA, Niehaus L, Hoffmann KT, Hengst S, Liebetanz D, Paulus W, Meyer BU. MRI study of human brain exposed to weak direct current stimulation of the frontal cortex. *Clinical Neurophysiology* 2004;**115**(10):2419–23.
12. Baudewig J, Nitsche MA, Paulus W, Frahm J. Regional modulation of BOLD MRI responses to human sensorimotor activation by transcranial direct current stimulation. *Magnetic Resonance in Medicine* 2001;**45**(2):196–201.
13. Kim CR, Kim DY, Kim LS, Chun MH, Kim SJ, Park CH. Modulation of cortical activity after anodal transcranial direct current stimulation of the lower limb motor cortex: a functional MRI study. *Brain Stimulation* 2012;**5**(4):462–7.
14. Stagg CJ, O'Shea J, Kincses ZT, Woolrich M, Matthews PM, Johansen-Berg H. Modulation of movement-associated cortical activation by transcranial direct current stimulation. *European Journal of Neuroscience* 2009;**30**(7):1412–23.
15. Antal A, Kovács G, Chaieb L, Cziraki C, Paulus W, Greenlee MW. Cathodal stimulation of human MT+ leads to elevated fMRI signal: a tDCS-fMRI study. *Restorative Neurology and Neuroscience* 2012;**30**(3):255–63.
16. Ellison A, Ball KL, Moseley P, Dowsett J, Smith DT, Weis S, Lane AR. Functional interaction between right parietal and bilateral frontal cortices during visual search tasks revealed using functional magnetic imaging and transcranial direct current stimulation. *PLoS One* 2014;**9**(4):e93767.
17. Kwon YH, Jang SH. The enhanced cortical activation induced by transcranial direct current stimulation during hand movements. *Neuroscience Letters* 2011;**492**(2):105–108.
17a. Falcone B, Wada A, Parasuraman R, Callan DE. Individual differences in learning correlate with modulation of brain activity induced by transcranial direct current stimulation. *PLoS one* 2018;**13**(5):e0197192.
18. Kwon YH, Ko MH, Ahn SH, Kim YH, Song JC, Lee CH, Jang SH. Primary motor cortex activation by transcranial direct current stimulation in the human brain. *Neuroscience Letters* 2008;**435**(1):56–9.
19. Antal A, Polania R, Schmidt-Samoa C, Dechent P, Paulus W. Transcranial direct current stimulation over the primary motor cortex during fMRI. *Neuroimage* 2011;**55**(2):590–6.
20. Holland R, Leff AP, Josephs O, Galea JM, Desikan M, Price CJ, Crinion J. Speech facilitation by left inferior frontal cortex stimulation. *Current Biology* 2011;**21**(16):1403–7.
21. Biswal B, Yetkin FZ, Haughton VM, Hyde JS. Functional connectivity in the motor cortex of resting human brain using echo-planar MRI. *Magnetic Resonance in Medicine* 1995;**34**(4):537–41.
22. Greicius MD, Krasnow B, Reiss AL, Menon V. Functional connectivity in the resting brain: a network analysis of the default mode hypothesis. *Proceedings of the National Academy of Sciences of the United States of America* 2003;**100**(1):253–8.
23. Keeser D, Meindl T, Bor J, Palm U, Pogarell O, Mulert C, Padberg F. Prefrontal transcranial direct current stimulation changes connectivity of resting-state networks during fMRI. *The Journal of Neuroscience* 2011;**31**(43):15284–93.
24. Fransson P. Spontaneous low-frequency BOLD signal fluctuations: an fMRI investigation of the resting-state default mode of brain function hypothesis. *Human Brain Mapping* 2005;**26**(1):15–29.
25. Callan DE, Falcone B, Wada A, Parasuraman R. Simultaneous tDCS-fMRI identifies resting state networks correlated with visual search enhancement. *Frontiers in Human Neuroscience* 2016;**10**.
26. Matsumoto M, Takada M. Distinct representations of cognitive and motivational signals in midbrain dopamine neurons. *Neuron* 2013;**79**:1011–24.
27. Montague P, Dayan P, Sejnowski T. A framework for mesencephalic dopamine systems based on predictive Habbian learning. *Journal of Neuroscience* 1996;**16**:1936–47.
28. Schultz W. Predictive reward signal of dopamine neurons. *Journal of Neurophysiology* 1998;**80**:1–27.
29. Doya K. Metalearning and neuromodulation. *Neural Networks* 2002;**15**:495–506.
30. Callan D, Schweighofer N. Positive and negative modulation of word learning by reward anticipation. *Human Brain Mapping* 2008;**29**:237–49.

Chapter 24

Neurophysiological Correlates of tDCS-Induced Modulation of Cortical Sensorimotor Networks: A Simultaneous fNIRS–EEG Study

Makii Muthalib[1], Pierre Besson[2], Anirban Dutta[2], Mitsuhiro Hayashibe[2], Stephane Perrey[2]

[1]Silverline Research, Brisbane, Australia; [2]EuroMov, University of Montpellier, Montpellier, France

INTRODUCTION

As noninvasive brain stimulation research and technology evolve, the neuroergonomics community should endeavor to apply brain stimulation techniques for the study of brain and behavior in natural settings. Transcranial direct current stimulation (tDCS) is particularly useful due to its portability. While a considerable amount of research remains to be done, tDCS is an exciting potential method for positively augmenting the human operator to improve efficiency and enhance a variety of cognitive/motor skills that should ultimately improve performance at work Parasuraman and McKinley, 2014[1d]. The commonly hypothesized neuromodulatory effect of tDCS is subthreshold membrane polarization with subsequent synaptic plasticity, such that anodal tDCS (atDCS) upregulates cortical excitability while cathodal tDCS (ctDCS) downregulates it (Nitsche et al., 2003)[1c].

The conventional method of atDCS application to the sensorimotor cortex (SMC) region is placing a large sponge rubber electrode (35 cm^2) on the scalp of the target primary motor cortex (M1) with the return electrode on the contralateral supraorbital region. However, modeling studies have shown that the currents delivered to the target M1 is not directly under the stimulating electrode but rather spread in the region between the anode and cathode electrode. More recently, focal methods of tDCS have been developed using modeling simulations of current flow. For instance, high-definition tDCS (HD-tDCS), which uses an anode centered over the M1 and four surrounding return electrodes (anodal 4 × 1 HD montage), is thought to target a brain region with focal current delivery (Datta et al., 2009)[1a]. Computational models of current flow based upon individual magnetic resonance imaging (MRI) scans confirmed prediction of increased current focality with such a 4 × 1 HD-atDCS montage by experimental measurements of evoked motor responses.[1] While such an individualized modeling approach of current flow distribution can be accessible to some tDCS users, it is not easily applied to routine tDCS use, particularly to gain some idea of the individual current dosage. For HD-atDCS to be applied optimally to stimulate a target cortical region, a neurophysiological correlate of the strength of the applied electric field should be measured during the stimulation.

In this context, neuroimaging measures have been proposed to guide tDCS, providing information about the brain-tissue effects of the stimulation. Imaging techniques such as positron emission tomography (PET) offer the possibility of measuring the regional cerebral blood flow (rCBF). With PET and conventional tDCS montage, widespread increases in rCBF for atDCS have been reported in cortical and subcortical areas Lang et al., 2005[1b]. With arterial spin labeling and using functional MRI, Zheng et al.[2] showed that tDCS modulates rCBF in the brain region under and in close proximity to the stimulating electrodes differentially depending on the polarity and, to a lesser degree, the strength of the stimulation. Due to safety concerns (presence of wires and electrodes) in the MRI environment in the study of primary tDCS effects, an alternative neuroimaging technique for tracking regional changes evoked by tDCS is warranted.

The temporal and spatial changes of cortical neurovascular dynamics during tDCS can be measured noninvasively using fNIRS (hemodynamics) and EEG (neuronal activity) neuroimaging methods. While EEG measures functional brain activity directly by detecting the variations of electrical fields, fNIRS measures functional brain activity indirectly via changes in the oxygenated (O$_2$Hb) and deoxygenated (HHb) hemoglobin concentrations. These hemodynamic measurements are related to an increase in rCBF subsequent to increased neuronal activity (i.e., a consequence of neurovascular coupling).[3] Our previous

Neuroergonomics. https://doi.org/10.1016/B978-0-12-811926-6.00024-5
Copyright © 2019 Elsevier Inc. All rights reserved.

study[4] using a 4×1 HD-atDCS montage on the left M1 and two selected fNIRS channels, one on the left and one on the right SMC, showed an increase in O_2Hb within the vicinity of the anode of a 4×1 HD montage with no changes seen on the right hemisphere channel. This suggests a direct tDCS-induced current effect on the stimulated left hemisphere, but whether the time course of cerebral hemodynamic changes are spatially localized within the border of the 4×1 HD montage is not known. Thus the primary aim of the present study was to utilize a multichannel fNIRS setup with four channels covering inside and outside the 4×1 HD montage border to determine the spatial specificity of the hemodynamic changes. To determine how well the EEG can predict the fNIRS time course, we developed and experimentally tested a computational autoregressive (ARX) model for online parameter estimation with a Kalman filter to track resting-state transient coupling relations between $\log_{10}$ mean-power EEG band (0.5–11.25 Hz) and fNIRS O_2Hb signals (≤0.1 Hz) acquired simultaneously from the left SMC during anodal HD-tDCS. These latter findings have recently been published and are briefly discussed here.

METHODS

Design and Protocol

In a three-session cross-over study design, 14 subjects were randomly allocated to receive initially either sham or real (tDCS-1 or tDCS-2) atDCS (2 mA) targeting the left M1 via a 4×1 high-definition electrode montage with 1 week separating each session. In the real tDCS sessions, atDCS was applied with 30 s ramp up to 2 mA and maintained at this level for 10 min. In sham, the atDCS was applied with 30 s ramp up, 30 s at 2 mA, and 30 s ramp down followed by no stimulation for the remaining 8.5 min.

For each session, after the experimental setup fNIRS and EEG measurements were initiated with 3 min of baseline followed by the specific stimulation session, in which participants were asked to relax for the duration of the session. During each tDCS session, fNIRS (16 channels) and EEG (23 channels) were used simultaneously to measure changes in hemodynamics and neuronal activity, respectively from the scalp of the left (stimulated) and right (unstimulated) hemispheres (see Fig. 24.1).

tDCS

A Startstim tDCS system (Neuroelectrics, Barcelona, Spain) was used to deliver a constant direct current to the left M1 via an anodal 4×1 HD-tDCS electrode montage. The active anode electrode was positioned on the scalp (C3) surrounded by

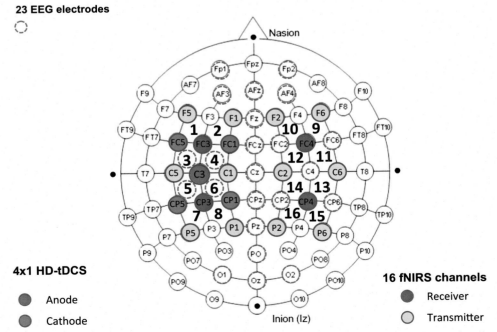

FIGURE 24.1 Locations of the 4×1 HD-atDCS electrodes (anode, in red; cathode, in blue), 16 fNIRS channels (transmitter, in yellow; receiver, in green; channel numbered), and 23 EEG electrodes (*dashed red circles*) on a 10/10 EEG layout.

four return electrodes each at a distance of ~3.5 cm from the anode electrode at FC1, FC5, CP5, and CP1 based on the 10/10 EEG system. The five electrodes ($3.14 \, cm^2$ AgCl electrodes) were secured on the scalp using conductive paste (Ten20, Weaver and Company, USA) and a specially designed plastic headgear to hold the HD-tDCS/EEG electrodes and fNIRS probes in place on the head (see Fig. 24.1).

fNIRS

Measurements of changes in O_2Hb and HHb concentrations were made from a 16-channel continuous-wave NIRS system (Oxymon MkIII, Artinis, Zetten, Netherlands) at a sampling frequency of 10 Hz. A receiver–transmitter distance of 3 cm was used, with the receivers placed on FC3 and CP3 for the left hemisphere and FC4 and CP4 for the right hemisphere (see Fig. 24.1). Transmitters were placed diagonally, i.e., at P1, P5, C1, C5, F5, and F1 for the left hemisphere and at P6, P2, C6, C2, F2, and F6 for the right hemisphere, as shown in Fig. 24.1.

EEG

Twenty-three channels of raw unreferenced (active electrode) EEG signals were recorded (ActiveTwo, Biosemi BV, Netherlands). The 23 EEG channels were AF3, AFz, AF4, Fz, FCz, FC2, FC6, FCC5h, FCC3h, Cz, C4, CCP5h, CCP3h, CPz, CP2, CP6, Pz, POz, O1, Oz, O2, Fp1, and Fp2 in the 10/10 system (see Fig. 24.1).

fNIRS and EEG Analysis

For fNIRS, the changes in O_2Hb and HHb concentrations (expressed in μM) were calculated according to a modified Beer–Lambert law that included an age-dependent constant differential pathlength factor ($4.99 + 0.067 \times Age^{0.814}$).[5] The time course of changes in O_2Hb and HHb concentrations for each of the 16 channels were first low-pass filtered at 0.1 Hz to attenuate cardiac signal, respiration, and Mayer-wave systemic oscillations,[5] and then offset to zero at the start of the stimulation. Since the time course of HHb signals were more variable between subjects, only O_2Hb signals are reported.

For the EEG analysis and ARX model development see Ref. 6.

RESULTS

In general the O_2Hb time course showed a larger increase in the fNIRS channels surrounding the anode (channels 3, 4, 5, 6) in the stimulated left hemisphere during real HD-atDCS sessions (tDCS-1 and tDCS-2) than in the sham session (see Fig. 24.2). Although there were some increases in O_2Hb concentration in the fNIRS channels outside the perimeter of the return electrodes (channels 1, 2, 7, 8) for only the real tDCS sessions, these changes were of much smaller magnitude (data not shown). The contralateral channels (channels 9–16) showed no important changes from baseline for any of the stimulation sessions (data not shown).

In a subsample of five subjects, a Kalman filter using an ARX model was able to track O_2Hb signals (<0.1 Hz) appropriately using EEG band-power signals ($\log_{10}$ EEG band 0.5–11.25 Hz).

DISCUSSION

In this study we used a 4×1 HD-tDCS montage to constrain the electric field between the active anode electrode at the center and the four surrounding return electrodes,[1] and then used multichannel fNIRS to determine the time course of tDCS-induced changes in cortical hemodynamics in the stimulated left and unstimulated right hemispheres. We showed that fNIRS channels located close to the anode electrode of the 4×1 HD-tDCS montage of the real tDCS sessions (tDCS-1 and tDCS-2) had a larger increase in O_2Hb compared to the sham. Since there were no obvious changes in O_2Hb in the contralateral channels for any of the stimulation sessions, we consider that the specific O_2Hb time course changes in the stimulated left SMC region represent some aspect of the induced electric field. Our previous study[4] showed that HD-tDCS induced an increase in O_2Hb only in the stimulated left hemisphere, but since we did not include a sham tDCS condition we were not able to determine if these hemodynamic changes were specific to the tDCS electric field. In the present study we confirm that the O_2Hb changes represent a real HD-tDCS-induced effect, since the channels in the vicinity of the anode showed a relatively large increase in O_2Hb that was maintained throughout the stimulation period, while the sham stimulation induced a small increase in O_2Hb that reduced soon after the current was terminated after 1.5 min. For further details and discussion of these results see the full paper by Muthalib et al. 2018.[6a]

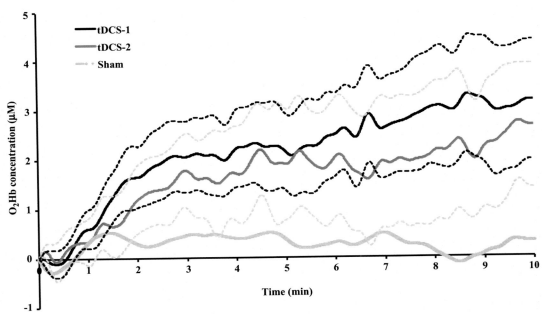

FIGURE 24.2 Group mean (±95% confidence interval) time course of oxygenated hemoglobin (O_2Hb) concentration changes during the real (tDCS-1 and tDCS-2) and sham HD-atDCS sessions. The O_2Hb values correspond to the four channels (channels 3, 4, 5, 6) surrounding the anode electrode (see Fig. 24.1 for layout).

In a subsample of five subjects we developed and tested a ARX model-based tracking method to monitor continuously the transient coupling between the electrophysiological (EEG) and the hemodynamic (fNIRS) signals during HD-atDCS. We presented a Kalman-filter-based online parameter estimation of an ARX model with four poles, six zeros, and a constant dead time of 0.5 sec that was found suitable to capture the transfer function from $\log_{10}$ EEG band (0.5–11.25 Hz) power alterations to fNIRS O_2Hb signal changes in the slow frequency range (≤ 0.1 Hz) during HD-tDCS. For more details of the ARX method and results see the full paper by Sood et al.[6]

Previous studies have shown that atDCS increases rCBF[2] and cortical hemodynamics.[4] The driving mechanisms responsible for these hemodynamic changes is unclear, but Nikulin et al.[7] showed that monochromatic ultraslow oscillations around 0.1 Hz in the EEG signals were related to the NIRS O_2Hb signals, and the authors hypothesized that EEG ultraslow oscillations represent an electric counterpart of the hemodynamic responses. In the present study, online tracking of the transient coupling relation between EEG ($\log_{10}$ mean-power EEG band 0.5–11.25 Hz) and fNIRS (<0.1 Hz) during tDCS could provide information on the momentary state of cortical excitability and bidirectional interactions within the neurovascular unit that would be important for determining the dose of tDCS neuromodulation.

CONCLUSION

The temporal and spatial increase of O_2Hb in the stimulated left hemisphere by HD-atDCS could represent the strength of the induced electrical field, and thus provide an indication of the dose of cortical neuromodulation. The ARX model using neuronal (EEG) and hemodynamic (fNIRS) responses can lead to closed-loop control of HD-atDCS for optimized neuromodulation in various neuroergonomic applications.

ACKNOWLEDGMENTS

This work was supported by LabEx NUMEV (ANR-10-LABX-20).

REFERENCES

1. Edwards D, Cortes M, Datta A, Minhas P, Wassermann EM, Bikson M. Physiological and modeling evidence for focal transcranial electrical brain stimulation in humans: a basis for high-definition tDCS. *Neuroimage* 2013;**74**:266–275.

1a. Datta A, Bansal V, Diaz J, Patel J, Reato D, Bikson M. Gyri-precise head model of transcranial direct current stimulation: improved spatial focality using a ring electrode versus conventional rectangular pad. *Brain Stimul* 2009;**2**(4):201–207. 207 e201.

1b. Lang N, Siebner HR, Ward NS, Lee L, Nitsche MA, Paulus W, Frackowiak RS. How does transcranial DC stimulation of the primary motor cortex alter regional neuronal activity in the human brain? *Eur J Neurosci* 2005;**22**(2):495–504.

1c. Nitsche MA, Fricke K, Henschke U, Schlitterlau A, Liebetanz D, Lang N, Paulus W. Pharmacological modulation of cortical excitability shifts induced by transcranial direct current stimulation in humans. *Journal of Physiology* 2003;**553**(Pt 1):293–301. https://doi.org/10.1113/jphysiol.2003.049916.

1d. Parasuraman R, McKinley RA. Using noninvasive brain stimulation to accelerate learning and enhance human performance. *Hum Factors* 2014;**56**(5):816–824.

2. Zheng X, Alsop DC, Schlaug G. Effects of transcranial direct current stimulation (tDCS) on human regional cerebral blood flow. *Neuroimage* 2011;**58**(1):26–33.

3. Attwell D, Iadecola C. The neural basis of functional brain imaging signals. *Trends in Neurosciences* 2002;**25**(12):621–5.

4. Muthalib M, Re R, Besson P, Perrey S, Rothwell J, Contini D, Torricelli A. Transcranial direct current stimulation induced modulation of cortical haemodynamics: a comparison between time-domain and continuous-wave functional near-infrared spectroscopy. *Brain Stimulation* 2015;**8**:392.

5. Basso Moro S, Bisconti S, Muthalib M, Spezialetti M, Cutini S, Ferrari M, Quaresima V. A semi-immersive virtual reality incremental swing balance task activates prefrontal cortex: a functional near-infrared spectroscopy study. *Neuroimage* 2014;**85**(Pt 1):451–60.

6. Sood M, Besson P, Muthalib M, Jindal U, Perrey S, Dutta A, Hayashibe M. NIRS-EEG joint imaging during transcranial direct current stimulation: online parameter estimation with an autoregressive model. *Journal of Neuroscience Methods* 2016;**274**:71–80.

6a. Muthalib M, Besson P, Rothwell J, Perrey S. Focal hemodynamic responses in the stimulated hemisphere during high-definition transcranial direct current stimulation. *Neuromodulation* 2018;**21**(4):348–354. https://doi.org/10.1111/ner.12632.

7. Nikulin VV, Fedele T, Mehnert J, Lipp A, Noack C, Steinbrink J, Curio G. Monochromatic ultra-slow (~0.1 Hz) oscillations in the human electroencephalogram and their relation to hemodynamics. *Neuroimage* 2014;**97**:71–80.

Chapter 25

Opinion: The Use of Online/Offline Terminology for Transcranial Direct Current Stimulation Can Bring Confusion

Pierre Besson[1], Vincent Cabibel[1], Mark Muthalib[1,2], Stephane Perrey[1]

[1]*EuroMov, University of Montpellier, Montpellier, France;* [2]*Deakin University, Melbourne, Australia*

Neuroergonomics aggregates different fundamental fields necessary to understand how the human brain functions to operate in everyday life. Beyond this motivation, there is a desire for neuroenhancement through noninvasive brain stimulation techniques, such as transcranial direct current stimulation (tDCS).[1] Gaining the enthusiasm of the general public requires research to prove the value of this technique. To achieve this goal, neuroimaging and tDCS techniques have been combined to offer complementary information on how interactions between different brain regions underlie complex brain functions, and these specific regions have been stimulated with tDCS to enhance cognitive or motor functions. Motor coordination in healthy people and in patients with neurological diseases, including poststroke dysfunctions, movement disorders, and other pathological conditions, has been investigated using these combined brain stimulation and imaging technologies.[2]

tDCS gives the potential to modulate cortical excitability, thus enabling researchers to target a cortical region in an effort to enhance performance across a various range of cognitive and motor tasks.[3] Although the exact mechanism(s) of tDCS remains partially unclear, recent studies have reported large variability among subjects in response to tDCS (reviewed elseshere[4,5]). To fix the latter issue partly in an effort to increase the effectiveness of tDCS application as a therapeutic adjuvant, new study designs are needed and a key issue is the timing of tDCS application in relation to a cognitive or motor task. Here we highlight two main types of study design encountered in the field of tDCS that could benefit the neuroergonomics community: "online" versus "offline" tDCS.

Some studies have reported enhanced performance when tDCS is applied concurrently with a task.[6–9] It is one way to potentiate the impact of tDCS, as the neurons involved in the task are thought to be more sensitive to the polarizing effects of tDCS.[10] In this context, the concurrent usage of tDCS and task is usually termed "online," whereas the term "offline" is employed when the task is performed before or after tDCS in the protocol design. But online/offline terminology is also used to indicate if the effects of tDCS are produced during or after the stimulation period, without any reference to a task. Furthermore, since tDCS is applied to enhance learning,[11] the terms online/offline are also used to signify performance gains during and after the task training period without reference to tDCS. Among neuroscientists, such important terms are used in different ways, and this can be confusing to researchers in the field if it is not explicitly stated in the protocol design. In our opinion, the use of generic terms such as online/offline has added to the confusion.

In a recent article,[12] the authors took an interest to one of our publications[13] that utilized tDCS with neuroimaging to determine the online and offline effects of tDCS and task. The authors reported: "The study that measured the blood flow after the stimulation (offline) showed an increase in task related brain activations,[14] whereas the other study measured the blood flow during the stimulation (online) showed a decrease."[13] The reading is not fluent due to the online/offline terminology that represents at the same time the type of stimulation and the moment of assessment. Brunoni et al.[15] emphasized the multiple meanings for the "online" term in the tDCS field: "When tDCS and the main outcome are coincident in time (i.e., when the variable is collected during tDCS application) the experiment is said to test the 'online' effects of tDCS. The concept is also used when another intervention (usually having a similar time span than tDCS such as physical therapy or a given task) and tDCS are applied simultaneously." But the authors did not propose clarifying their statement or providing readers with a solution to fix the ambiguity of terminology. Some other researchers used quotes for "online" and constantly recall the meaning of this term by using during, combined, and even concurrent.[16] Since "combined" is already widely used in the field of neuroimaging,[17] this could create confusion. In the field of motor learning and cognitive enhancement studies, simultaneous task and tDCS intervention have been termed "concurrent."[18,19] This formulation appears clearer and less

Neuroergonomics. https://doi.org/10.1016/B978-0-12-811926-6.00025-7
Copyright © 2019 Elsevier Inc. All rights reserved.

ambiguous, and could be replicated by all tDCS users whatever the nature of the task. Thus we propose to use this terminology. Concerning offline terminology, however, we do not know any clear formulation, so our proposal is to use the qualifier "sequential," as the task and tDCS interventions are placed contiguously (before and after). This also requires more emphasis on the detailed description of methods. Concerning the expected online/offline effects, as proposed by Woods et al.[20] "online" identifies the ongoing changes occurring during tDCS while "offline" identifies short- and long-term aftereffects of the tDCS. Precision about the timing of the offline evaluation (i.e., time point after tDCS) can be specified by "post" tDCS, which should go along with these qualifiers. This is suitable for both learning[11] and neuroimaging[21] fields.

To summarize, qualifiers are needed to shed light on the use of "online" and "offline" terminology in tDCS when related to a task, learning, and neuroimaging. If a cognitive or motor task is performed simultaneously or delayed with regard to the tDCS application, "concurrent" or "before/after" terms can be employed, respectively. If authors aim to identify the evaluations of the effects of the protocols, "online" or "offline" terms can be employed alongside timing precision. With the diversification of the stimulation protocols in the field of neuroscience, these simple clarifications will assure a basis and a common method that can benefit everyone.

ACKNOWLEDGMENTS

The authors declare that there is no conflict of interest.

REFERENCES

1. Dubljević V, Saigle V, Racine E. The rising tide of tDCS in the media and academic literature. *Neuron* 2014;**82**:731–6.
2. Shafi MM, Westover MB, Fox MD, Pascual-Leone A. Exploration and modulation of brain network interactions with noninvasive brain stimulation in combination with neuroimaging. *European Journal of Neuroscience* 2012;**35**:805–25.
3. Zhou J, Hao Y, Wang Y, Jor'dan A, Pascual-Leone A, Zhang J, Fang J, Manor B. Transcranial direct current stimulation reduces the cost of performing a cognitive task on gait and postural control. *European Journal of Neuroscience* 2014;**39**:1343–8.
4. López-Alonso V, Cheeran B, Río-Rodríguez D, Fernández-del-Olmo M. Inter-individual variability in response to non-invasive brain stimulation paradigms. *Brain Stimulation* 2014;**7**:372–80.
5. Wiethoff S, Hamada M, Rothwell JC. Variability in response to transcranial direct current stimulation of the motor cortex. *Brain Stimulation* 2014;**7**:468–75.
6. Boggio PS, Castro LO, Savagim EA, Braite R, Cruz VC, Rocha RR, Rigonatti SP, Silva MT, Fregni F. Enhancement of non-dominant hand motor function by anodal transcranial direct current stimulation. *Neuroscience Letters* 2006;**404**:232–6.
7. Galea JM, Celnik P. Brain polarization enhances the formation and retention of motor memories. *Journal of Neurophysiology* 2009;**102**:294–301.
8. Nitsche MA, Schauenburg A, Lang N, Liebetanz D, Exner C, Paulus W, Tergau F. Facilitation of implicit motor learning by weak transcranial direct current stimulation of the primary motor cortex in the human. *Journal of Cognitive Neuroscience* 2003;**15**:619–26.
9. Stagg CJ, Jayaram G, Pastor D, Kincses ZT, Matthews PM, Johansen-Berg H. Polarity and timing-dependent effects of transcranial direct current stimulation in explicit motor learning. *Neuropsychologia* 2011;**49**:800–4.
10. Bikson M, Name A, Rahman A. Origins of specificity during tDCS: anatomical, activity-selective, and input-bias mechanisms. *Frontiers in Human Neuroscience* 2013;**7**:688.
11. Reis J, Schambra HM, Cohen LG, Buch ER, Fritsch B, Zarahn E, Celnik PA, Krakauer JW. Noninvasive cortical stimulation enhances motor skill acquisition over multiple days through an effect on consolidation. *Proceedings of the National Academy of Sciences* 2009;**106**:1590–5.
12. Gözenman F, Berryhill ME. Working memory capacity differentially influences responses to tDCS and HD-tDCS in a retro-cue task. *Neuroscience Letters* 2016;**629**:105–9.
13. Muthalib M, Besson P, Rothwell J, Ward T, Perrey S. Effects of anodal high-definition transcranial direct current stimulation on bilateral sensorimotor cortex activation during sequential finger movements: an fNIRS study. *Advances in Experimental Medicine and Biology* 2016;**876**:351–9.
14. Jones KT, Gözenman F, Berryhill ME. The strategy and motivational influences on the beneficial effect of neurostimulation: a tDCS and fNIRS study. *Neuroimage* 2015;**105**:238–47.
15. Brunoni AR, Nitsche MA, Bolognini N, Bikson M, Wagner T, Merabet L, Edwards DJ, Valero-Cabre A, Rotenberg A, Pascual-Leone A, Ferrucci R, Priori A, Boggio PS, Fregni F. Clinical research with transcranial direct current stimulation (tDCS): challenges and future directions. *Brain Stimulation* 2012;**5**:175–95.
16. Martin DM, Liu R, Alonzo A, Green M, Loo CK. Use of transcranial direct current stimulation (tDCS) to enhance cognitive training: effect of timing of stimulation. *Experimental Brain Research* 2014;**232**:3345–51.
17. Devlin JT, Matthews PM, Rushworth MFS. Semantic processing in the left inferior prefrontal cortex: a combined functional magnetic resonance imaging and transcranial magnetic stimulation study. *Journal of Cognitive Neuroscience* 2003;**15**:71–84.
18. Karok S, Witney AG. Enhanced motor learning following task-concurrent dual transcranial direct current stimulation. *PLoS One* 2013;**8**:e85693.
19. Segrave RA, Arnold S, Hoy K, Fitzgerald PB. Concurrent cognitive control training augments the antidepressant efficacy of tDCS: a pilot study. *Brain Stimulation* 2014;**7**:325–31.

20. Woods AJ, Antal A, Bikson M, Boggio PS, Brunoni AR, Celnik P, Cohen LG, Fregni F, Herrmann CS, Kappenman ES, Knotkova H, Liebetanz D, Miniussi C, Miranda PC, Paulus W, Priori A, Reato D, Stagg C, Wenderoth N, Nitsche MA. A technical guide to tDCS, and related non-invasive brain stimulation tools. *Clinical Neurophysiology* 2015;**127**:1031–48.
21. Obeso I, Robles N, Marrón EM, Redolar-Ripoll D. Dissociating the role of the pre-SMA in response inhibition and switching: a combined Online and Offline TMS approach. *Frontiers in Human Neuroscience* 2013;**7**:150.

Emerging Applications in Decision-Making, Usability, Trust & Emotions

Neural Signatures of Advice Utilization During Human–Machine Agent Interactions: Functional Magnetic Resonance Imaging and Effective Connectivity Evidence

Kimberly Goodyear[1,2], Frank Krueger[3]
[1]*Brown University, Providence, RI, United States;* [2]*National Institutes of Health, Bethesda, MD, United States;* [3]*George Mason University, Fairfax, VA, United States*

INTRODUCTION

On July 24, 2013, an Alvia high-speed train derailed in Santiago de Compostela, Spain, killing 80 and injuring 144 people.[1] After a thorough investigation it was found that the train was going more than twice the speed limit (approximately 111–118 miles per hour), and the train conductor was distracted by a phone call and failed to respond to warnings to slow the train down. This disaster was completely preventable. As technology advances exponentially, understanding the contributing factors involved in such tragedies is becoming progressively more pertinent. As individuals become more reliant on machines and job roles shift from humans to automation, how individuals decide to utilize advice has become significant. The prevalence of new technology in our society has created an emergent need to understand all the behavioral factors and neural signatures involved in these complex human–machine interactions.

This chapter gives a background on factors that can affect decisions during advice taking from humans and machines, and the neural mechanisms involved during those complex human–machine interactions. Empirical behavioral and neural evidence from two functional magnetic resonance imaging (fMRI) studies is presented, examining the impact of misses and false alarms when utilizing advice from human and machine agents. Lastly, a summary presents conclusions and future directions for the emerging field of social neuroergonomics.

BACKGROUND

Various factors influence how a person responds to advice from different sources and agents. For instance, variables such as source credibility (expert versus novice) or type of advice (accurate versus inaccurate) can change how an individual responds to such advice. A person may decide to use or discount advice based on these different factors; studies have shown that poor (inaccurate) advice is discounted more than good (accurate) advice,[2] and expert advice is utilized more than novice advice.[3] In addition, trust decisions made while taking advice from humans and machines can vary depending on the pedigree (novice, expert) of the decision aids.[4] These differences in strategies for using advice from humans and machines may be due to dispositional credibility and/or high expectations of reliable advice. Another factor influencing a decision to accept or reject advice is reliability of the source. For instance, responses to imperfect automation (e.g., errors) may be affected by automation characteristics such as reliability, predictability, and ability.[5] Initial expectations of reliable advice can be altered when disconfirmation evidence is revealed, which ultimately impacts decision-making behaviors.[6] Initial confirmatory experiences can be explained by the expectation disconfirmation theory, a cognitive theory that explains satisfaction as a function of expectations (e.g., attributes that a person anticipates), perceived performance (e.g., perceptions of actual performance), and disconfirmation of beliefs (e.g., evaluations based on original expectations).[7]

Neuroergonomics. https://doi.org/10.1016/B978-0-12-811926-6.00026-9
Copyright © 2019 Elsevier Inc. All rights reserved.

Signal detection theory provides a means for understanding the circumstances in which a person makes decisions under uncertainty (e.g., unreliable or reliable advice).[8] During decision-making, responses to stimuli can be broken up into different categories: signal absent (correct rejection [correct nonalert], false alarm [incorrect alert]) and signal present (hit [correct alert], miss [incorrect nonalert]). The incorrect stimuli categories (i.e., false alarms and misses) can provide useful information to discern how individuals respond during advice taking. For instance, false alarms have been shown to hurt overall performance, operator compliance, and operator reliance.[9,10] In addition, reoccurring false alarms may cause a "cry wolf effect," in which operators may not respond to alerts at all.[11] Similar to false alarms, misses may affect operator reliance and also monitoring strategies, which can cause a shift in attention allocation during nonalarm periods.[12] Furthermore, aids with high reliability (high accuracy [low errors]) or low reliability (low accuracy [high errors]) can differentially impact decision-making behaviors. For instance, if an aid has high reliability this can lead to misuse, or over-reliance on the aid, and if an aid has low reliability this can lead to disuse, ignoring aid alerts, or underutilizing the aid.[13] Studies show that varying reliability may disrupt complacency,[14] and complacent behaviors may be due to conditions under a multiple-task load.[15] Lastly, the importance of reliable advice is emphasized by a study demonstrating that aid reliability below 70% can impair an individual's performance.[16]

Numerous neuroimaging studies have investigated advice taking,[17] personal traits and dispositions,[18] and human–robot interactions.[19] Brain regions associated with three large-scale brain networks (salience, default mode, and central executive networks) have been implicated in these different fMRI studies. For instance, the salience network (dorsal anterior cingulate cortex, anterior insula) has been implicated in saliency detection of internal and external events; the default-mode network (posterior cingulate cortex and precuneus) is associated with self-processing cognitions and evaluations; and the central executive network (dorsolateral prefrontal cortex) is used in higher-order executive functions.[20] These large-scale brain networks are further implicated in neuroimaging studies investigating tracking of expertise during estimates of human and algorithm abilities,[21] disobedience during expert advice,[22] inferring mental states of humans and machines,[23] and attribution of personal traits and characteristics.[24] The considerable overlap between these brain networks provides evidence that different factors need to be considered when evaluating how people perceive and decide to utilize advice from different sources and agents.

EMPIRICAL EVIDENCE

In previous work we investigated advice taking from humans and machines with false alarms[25] and misses[26] by combining fMRI with an X-ray luggage-screening task. We aimed to elucidate the neural bases of advice utilization from different agents (human, machine) and the corresponding brain activity and effective (or directional) connectivity. Stimuli for the luggage-screening task included X-ray luggage images that contained everyday objects (e.g., hairdryers, clothes) and a possible target (five different knives, with one possible per image).[27] Participants were asked to search for the presence or absence of a knife after being randomly assigned to receive advice from either a human agent or a machine agent framed as experts (Fig. 26.1). A reliability of 60% for the agents was chosen based on earlier work demonstrating differences between false alarms and misses:[9] study 1 (24 subjects) investigated false alarms (good advice [advice congruent]: 50% hits, 10% correct rejections; bad advice [advice incongruent]: 40% false alarms) and study 2 (24 subjects) investigated misses (good advice [advice congruent]: 50% hits, 10% correct rejections; bad advice [advice incongruent]: 40% misses). During each trial, participants would first see a fixation cross, then advice from one of the agents to "search" or "clear" the bag, an X-ray image of the luggage bag, a decision to accept or reject the advice of the agent to search or clear the bag (i.e.,

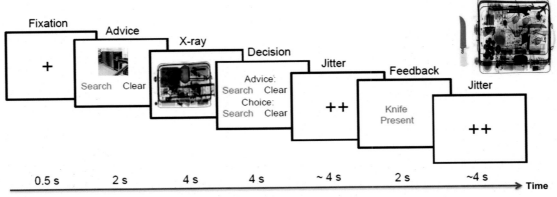

FIGURE 26.1 X-ray luggage screening task. (*Figure adjusted and reprinted from Goodyear K, Parasuraman R, Chernyak S, Madhavan P, Deshpande G, Krueger F. Advice taking from humans and machines: an fMRI and effective connectivity study. Frontiers in Human Neuroscience 2016b;10.*)

decision phase), fixation crosses, feedback indicating if their decision was correct or incorrect, and, lastly, fixation crosses. Participants completed two runs (each of about 13 min); they were given an initial endowment of $40, and each incorrect answer resulted in a deduction of $0.30 from the total.

In each study we demonstrated distinct behavioral patterns associated with different error types. Participants used bad advice in study 1 (false alarms) more than in study 2 (misses) (Fig. 26.2A), decreasing their overall performance (% correct) in the X-ray luggage-screening task (Fig. 26.2B). For false alarms, advice utilization during bad advice degraded more over time when participants interacted with the human agent compared to the machine agent (Fig. 26.2C). Because false alarms are more of a nuisance and not necessarily detrimental, participants might have focused more on the dispositional credibility of the agent, leading to less advice utilization for the human agent (evaluation-based strategy).[4] For misses, advice utilization degraded more when participants interacted with a machine agent compared to the human agent independent of advice type (good, bad) and time (Run1, Run2) (Fig. 26.2D). Because misses have costly consequences and may be fatal, participants might have focused more on the perceptional credibility (i.e., perfect automation schema) of the agent, leading to less advice utilization for the machine agent (perception-based strategy).[29]

Differences in behavioral responses to error types were reflected in distinct brain activation patterns. For false alarms the fMRI results showed that differences in advice utilization due to a more evaluation-based strategy engaged brain regions associated with attribution of personal traits and depositions (precuneus, posterior cingulate cortex, temporo–parietal junction)[18,30] and interoception (i.e., recruitment of physiological responses to environmental cues) (posterior insula)[31] (Fig. 26.3A). The effective connectivity results revealed that the right posterior insula and left precuneus (reciprocally connected to each other) were the drivers of the advice utilization network and projected to all other brain regions. These activation and connectivity patterns are in accordance with behavioral findings that false alarms may cause operators to reevaluate a human agent's dispositional credibility, leading to a change in expectations and ultimately a change in advice utilization behavior.[11] For misses, the fMRI results showed that differences due to a more perception-based strategy involved brain regions associated with sensory processing (e.g., cuneus, lingual gyrus)[32] and error monitoring (e.g., anterior cingulate cortex)[33] (Fig. 26.3B). The effective connectivity results revealed that the lingual gyrus was the driver of the advice utilization network and projected to all other brain regions (anterior cingulate cortex, anterior precuneus, and cuneus). These findings align with the concept that misses may cause changes in operator strategies during nonalarm periods, causing a reallocation of perception.[12] Thus participants interacting with the machine agent could have felt more

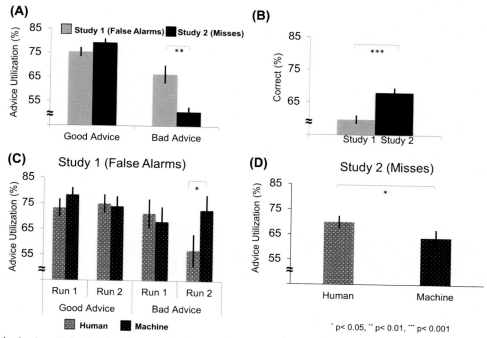

FIGURE 26.2 Behavioral results for decision phase. **A) Overall Advice Utilization.** Bad advice in study 1 (false alarms) was utilized more than in study 2 (misses). **B) Task Performance.** Performance was lower in study 1 compared to study 2. C) **Advice Utilization in Study 1.** Advice utilization was lower for the human agent compared to the machine agent during run 2. D) **Advice Utilization in Study 2.** Advice utilization was lower for the machine agent compared to the human agent. *(Figure adjusted and reprinted from Goodyear KS. The neural basis of advice utilization during human and machine agent interactions. [Doctoral dissertation]: George Mason University; 2016.)*

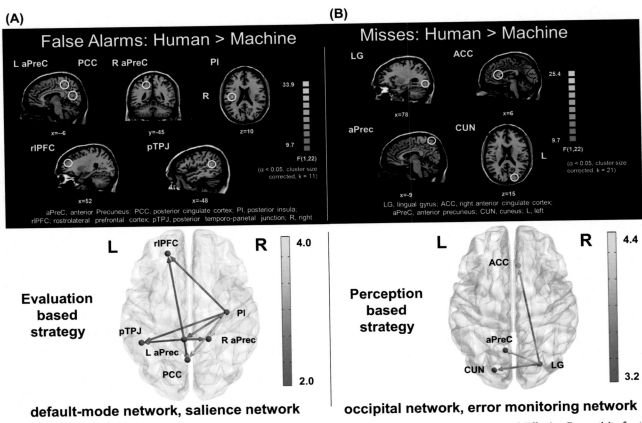

FIGURE 26.3 Brain activations and effective connectivity for decision phase. ((A)*False Alarm Brain Activations and Effective Connectivity for the Decision Phase. An evaluation based strategy was found for false alarms with activation in the right posterior insula (PI), right anterior precuneus (aPreC), left aPreC, left posterior cingulate cortex (PCC), left rostrolateral prefrontal cortex (rlPFC) and left posterior temporoparietal junction (pTPJ). The effective connectivity network showed that the PI (posterior insula) and L aPreC (anterior precuneus) were drivers of the network and also the source ROIs for all other target ROIs (R aPreC, PCC (posterior cingulate cortex), rlPFC (rostrolateral prefontal cortex) and pTPJ (posterior temporoparietal junction). Adjusted and reprinted from Goodyear K, Parasuraman R, Chernyak S, Madhavan P, Deshpande G, Krueger F. Advice taking from humans and machines: an fMRI and effective connectivity study. Frontiers in Human Neuroscience 2016b;10; (B) Miss Brain Activations and Effective Connectivity for the Decision Phase. A perception based strategy was found for misses with activation in the right lingual gyrus (LG), right anterior cingulate cortex (ACC), left anterior precuneus (aPreC) and left cuneus (CUN). The effective connectivity network showed that the LG (lingual gyrus) was the driver of the network and source ROI, sending outputs to all target ROIs (ACC (anterior cingulate cortex), aPreC (anterior precuneus), and CUN (cuneus).Adjusted and reprinted by permission from Taylor & Francis Ltd.: http://www.tandfonline.com. Goodyear K, Parasuraman R, Chernyak S, de Visser E, Madhavan P, Deshpande G, Krueger F. An fMRI and effective connectivity study investigating miss errors during advice utilization from human and machine agents. Social Neuroscience 2016a:1–12.)*

accountable for their actions due to differences in perceptions involved with the perfect automation schema leading to a change in expectations, and ultimately a shift in advice utilization behaviors.

SUMMARY

The findings of our previous work provide insight into the behavioral and neural differences between error types when taking advice from humans and machines, and can ultimately serve to optimize our understanding of how individuals decide to use or discount advice. The results of our studies can provide a foundation for further research in social neuroergonomics, human factors, and other related fields to expand and confirm our findings. For instance, future studies could investigate different factors contributing to advice utilization, such as agent etiquette or timing of advice. Our laboratory findings can be expanded to field studies and additional neuroimaging techniques such as transcranial magnetic stimulation that can be applied to establish causality between brain activation and different types of cognitive functions. A better understanding of behavioral metrics and neurobiological mechanisms involved during advice can ultimately allow development of safety measures. The long-term goal is a comprehensive understanding of all of the components involved with human–machine interactions, which can ultimately serve to prevent catastrophic disasters such as the train derailment in Santiago de Compostela.

ACKNOWLEDGMENTS

We would like to acknowledge the memory of our dear colleague and friend, Raja Parasuraman, who was a valuable contributor to this research. Direct portions of this chapter have been adjusted and reprinted from: Goodyear, K.S. (2016). *The Neural Basis of Advice Utilization During Human and Machine Agent Interactions* (Doctoral dissertation, George Mason University).

REFERENCES

1. Shultz JM, Garcia-Vera MP, Santos CG, Sanz J, Bibel G, Schulman C, Rechkemmer A. Disaster complexity and the Santiago de Compostela train derailment. *Disaster Health* 2016;**3**(1):11–31.
2. Yaniv I, Kleinberger E. Advice taking in decision making: egocentric discounting and reputation formation. *Organizational Behavior and Human Decision Processes* 2000;**83**(2):260–81.
3. Sniezek JA, Schrah GE, Dalal RS. Improving judgement with prepaid expert advice. *Journal of Behavioral Decision Making* 2004;**17**(3):173–90.
4. Madhavan P, Wiegmann DA. Effects of information source, pedigree, and reliability on operator interaction with decision support systems. *Human Factors: The Journal of the Human Factors and Ergonomics Society* 2007;**49**(5):773–85.
5. Lee JD, See KA. Trust in automation: designing for appropriate reliance. *Human Factors* 2004;**46**(1):50–80.
6. Staudinger MR, Buchel C. How initial confirmatory experience potentiates the detrimental influence of bad advice. *NeuroImage* 2013;**76**:125–33.
7. Oliver RL. A cognitive model of the antecedents and consequences of satisfaction decisions. *Journal of Marketing Research* 1980;**17**(4):460–9.
8. Tanner Jr WP, Swets JA. A decision-making theory of visual detection. *Psychological Review* 1954;**61**(6):401–9.
9. Dixon SR, Wickens CD, McCarley JS. On the independence of compliance and reliance: are automation false alarms worse than misses? *Human Factors: The Journal of the Human Factors and Ergonomics Society* 2007;**49**(4):564–72.
10. Meyer J. Conceptual issues in the study of dynamic hazard warnings. *Human Factors* 2004;**46**(2):196–204.
11. Breznitz S. *Cry wolf: the psychology of false alarms*. Psychology Press; 2013.
12. Onnasch L, Ruff S, Manzey D. Operators' adaptation to imperfect automation – impact of miss-prone alarm systems on attention allocation and performance. *International Journal of Human-Computer Studies* 2014;**72**(10–11):772–82.
13. Parasuraman R, Riley V. Human and automation: use, misuse, disuse, abuse. *Human Factors* 1997;**39**(2):230–53.
14. McBride SE, Rogers WA, Fisk AD. Understanding human management of automation errors. *Theoretical Issues in Ergonomics Science* 2014;**15**(6):545–77.
15. Parasuraman R, Molloy R, Singh IL. Performance consequences of automation-induced 'complacency'. *The International Journal of Aviation Psychology* 1993;**3**(1):1–23.
16. Wickens CD, Dixon SR. The benefits of imperfect diagnostic automation: a synthesis of the literature. *Theoretical Issues in Ergonomics Science* 2007;**8**(3):201–12.
17. Biele G, Rieskamp J, Krugel LK, Heekeren HR. The neural basis of following advice. *PLoS Biology* 2011;**9**(6):e1001089.
18. Brosch T, Schiller D, Mojdehbakhsh R, Uleman JS, Phelps EA. Neural mechanisms underlying the integration of situational information into attribution outcomes. *Social Cognitive and Affective* 2013;**8**(6):640–6.
19. Krach S, Hegel F, Wrede B, Sagerer G, Binkofski F, Kircher T. Can machines Think? Interaction and perspective taking with robots investigated via fMRI. *PLoS One* 2008;**3**(7):e2597.
20. Menon V. Large-scale brain networks and psychopathology: a unifying triple network model. *Trends in Cognitive Sciences* 2011;**15**(10):483–506.
21. Boorman ED, O'Doherty JP, Adolphs R, Rangel A. The behavioral and neural mechanisms underlying the tracking of expertise. *Neuron* 2013;**80**(6):1558–71.
22. Suen VYM, Brown MRG, Morck RK, Silverstone PH. Regional brain changes occurring during disobedience to "experts" in financial decision-making. *PLoS One* 2014;**9**(1):e87321.
23. Chaminade T, Rosset D, Da Fonseca D, Nazarian B, Lutcher E, Cheng G, Deruelle C. How do we think machines think? An fMRI study of alleged competition with an artificial intelligence. *Frontiers in Human Neuroscience* 2012;**6**.
24. Cabanis M, Pyka M, Mehl S, Muller BW, Loos-Jankowiak S, Winterer G, Kircher T. The precuneus and the insula in self-attributional processes. *Cognitive, Affective and Behavioral Neuroscience* 2013;**13**(2):330–45.
25. Goodyear K, Parasuraman R, Chernyak S, Madhavan P, Deshpande G, Krueger F. Advice taking from humans and machines: an fMRI and effective connectivity study. *Frontiers in Human Neuroscience* 2016b;**10**.
26. Goodyear K, Parasuraman R, Chernyak S, de Visser E, Madhavan P, Deshpande G, Krueger F. An fMRI and effective connectivity study investigating miss errors during advice utilization from human and machine agents. *Social Neuroscience* 2016a:1–12.
27. Madhavan P, Gonzalez C. Effects of sensitivity, criterion shifts, and subjective confidence on the development of automaticity in airline luggage screening. In: *Paper presented at the Proceedings of the human factors and Ergonomics society Annual Meeting*. 2006.
28. Deleted in review.
29. Mosier KL, Skitka LJ, Heers S, Burdick M. Automation bias: decision making and performance in high-tech cockpits. *The International Journal of Aviation Psychology* 1998;**8**(1):47–63.
30. Harris LT, Todorov A, Fiske ST. Attributions on the brain: neuro-imaging dispositional inferences, beyond theory of mind. *Neuroimage* 2005;**28**(4):763–9.

31. Kurth F, Eickhoff SB, Schleicher A, Hoemke L, Zilles K, Amunts K. Cytoarchitecture and probabilistic maps of the human posterior insular cortex. *Cerebral Cortex* 2010;**20**(6):1448–61.

32. Schilbach L, Eickhoff SB, Schultze T, Mojzisch A, Vogeley K. To you I am listening: perceived competence of advisors influences judgment and decision-making via recruitment of the amygdala. *Social Neuroscience* 2013;**8**(3):189–202.

33. Crottaz-Herbette S, Menon V. Where and when the anterior cingulate cortex modulates attentional response: combined fMRI and ERP evidence. *Journal of Cognitive Neuroscience* 2006;**18**(5):766–80.

Chapter 27

Psychophysical Equivalence of Static Versus Dynamic Stimuli in a Two-Alternative Forced-Choice Detection Task

Gabriella M. Hancock

California State University, Long Beach, Long Beach, CA, United States

INTRODUCTION

Vigilance is typically defined as the mental capacity to sustain attention over extended periods of time.[1–3] Research often operationalizes this phenomenon as the ability to detect rare and critical signals over the course of multiple periods on watch, spanning minutes[4] to several hours in duration.[5,6] Due to inherent limitations in human physical and cognitive capacities, the ability to maintain attention wanes as a function of time on task, as does performance, a phenomenon noted as the vigilance decrement function.[5] Successful vigilance performance is of vital importance to a host of operational tasks that directly impact health and safety. Real-world operational domains that necessitate observers discriminating signals from extraneous noise include civil aviation,[7] air traffic control,[8] baggage screening,[9] diagnostic medical screening,[10] nuclear power plant operation,[11] military surveillance,[12] and security operations,[13] among others.

Most of the monitoring tasks inherent to these performance domains utilize dynamic displays to present information pertaining to system state. However, over the past six decades vigilance research has predominantly used static stimuli in experimental protocols designed to evoke and examine the vigilance decrement.[8,14] Such a divorce between laboratory-based simulations and real-world operational task demands can significantly hamper the validity and generalizability of vigilance research findings.[15] To address this issue, studies in recent years have attempted to build and validate a new vigilance task that employs a videogame-based protocol (Virtual Battlespace 2, VBS2). This task seeks to enhance ecological validity by presenting both signal and nonsignal stimuli in a dynamic, continuous fashion using motion with a first-person perspective.[16] To establish the task parameters of this new protocol and ensure that any changes in performance are due to the task demand of monitoring, rather than signal discrimination, the study investigated the effect of stimulus type (i.e., static versus dynamic) on detection performance using a videogame-based platform under alerted conditions.

METHODS

This study is framed as an improvised explosive device (IED) detection task. The videogame protocol presents a virtual environment depicting an Afghan village complete with structures, inhabitants, and domestic animals. Within these scenes were randomly placed ecologically valid signals that indicate the potential presence of an IED, such as fuel cans, trash bags, motorcycle batteries, and wooden planks sunk into the ground. The dynamic video clips and static images were excised from this VBS2 software platform and presented via Qualtrics.

To determine effects of stimulus type on detection performance, the dependent variables of interest were correct detections and participants' cognitive and affective attributes, specifically task engagement, distress, worry, and global workload. Correct detections (instances where participants identified the presence of a signal when one was indeed presented) were tabulated by Qualtrics software for offline analysis. Cognitive and affective states were gauged via the administration of validated questionnaires, including the Dundee Stress State Questionnaire (DSSQ; 30-item short version[17–19]) and the NASA Task Load Index (NASA-TLX[20]).

Participants first signed an informed consent form and demographics data were collected (e.g., age, gender, previous exposure to virtual environments, videogame-playing experience). The pre-task DSSQ was then administered to assess individuals' cognitive and affective states prior to completing the task. The questionnaire was completed at this time so that difference scores (post-task scores versus pre-task scores) could later be calculated for analysis.

Neuroergonomics. https://doi.org/10.1016/B978-0-12-811926-6.00027-0
Copyright © 2019 Elsevier Inc. All rights reserved.

Participants then completed a two-alternative forced-choice task (2AFCT). The experiment consisted of two counterbalanced task conditions: one composed of static pictures and one containing dynamic video clips. Each trial comprised one signal-present stimulus and one signal-absent stimulus shown sequentially. After both stimuli had been shown, participants were instructed to identify via a mouse click which stimulus included the signal. Duration was manipulated with a total of three durations: 5, 3, and 1 s stimulus presentations. The order of presentation for the trials (i.e., 10 signal-first trials, 10 nonsignal-first trials) for each duration time was randomized throughout each task condition. The task thus comprised a total of 60 trials per task condition (20 five-second trials, 20 three-second trials, and 20 one-second trials).

RESULTS

Twenty-nine participants were recruited to take part in this study; as they were university students, they were remunerated with extra credit for their participation. The data of three participants were excluded from the final analysis due to one incident of equipment failure, one participant with a self-reported history of simulator sickness, and one individual with a congenital color perception deficiency. Data from a total of 26 participants (15 females, 11 males) with an average age of 21.12 years was therefore analyzed.

Performance outcomes were analyzed via a 2 (Stimulus Type: Static image versus dynamic clip) × 2 (Task Order: Images first versus clips first) × 3 (Duration: 1, 3, 5 s) mixed-model ANOVA with repeated measures on the first and third factors. DSSQ difference scores and NASA-TLX workload scores were analyzed via a between-subjects (Task Order: Clips first versus images first) ANOVA.

Analyses revealed a statistically significant main effect of stimulus type on detection performance ($F(1, 24) = 25.333$, $p < .0001$, $\eta_P^2 = 0.514$). Pairwise comparisons indicated that participants made significantly fewer correct detections when the stimuli were presented via dynamic clips as opposed to static images (mean difference = 1.462, $p < .0001$). This main effect of stimulus type is presented in Fig. 27.1.

A statistically significant main effect for duration on detection performance was also observed ($F(1.545, 38.625) = 18.616$, $p < .0001$, $\eta_P^2 = 0.437$). As Mauchly's test identified a violation in the sphericity assumption for duration ($\chi^2 (2) = 8.028$, $p = .018$, $\varepsilon = 0.772$), the Greenhouse–Geisser adjusted degrees of freedom are reported. Pairwise comparisons revealed that participants exhibited greater correct detection rates when the stimuli were presented over 5 s compared to 3 s (mean difference = 1.115 correct detections, $p < .0001$) or 1 s (mean difference = 1.846 correct detections, $p < .0001$). These comparisons are shown in Fig. 27.2.

Finally, there was a statistically significant stimulus type by duration interaction ($F(1.897, 47.425) = 9.390$, $p < .0001$, $\eta_P^2 = 0.281$). Both conditions exhibited the same general trend of correct detections decreasing as duration decreased. However, the decline appears much steeper in the dynamic condition as illustrated in Fig. 27.3.

Two-tailed t-tests conducted on DSSQ difference scores (post-scores minus pre-scores) showed that participants reported lower task engagement ($t(25) = -2.736$, $p = .011$) and greater distress ($t(25) = 3.682$, $p = .001$) as a result of completing the detection task. Worry was not significantly affected ($t(25) = -1.873$, $p = .073$).

Unweighted global workload scores, as gauged by the NASA-TLX, were analyzed via two-tailed t-tests. Analyses revealed that the task imposed significant workload ($t(25) = 14.668$, $p < .001$). The overall average workload score (46.42) was commensurate with the averages of other classification (46.00) and cognitive (46.00) tasks.[21]

DISCUSSION

The current study sought to determine a possible statistical psychophysical equivalency of the static and dynamic visual stimuli. Participants correctly detected the critical signals equally well (static mean out of 20 = 16.923, dynamic mean out of 20 = 16.692) under alerted conditions when the stimuli were presented over a longer duration (5 s) versus shorter durations of 3 s and 1 s. Moreover, the difference between the static average and dynamic average of correct detections at 5 s yielded the smallest effect size (d = 0.046) of any of the collective comparisons.

To be considered truly "equivalent," the distributions of both the static and the dynamic conditions should be Gaussian. However, when comparing across the two stimulus types and all three durations, and having failed to reject the null hypothesis, the 5 s duration stimuli were considered equivalent between the static and dynamic conditions. As a result, the 5 s images and clips were designated as equally detectable. These findings were of import, as they were used to establish task parameters of a subsequent study to test the effects of stimulus type (i.e., static versus dynamic) on detection performance in a vigilance task.

According to subjective reports, the task imposed significant workload and stress on the participants. DSSQ subscales (t-test results) indicated that the 2AFCT imposed much the same stress profile as traditional vigilance protocols: lower

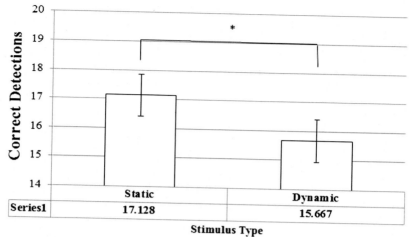

FIGURE 27.1 Main effect of stimulus type on detection performance. Error bars are standard errors. The asterisk denotes a statistically significant difference.

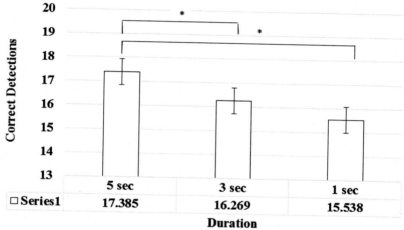

FIGURE 27.2 Main effect of duration on detection performance. Error bars are standard errors. The asterisk denotes a statistically significant difference.

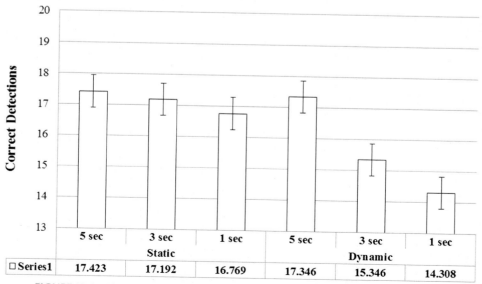

FIGURE 27.3 Significant stimulus type by duration interaction. Error bars are standard errors.

task engagement, higher distress, and no effect on worry. The 5 s duration was thus considered to be the length of stimulus presentation wherein the signals were most detectable, as indexed by both performance outcomes and subjective reports.

Empirical studies designed to investigate detection performance under alerted conditions and over extended periods of time (i.e., vigilance) have long utilized experimental protocols with exclusively static stimuli. The findings of the present work support the hypothesis that the nature of stimulus presentation can have a significant effect on detection performance under alerted conditions. Results can therefore help to inform and refine the design of future experimental protocols intended to investigate naturalistic detection in vigilance-type tasks.

REFERENCES

1. Davies DR, Parasuraman R. *The psychology of vigilance*. London (UK): Academic; 1982.
2. Warm JS, editor. *Sustained attention in human performance*. Chichester (UK): Wiley; 1984.
3. Parasuraman R. Memory load and event rate control sensitivity decrements in sustained attention. *Science* 1979;**205**:924–7.
4. Nuechterlein KH, Parasuraman R, Jiang Q. Visual sustained attention: Image degradation produces rapid sensitivity decrement over time. *Science* 1983;**220**:327–9.
5. Mackworth NH. The breakdown of vigilance during prolonged visual search. *Quarterly Journal of Experimental Psychology* 1948;**1**(1):6–21.
6. O'Hanlon JF. Adrenaline and noradrenaline: relation to performance in a visual vigilance task. *Science* 1965;**150**(3695):507–9.
7. Wiggins MW. Vigilance decrement during a simulated general aviation flight. *Applied Cognitive Psychology* 2011;**25**(2):229–35.
8. Hitchcock EM, Dember WN, Warm JS, Moroney BW, See JE. Effects of cueing and knowledge of results on workload and boredom in sustained attention. *Human Factors* 1999;**41**(3):365–72.
9. Harris DH. How to really improve airport security. *Ergonomics in Design* 2002;**10**(1):17–22.
10. Gill GW. Vigilance in cytoscreening: looking without seeing. *Advance for Medical Laboratory Professionals* 1996;**8**:14–5.
11. Reinerman-Jones L, Matthews G, Mercado JE. Detection tasks in nuclear power plant operation: vigilance decrement and physiological workload monitoring. *Safety Science* 2016;**88**:97–107.
12. Gunn DV, Warm JS, Nelson WT, Bolia RS, Schumsky DA, Corcoran KJ. Target acquisition with UAVs: vigilance displays and advanced cuing interfaces. *Human Factors* 2005;**47**(3):488–97.
13. Hancock PA, Hart SG. What can human factors/ergonomics offer? *Ergonomics in Design* 2002;**10**(1):6–16.
14. Becker AB, Warm JS, Dember WN, Hancock PA. Effects of feedback on perceived workload in vigilance performance. In: *Proceedings of the human factors and ergonomics Society Annual Meeting*. September 1991. p. 20.
15. Hancock PA. In search of vigilance: the problem of iatrogenically created psychological phenomena. *American Psychologist* 2013;**68**(2):97–109.
16. Szalma JL, Schmidt TN, Teo GWL, Hancock PA. Vigilance on the move: video game-based measurement of sustained attention. *Ergonomics* 2014;**57**(9):1315–36.
17. Matthews G, Joyner L, Gilliland K, Campbell SE, Huggins J, Falconer S. Validation of a comprehensive stress state questionnaire: towards a state 'Big Three'?. In: Mervielde I, Deary IJ, De Fruyt F, Ostendorf F, editors. *Personality psychology in Europe*, vol. 7. Tilburg: Tilburg University Press; 1999.
18. Matthews G, Campbell SE, Falconer S, Joyner LA, Huggins J, Gilliland K, Grier R, Warm JS. Fundamental dimensions of subjective state in performance settings: task engagement, distress, and worry. *Emotion* 2002;**2**(4):315.
19. Matthews G, Szalma JL, Panganiban AR, Neubauer C, Warm JS. Profiling task stress with the Dundee stress state questionnaire. In: Cavalcanti I, Azevedo S, editors. *Psychology of stress: new research*. Hauppauge (NY): Nova; 2013. p. 49–90.
20. Hart SG, Staveland LE. Development of NASA-TLX (task load Index): results of empirical and theoretical research. In: Hancock PA, Meshkati N, editors. *Human mental workload*. Amsterdam: Elsevier; 1988. p. 139–83.
21. Grier RA. How high is high? A meta-analysis of NASA-TLX global workload scores. In: *Proceedings of the human factors and Ergonomics Society Annual Meeting*, vol. 59, No. 1. SAGE Publications; September 2015. p. 1727–31.

Chapter 28

Functional Near-Infrared Spectroscopy: Proof of Concept for Its Application in Social Neuroscience

Stefano I. Di Domenico[1,2], Achala H. Rodrigo[2], Mengxi Dong[2], Marc A. Fournier[2], Hasan Ayaz[3,5,6], Richard M. Ryan[1,4], Anthony C. Ruocco[2]

[1]Institute for Positive Psychology and Education, Australian Catholic University, North Sydney, NSW, Australia; [2]University of Toronto Scarborough, Toronto, ON, Canada; [3]Drexel University, Philadelphia, PA, United States; [4]University of Rochester; [5]University of Pennsylvania, Philadelphia, PA, United States; [6]Children's Hospital of Philadelphia, Philadelphia, PA, United States

The mission of social psychology is to understand "how the thoughts, feelings, and behaviour of individuals are influenced by the actual, imagined, or implied presence of others."[1] The mission of social neuroscience is to understand the neural and broader biological underpinnings of social psychological phenomena.[2] Like every scientific discipline, the success of social neuroscience rests on the effectiveness of its methods. Recent years have witnessed a proliferation of methods that hold promise for advancing social neuroscience. One such method is continuous-wave functional near-infrared spectroscopy (fNIRS), a functional neuroimaging technique that can be used to measure brain activity noninvasively.[a] In this chapter we demonstrate the utility of fNIRS for social neuroscience in identifying well-established patterns of prefrontal activity when people make self- and other-referential judgments.[3]

A BRIEF INTRODUCTION TO FNIRS AND ITS POTENTIAL FOR SOCIAL NEUROSCIENCE

Detailed reviews of continuous-wave fNIRS methodology and instrumentation are available elsewhere,[4,5] so here we provide only a brief outline of fNIRS principles. Measurements of brain activity obtained by fNIRS are based on the hemodynamic response, or more specifically on the fact that neuronal activity is fueled by glucose metabolism in the presence of oxygen. The hemodynamic response is a homeostatic process that replenishes the nutrients used by biological tissues by adjusting blood flow to areas of focal activity. Increases in neuronal activity set off a series of vascular events that result in the flooding of neuronal tissues with oxygenated hemoglobin (oxy-Hb), the protein molecules that carry oxygen within the blood. During bouts of activity the rate of oxy-Hb delivery typically exceeds the rate of oxygen utilization, resulting in a temporary increase in the concentration of oxy-Hb and a decrease in the concentration of deoxygenated hemoglobin (deoxy-Hb).

Whereas most biological tissues are transparent to near-infrared (NIR) light, oxy-Hb and deoxy-Hb are known to absorb and scatter NIR light of slightly different wavelengths in the range of 700–1000 nm. Continuous-wave fNIRS capitalizes on this property of oxy-Hb and deoxy-Hb. Light emitters placed on the surface of the scalp radiate NIR light into the head. Given the differential absorption and backscattering of oxy-Hb and deoxy-Hb, a portion of this NIR light returns to the surface of the scalp, where it is measured with photodetectors. Spectroscopic methods may thus be used to detect changes in the concentrations of oxy-Hb and deoxy-Hb. Typical fNIRS sensor pads geometrically position emitters and photoreceptors so that activity at the outer surface of the cortex may be measured with a spatial resolution in the order of square centimeters.

There are three main reasons why fNIRS is a promising neuroimaging technique for social neuroscience. First, compared to traditional neuroimaging systems such as functional magnetic resonance imaging (fMRI), which carry initial costs of up to several million dollars, continuous-wave fNIRS systems are relatively inexpensive—some available for less than $100,000 USD. A second key advantage of fNIRS is the low monetary cost of running participants.[4] Whereas the cost of running participants in fMRI studies can reach several hundred dollars per head, the cost of running participants with fNIRS

---

a. We restrict our focus on continuous-wave fNIRS. "Time resolved" and "frequency domain" fNIRS systems are outside the purview of the current chapter.

Neuroergonomics. https://doi.org/10.1016/B978-0-12-811926-6.00028-2
Copyright © 2019 Elsevier Inc. All rights reserved.

is no different than the cost of administering a standard paper-and-pencil questionnaire. This allows social neuroscientists to enhance the statistical power of their research by recruiting larger samples. Indeed, an oft-cited challenge in social neuroscience is the curtailed levels of statistical power of individual studies, and fNIRS is well suited to help meet this challenge. Finally, fNIRS systems are relatively insensitive to participant motion and have a portable, compact, and increasingly miniaturized design. This means that fNIRS can be flexibly deployed in naturalistic settings for enhanced ecological validity.[7]

THE CURRENT STUDY

To illustrate the utility of fNIRS for social neuroscience, we attempted to identify and replicate well-established patterns of prefrontal activity when people make self- and other-referential judgments. One of the most robust findings of social neuroscience studies using fMRI is that regions within the medial prefrontal cortex (MFPC), particularly the frontal midline regions within Brodmann's area 10, play an important role in representing knowledge about the self, relative to knowledge about other people.[3,8] For example, Kelly et al.[9] asked people to reflect upon their own personality characteristics and those of the former United States president George W. Bush, and found that the MPFC was preferentially engaged when participants reflected upon their own characteristics. In the present research, we constructed a personality judgment task to test the role of the MPFC across self- and other-referential processing using fNIRS.

METHOD

Participants

The research involved 109 individuals (78 females, 31 males), who participated for course credit or monetary compensation ($45 CAD). They ranged in age from 18 to 27 ($M = 20.30$, $SD = 2.02$).

Personality Judgment Task

As part of a larger study examining neural correlates of social support and the representation of self-knowledge,[6] participants were instructed to nominate a friend with the understanding that they would then describe themselves and their nominated friend using a list of trait adjectives. They were asked to rate themselves and their nominated friend on 120 adjectival markers of the "big five" trait dimensions.[10] We focused on markers for two of the big five traits, conscientiousness and extraversion, because these traits have clear behavioral expressions and are the most accurately perceived of the big traits.[11] Sample adjectives for conscientiousness were *dependable*, *efficient*, and *orderly*. Sample adjectives for extraversion were *assertive*, *energetic*, and *playful*. The personality-reflection task was presented using a block design consisting of 24 blocks: 12 for each of the self and friend conditions. Each block featured 10 consecutive adjectives for either conscientiousness or extraversion.

Fig. 28.1 shows the flow of one trial in the self–other reflection task. Participants responded to these adjectives by sliding and clicking a computer mouse cursor with their right hand over the appropriate scale response. They were instructed to respond as quickly as possible to each trial. After each click the cursor disappeared and the selected response flashed on the screen. Participants were instructed to slide the mouse to its original position on the mouse pad after each response. At the start of each trial, the location of the mouse cursor was reset to its original position in the middle of the screen.

fNIRS Procedures and Signal Processing

Activity of the prefrontal cortex was monitored using the fNIR Imager 1000, a 16-channel continuous-wave fNIRS system (FNIR, Potomac, MD; www.fnirdevices.com). The system is composed of a sensor pad with a source–detector separation of 2.50 cm and a data acquisition control box running Cognitive Optical Brain Imaging Studio software. The sensor pad had a temporal resolution of 500 ms per scan, a penetration depth of 1.25 cm into the prefrontal cortex, and light sources with peak wavelengths at 730 and 850 nm. The sensor pad was secured in alignment with electrode positions F_7, F_{P1}, F_{P2}, and F_8 based on the international 10/20 system. This positioning corresponds to Brodmann areas 9, 10, 45, and 46. Fig. 28.2A displays the location of each channel on a standard MRI template.[12]

After acquisition, recorded light intensities were visually inspected by a trained experimenter. Saturated channels were excluded from analyses. Subsequently, signal and physiological artifacts were excluded with a low-pass filter consisting of a finite impulse response and a linear-phase filter with an order of 20 and a cut-off frequency of 0.1 Hz. A sliding-window rejection algorithm was applied to the filtered data to exclude motion artifacts. Activation segments were extracted using time-synchronization markers via a serial connection from the computer used to display the personality-reflection task. Relative changes in concentrations of oxy-Hb (Δoxy-Hb) for each activation segment were calculated using fNIRSoft Professional Edition.

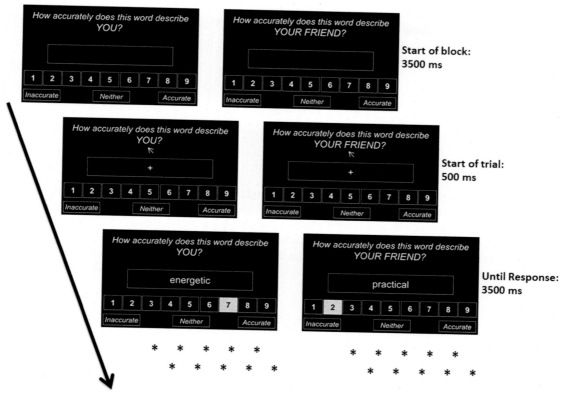

FIGURE 28.1 Exemplary flow of one trial in the personality judgment task.

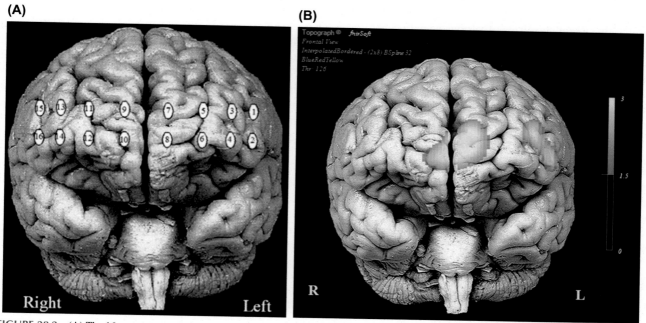

FIGURE 28.2 (A) The 16 measurement locations (channels) of the fNIRS sensor pad. The brain surface image is from the University of Washington, Digital Anatomist Project.[12] (B) Topographic images representing oxy-Hb contrasts across self- and other-referential processing.

TABLE 28.1 Multilevel Analyses Contrasting Prefrontal Activity Across the Self- and Friend-Referential Conditions

Channel	1	2	3	4	5	6	7	8	9	10	11	12	13	14	15	16
b	.06	.12	.10	.00	.00	.02	.14	.14	.07	.14	−.03	−.10	.04	−.11	−.02	.03
SE	.05	.05	.04	.09	.05	.08	.05	.06	.05	.06	.05	.10	.05	.08	.05	.05
t value	1.39	2.62**	2.76**	−.05	.02	.24	2.93**	2.32*	1.31	2.44*	−.57	−1.06	.87	−1.39	−.32	.56

Note: Unstandardized regression coefficients represented by bs and corresponding standard errors by SEs. **$P < .01$, *$P < .05$.

RESULTS

Data was analyzed with multilevel regression models.[13] We estimated random intercept models using the method of maximum likelihood, an unstructured covariance matrix, and the "between–within" method of estimating degrees of freedom. We estimated separate models for each channel across the fNIRS sensor pad. A dummy code contrasted activity in the self- and friend-referential conditions (friend=0, self=1). Results are shown in Table 28.1 and Fig. 28.2B. As expected, channels covering the anterior frontal pole and corresponding to the MPFC exhibited preferential activity in the self condition relative to the friend condition. Results also indicated preferential activity for the self condition in channels covering left lateral prefrontal regions.

DISCUSSION

The advancement of any scientific discipline depends in part upon the development of new methods. The present findings illustrate the utility of fNIRS for social neuroscience, showing its ability to recover well-established patterns of prefrontal activity when participants made self- and other-referential judgments. In keeping with previous fMRI studies,[3,8] participants exhibited greater activity in regions corresponding to the MPFC when rating trait adjectives with respect to themselves relative to adjectives with respect to their friends. Self-referential judgments were also associated with preferential activity in left lateral PFC regions. This finding is consistent with previous studies, suggesting the role of the left lateral PFC in the retrieval of autobiographical information during self-referential judgments.[8] These results thus provide initial evidence that fNIRS can be used to examine types of phenomena of interest to social neuroscientists.

Given these promising results, we can reflect upon the ways in which fNIRS may be best used to advance social neuroscience research. In addition to large-sample "brain-mapping" research (that is, traditional neuroimaging research in which the neural correlates of various social psychological phenomena are examined), fNIRS may be particularly useful for studies adopting a "brain-as-predictor" approach.[14] Studies adopting the brain-as-predictor approach have the goal of understanding how specific neurocognitive processes mediate ecologically valid outcomes, and accordingly use measures of task-related neural activity to predict real-world behaviors. For decisive results, however, large samples of participants are required, especially when task-related measures of neural activity are intended as variables that statistically mediate the relations between social psychological variables on the one hand and objective behavioral outcomes on the other. In this regard, fNIRS has distinct advantages relative to more expensive imaging methods.

Of course, like all methodologies fNIRS has clear limitations. Most saliently, it limits researchers to measuring neural activity at the surface of the cortex; but most, if not all, social psychological phenomena have key underpinnings in deeper cortical and subcortical structures. A related limitation of fNIRS is its lower spatial resolution relative to fMRI, which can assess neural activity in the order of square millimeters. For these reasons, fNIRS is optimally deployed in studies that target, on an a priori basis, specific operations of the outer cortex. Researchers can partially sidestep these limitations by coupling fNIRS technology with other methods, especially in protocols that capitalize on the portability of some fNIRS systems for naturalistic recording of brain activity.[7] Notwithstanding these limitations, the present results provide support for the idea that fNIRS is a promising method that can be effectively used to examine key topics within social neuroscience.

REFERENCES

1. Allport GW. The historical background of modern social psychology. In: 2nd ed. Lindzey G, Aronson E, editors. *The handbook of social psychology*, Vol. 1. Reading (MA): Addison-Wesley; 1968. p. 1–80.
2. Harmon-Jones E, Inzlicht M. A brief overview of social neuroscience: biological perspectives on social psychology. In: Harmon-Jones E, Inzlicht M, editors. *Social neuroscience: biological approaches to social psychology*. New York (NY): Psychology Press; 2016. p. 1–9.
3. Wagner DD, Haxby JV, Heatherton TF. The representation of self and person knowledge in the medial prefrontal cortex. *Wiley Interdisciplinary Reviews: Cognitive Science* 2012;**3**:451–70. https://doi.org/10.1002/wcs.1183.

4. Irani F, Platek SM, Bunce S, Ruocco AC, Chute D. Functional near infrared spectroscopy (fNIRS): an emerging neuroimaging technology with important applications for the study of brain disorders. *The Clinical Neuropsychologist* 2007;**21**:9–37. https://doi.org/10.1080/13854040600910018.

5. Scholkmann F, Kleiser S, Metz AJ, Zimmerman R, Pavia JM, Wolf U, Wolf M. A review on continuous wave functional near-infrared spectroscopy and imaging instrumentation and methodology. *NeuroImage* 2014;**85**:6–27. https://doi.org/10.1016/j.neuroimage.2013.05.004.

6. Di Domenico SI, Fournier MA, Rodrigo AH, Dong M, Ayaz H, Ryan RM, Ruocco AC. *Medial prefrontal activity during self-other judgments is modulated by relationship need fulfillment: an optical neuroimaging study.* Institute for Positive Psychology and Education, Australian Catholic University; 2018 (under review).

7. McKendrick R, Parasuraman R, Ayaz H. Wearable functional near infrared spectroscopy (fNIRS) and transcranial direct current stimulation (tDCS): expanding vistas for neurocognitive augmentation. *Frontiers in Systems Neuroscience* 2015;**9**:27. https://doi.org/10.3389/fnsys.2015.00027.

8. Denny BT, Kober H, Wager TD, Ochsner KN. A meta-analysis of functional neuroimaging studies of self- and other judgments reveals a spatial gradient for mentalizing in medial prefrontal cortex. *Journal of Cognitive Neuroscience* 2012;**24**:1742–52. https://doi.org/10.1162/jocn_a_00233.

9. Kelley WM, Macrae CN, Wyland CL, Caglar S, Inati S, Heatherton TF. Finding the self? An event-related fMRI study. *Journal of Cognitive Neuroscience* 2002;**14**:785–94. https://doi.org/10.1162/08989290260138672.

10. Saucier G, Goldberg LR. Evidence for the Big Five in analyses of familiar English personality adjectives. *European Journal of Personality* 1996;**10**:61–77. https://doi.org/10.1002/(SICI)1099-0984(199603)10:1<61::AID-PER246>3.0.CO;2-D.

11. Connelly BS, Ones DZ. An other perspective on personality: meta-analytic integration of observers' accuracy and predictive validity. *Psychological Bulletin* 2010;**136**:1092–122. https://doi.org/10.1037/a0021212.

12. Ayaz H, Shewokis PA, Bunce S, Izzetoglu K, Willems B, Onaral B. Optical brain monitoring for operator training and mental workload assessment. *NeuroImage* 2012;**59**:36–47. https://doi.org/10.1016/j.neuroimage.2011.06.023.

13. Snijders TAB, Bosker RJ. *Multilevel analysis: an introduction to basic and advanced multilevel modeling.* 2nd ed. London (England): Sage; 2012.

14. Berkman ET, Falk EB. Beyond brain mapping: using neural measures to predict real-world outcomes. *Current Directions in Psychological Science* 2013;**22**:45–50. https://doi.org/10.1177/0963721412469394.

Chapter 29

Quantifying Brain Hemodynamics During Neuromuscular Fatigue

Joohyun Rhee, Ranjana K. Mehta
Texas A&M University, College Station, TX, United States

INTRODUCTION

Neuromuscular function, essential for performing activities of daily living and preventing falls, involves a cooperative integration of the musculoskeletal and central nervous systems that is compromised in older adults. Specifically, neuromuscular fatigue, defined as exercise-induced loss of force-generating capacity, is a common disabling problem that substantially decreases quality of life in older adults. It has been demonstrated that older adults perform activities of daily living near their maximal capabilities. Additionally, older adults are more vulnerable to fatigue-related functional deficits because of age-related declines in both motor and cognitive functions.[1,2] The major contributors to neuromuscular fatigue include central and peripheral fatigue. In particular, central fatigue involves exhaustion of the central nervous system, thus it is necessary to have a better understanding of neural activation changes during fatigue development.[3] Substantial work has been done to understand the neural signatures of fatigue, but they are limited to the testing of smaller and more distal body parts (e.g., hand and ankle). Neural activation changes during fatigue of larger muscles, particularly those of the lower extremities, are important to establish, as these muscles play a major role in balance and locomotive functions that affect the quality of life when impaired.[4] Due to the physical constraints of neuroimaging devices[5] and inherent motion artifacts during larger muscle group exercises, few studies have investigated changes in neural activation patterns of the quadriceps, i.e., the thigh muscles.

Near-infrared spectroscopy (NIRS) is a noninvasive portable neuroimaging device that provides information in good temporal and spatial resolution with a flexibility to measure the area of interest during actual movement involving larger muscles.[6] However, the procedures for examining neuromuscular fatigue development in larger muscles using functional NIRS (fNIRS) have not been established. In this chapter we compare various NIRS processing methods to account for motion artifacts and systemic physiological changes that can be employed during submaximal fatiguing exercises of the quadriceps. We also demonstrate fatigue-related spatiotemporal brain activation changes across the different motor-function-related cortical regions in older adults.

METHODS

Participants

Fourteen sedentary older females with no self-reported musculoskeletal injuries or disorders were recruited from the local community. Mean (SD) age, height, weight, and body mass index of the participants were 72 (4.8) years, 1.64 (0.06) m, 61.3 (6.8) kg, and 22.7 (1.8) kg/m², respectively. Informed consent, approved by the Texas A&M University institutional review board, was obtained prior to the experiment.

Procedures

Upon consent, participants were seated upright in a commercial dynamometer (Humac NORM, Computer Sports Medicine, Stoughton, MA, USA) with the hip and knee flexed at 90 degrees. The right leg was supported by a fixture attached to the shin area that transmitted knee-joint force to a torque converter. Three isometric knee extension maximum voluntary contractions (MVCs) were measured with 2 min of rest in between each, and the maximum strength value was used to determine the target force level of 30% for the fatiguing task. Participants performed training trials of the fatiguing exercise.

Neuroergonomics. https://doi.org/10.1016/B978-0-12-811926-6.00029-4
Copyright © 2019 Elsevier Inc. All rights reserved.

175

After adequate rest, participants began the submaximal knee extension fatiguing task. Each trial in the fatiguing exercises was performed for 15 s with 15 s rest between each trial, and participants performed these trials until voluntary exhaustion. They were instructed to control their generated force against the target load level as closely as possible based on real-time visual feedback presented at eye height. The exercise was terminated upon a participant's failure to maintain the target force level or a participant's decision to stop based on self-reported exhaustion, and a postMVC trial was performed immediately after exhaustion.

Measurements

Neural hemodynamic responses of the anterior prefrontal cortex (PFC), primary motor area, and sensory area (Fig. 29.1) were recorded at a sampling rate of 50 Hz using a continuous-wave functional NIRS system (Techen Inc., MA, USA, CW6 system) during the entire session to investigate spatiotemporal neural activation patterns of fatigue development during the fatiguing exercise. Recorded hemodynamic responses were processed and analyzed using HomER2 (Center for Functional Neuroimaging Technologies, Massachusetts General Hospital East, MA, USA).[7]

NIRS ANALYSES

Prior to any NIRS analysis, the signal of each channel was evaluated for light saturation (i.e., signal too strong or weak). A saturated signal shows a flat line without fluctuation or waveforms with a fixed frequency. When the signal is too weak, with a small signal-to-noise ratio, the strong noise overrides hemodynamic response changes. If the signals show fluctuations reflecting cardiac perfusion, the channels can be considered to be of good quality. Based on these criteria, light saturated and noisy channels were excluded from the analyses. The light intensity was then converted into optical density changes by calculating the logarithm of the signal and low pass filtered with a 3 Hz cutoff frequency to reduce high-frequency noise.

Motion artifacts caused by sudden head movements, particularly when participants perform knee extensions, need to be corrected to have reliable results. Motion artifacts can be detected and excluded manually from the analysis, but this approach usually results in inconsistencies in outcomes, largely due to the analyst's prior experiences and biases. Several motion artifact detection and removal algorithms exist that can automate the process, including principal component analysis (PCA), spline interpolation, wavelet filtering, and kurtosis-based wavelet filtering. PCA works efficiently when motion is the main source of variance but requires multiple channels to process.[8] Spline interpolation is a simple and fast approach that can correct signal offsets, but requires a reliable technique to identify motion artifact segments.[9] Wavelet filtering is better at maintaining frequency content to keep physiological signals such as cardiac and respiration,[10] but it tends to reduce the signal amplitude when the signal to noise ratio is high. Kurtosis-based wavelet filtering removes the outlier wavelets from the filtering.[11] Since each method offers different strengths and weaknesses, the choice of a proper algorithm that fits

FIGURE 29.1 fNIRS probe design to measure neural activation during quadriceps fatigue. *Red dots* are emitters and *blue dots* are detectors. Neural activation was measured at the channels (*green lines*) between the emitters and detectors. The cortical mapping shows estimated spatial location of the measurement on the surface of the brain using the current probe design.

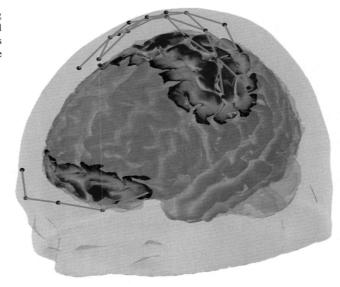

each study data type is important to acquire reliable results. Thus evaluation of corrected signals is necessary to choose the motion artifact removal algorithm that works best based on how it retains the integrity of the data. In this study motion artifacts were evaluated using all these algorithms (Fig. 29.2); based on the best fit, the motion artifacts were corrected using the kurtosis-based wavelet algorithm. The chosen algorithm was found to remove motion artifacts efficiently while retaining the characteristics of the original signal.

After the evaluation and correction of motion artifacts, the signal was bandpass filtered with a 0.5–0.016 Hz cutoff frequency range to remove systemic responses caused by heartbeat and slow wave drift caused by the NIRS system.[12] Processed optical density changes were converted into oxygenated (HbO) and deoxygenated (HbR) hemoglobin concentration using a modified Beer–Lambert law. HbO and HbR levels during each trial were compensated using the local baseline, which was computed as the average HbO and HbR levels of a 2 s window prior to event stimulation of each trial. The neural hemodynamic responses can be calculated by averaging the HbO activation or averaging the difference between HbO and HbR, i.e., Oxy.[13]

Quantification of calculated HbO or Oxy within each trial can be done by averaging the activation around maximum HbO value (HbO_{max}), or averaging the activation during a given time window (HbO_{mean}) depending on the task characteristics (Fig. 29.3). If the task duration is very short, HbO_{max} and HbO_{mean} values will be similar. If the task duration is longer, HbO_{mean} has been used to assess the activation during cognitive tasks. However, if perceptual or motor processes are involved in the task, such as the knee extension task in this study, the amplitude of neural activation may change across trials or the activation may decrease within trial due to habituation. In this case, HbO_{max} would be a better representation of task-related neural activation than HbO_{mean}.[14] The task in the current study was intermittent knee extension repeated until

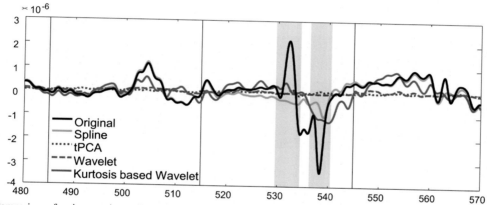

FIGURE 29.2 Comparison of various motion artifact correction algorithms. The *black line* shows the original signal of oxygenated hemoglobin (HbO) concentration of one channel, and *shaded areas* indicate the motion artifact caused by sudden head movement. The *colored lines* show the result of each motion artifact correction algorithm. The data is from one sample participant.

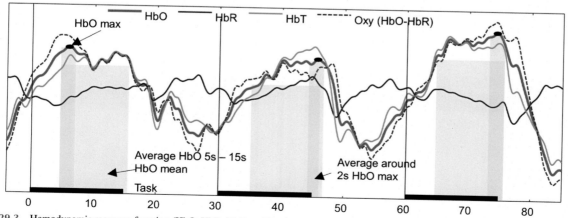

FIGURE 29.3 Hemodynamic response function (HbO, HbR, HbT, and Oxy) over the course of three knee extension trials (i.e., three work and rest periods). *Black dots* depict maximum HbO (HbO_{max}) of each trial. Neural activation can be calculated by averaging 2 s windows around HbO_{max} or averaging 5–15 s windows for each trial (HbO_{mean}).

exhaustion, and the amplitude and temporal activation pattern within each trial were expected to change over time. Thus both HbO_{max} and the HbO_{mean} were used to capture spatial and temporal activation pattern changes to explore their suitability as hemodynamic metrics. Because the emphasis here is on examining the various analysis techniques to process and present NIRS data associated with neuromuscular fatigue, statistical analysis of the fNIRS signal is not within the scope of this chapter. It should also be noted that procedures to evaluate statistical changes of neural spatiotemporal activation during fatigue development are not well established, unlike those used for functional magnetic resonance imaging. Potential approaches used in mobile NIRS studies include comparisons of channels using false discovery rate, a linear mixed model, and statistical parametric mapping.

DATA PRESENTATION

In this study the number of trials differed for each individual, since fatigue resistance varies between individuals. To quantify fatigue development from the data, trials of each participant were separated into three blocks based on the temporal occurrence. Each one-third of the trials were labeled as early, middle, or late phase. Trials within each phase were averaged for each participant, and used for group averages for the study sample. This averaged activation can be presented in several ways to display the spatiotemporal neural changes with fatigue development. First, cortical activation maps within each phase (i.e., spatial activation) and between each phase (i.e., temporal activation) are shown in Fig. 29.4A; these maps were created using the HbO_{max} and HbO_{mean} metrics. Second, hemodynamic response function curves based on HbO_{mean} for each channel (with the phase embedded inside each channel) are shown in Fig. 29.4B.

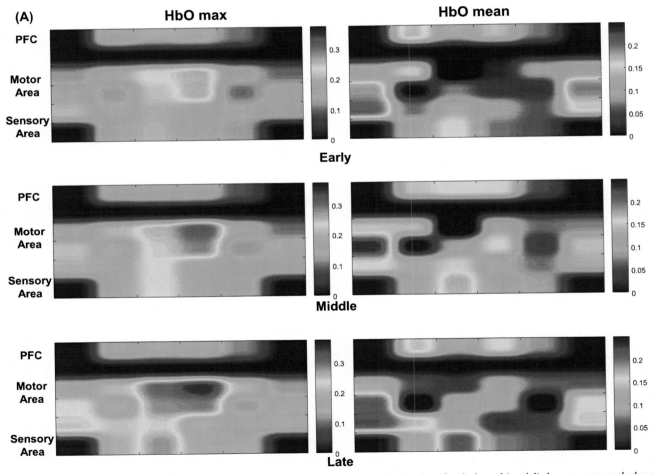

FIGURE 29.4 Cortical activation maps (A) for HbO_{max} and HbO_{mean} show how the neural activation of each channel (spatial) changes across each phase (temporal). Hemodynamic response function curves (B) shows HbO_{mean} changes within each phase for each channel.

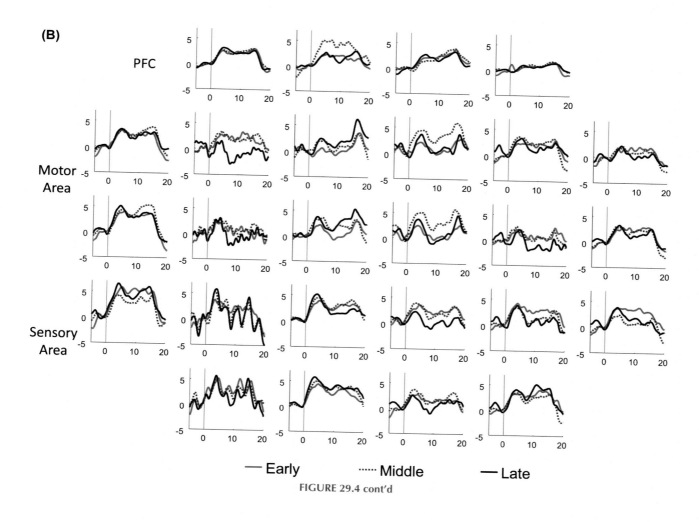

— Early ⋯⋯ Middle — Late

FIGURE 29.4 cont'd

RESULTS AND DISCUSSION

Based on the HbO_{max} metric, the medial motor region was found to have higher activation than other regions during the knee extension trials. This region is known to play a major role in neural control of the lower extremity. As participants continued the exercise (i.e., in middle and late phases), this region was found to have increased activation. This temporal pattern of increased activation over time indicates that the participants increased their effort to maintain their force levels at the target level. In the late phase the left medial sensory region was also found to contribute to the neural activation pattern, thereby demonstrating the spatiotemporal neural changes in quadriceps fatigue development. On the other hand, based on the HbO_{mean} metric, increased activation in the lateral motor and left medial sensory region was found, which strengthened in activation over time. Since HbO_{mean} is a more conservative metric, i.e., average for 10 s, it relays distinctly different information than the HbO_{max} metric. It should be noted that in the cortical map focused on HbO_{max} the lateral motor and left medial sensory regions were found to have a slight increase in activation, thus it can be argued that the HbO_{max} metric provides a more comprehensive understanding of the spatiotemporal changes in neural patterns during quadriceps fatigue.

In summary, this chapter presents a methodological approach to quantifying neural patterns of neuromuscular fatigue development using fNIRS, and proposes reliable procedures to process fNIRS signals during lower-extremity exercises. Neural activation scores were grouped and averaged based on the temporal location of the task during the fatiguing exercise, and spatiotemporal cortical activation maps of fatigue development are presented. The observed spatiotemporal activation pattern was congruent with a previous study that reported brain activity with pedaling exercises.[15] This supports the robustness of fNIRS as a neuroimaging technique that can reliably monitor neural activations of larger lower-extremity muscles during fatiguing motor tasks.

REFERENCES

1. Lewis G, Wessely S. The epidemiology of fatigue: more questions than answers. *Journal of Epidemiology and Community Health* 1992;**46**:92–7. https://doi.org/10.1136/jech.46.2.92.

2. Vanden Noven ML, Pereira HM, Yoon T, Stevens AA, Nielson KA, Hunter SK. Motor variability during sustained contractions increases with cognitive demand in older adults. *Frontiers in Aging Neuroscience* 2014;**6**:1–14. https://doi.org/10.3389/fnagi.2014.00097.

3. Allman BL, Rice CL. Neuromuscular fatigue and aging: central and peripheral factors. *Muscle and Nerve* 2002;**25**:785–96. https://doi.org/10.1002/mus.10116.

4. Renfro M, Maring J, Bainbridge D, Blair M. Fall risk among older adult high-risk populations: a review of current screening and assessment tools. *Current Geriatrics Reports* 2016:1–12. https://doi.org/10.1007/s13670-016-0181-x.

5. Dobkin BH, Firestine A, West M, Saremi K, Woods R. Ankle dorsiflexion as an fMRI paradigm to assay motor control for walking during rehabilitation. *Neuroimage* 2004;**23**:370–81. https://doi.org/10.1016/j.neuroimage.2004.06.008.

6. Karim H, Schmidt B, Dart D, Beluk N, Huppert T. Functional near-infrared spectroscopy (fNIRS) of brain function during active balancing using a video game system. *Gait and Posture* 2012;**35**:367–72. https://doi.org/10.1016/j.gaitpost.2011.10.007.

7. Huppert TJ, Diamond SG, Franceschini MA, Boas DA. HomER: a review of time-series analysis methods for near-infrared spectroscopy of the brain. *Applied Optics* 2009;**48**:280–98. https://doi.org/10.1016/j.drugalcdep.2008.02.002.A.

8. Yücel MA, Selb J, Cooper RJ, Boas DA. Targeted principle component analysis: a new motion artifact correction approach for near-infrared spectroscopy. *Journal of Innovative Optical Health Sciences* 2014;**7**(02):1350066.

9. Scholkmann F, Spichtig S, Muehlemann T, Wolf M. How to detect and reduce movement artifacts in near-infrared imaging using moving standard deviation and spline interpolation. *Physiological Measurement* 2010;**31**:649–62. https://doi.org/10.1088/0967-3334/31/5/004.

10. Molavi B, Dumont GA. Wavelet-based motion artifact removal for functional near-infrared spectroscopy. *Physiological Measurement* 2012;**33**:259–70. https://doi.org/10.1088/0967-3334/33/2/259.

11. Chiarelli AM, Maclin EL, Fabiani M, Gratton G. A kurtosis-based wavelet algorithm for motion artifact correction of fNIRS data. *Neuroimage* 2015;**112**:128–37. https://doi.org/10.1016/j.neuroimage.2015.02.057.

12. Koenraadt KL, Roelofsen EG, Duysens J, Keijsers NL. Cortical control of normal gait and precision stepping: an fNIRS study. *Neuroimage* 2014;**85**:415–22. https://doi.org/10.1016/j.neuroimage.2013.04.070.

13. Hyodo K, Dan I, Kyutoku Y, Suwabe K, Byun K, Ochi G, Soya H. The association between aerobic fitness and cognitive function in older men mediated by frontal lateralization. *Neuroimage* 2016;**125**:291–300. https://doi.org/10.1016/j.neuroimage.2015.09.062.

14. Yücel MA, Aasted CM, Petkov MP, Borsook D, Boas DA, Becerra L. Specificity of hemodynamic brain responses to painful stimuli: a functional near-infrared spectroscopy study. *Scientific Reports* 2015;**5**:9469. https://doi.org/10.1038/srep09469.

15. Mehta JP, Verber MD, Wieser JA, Schmit BD, Schindler-Ivens SM. A novel technique for examining human brain activity associated with pedaling using fMRI. *Journal of Neuroscience Methods* 2009;**179**:230–9. https://doi.org/10.1016/j.jneumeth.2009.01.029.

Chapter 30

Web Usability Testing With Concurrent fNIRS and Eye Tracking

Siddharth Bhatt[1], Atahan Agrali[1], Kevin McCarthy[1], Rajneesh Suri[1], Hasan Ayaz[1,2,3]

[1]Drexel University, Philadelphia, PA, United States; [2]University of Pennsylvania, Philadelphia, PA, United States; [3]Children's Hospital of Philadelphia, Philadelphia, PA, United States

INTRODUCTION

Digitization in the financial sector has given rise to a new industry in its own right: fintech. Within financial services, wealth management has been one of the prime beneficiaries of the digitization wave. According to the consulting firm KPMG,[1] global investment in fintech companies increased from US$4 billion in 2012 to US$47 billion in 2015. Today every major player in the wealth management industry offers digital portfolio management services to its clientele. With all major players having digitized their services, the competitive advantage no longer lies in firms merely having digital capabilities; rather, the winners will be firms which can provide the best possible wealth management experience by designing web interfaces that give consumers multiple features on the website along with ease of use. Given customers' need to self-manage their portfolios, the quality of the web interface offered by financial firms is becoming decisive. Thus creating and maintaining web interfaces which allow customers to have a pleasant experience on the firm's website have become a top priority for wealth management firms.

Unfortunately, firms often learn about issues with their websites only after such websites are launched. Poorly designed websites often signal an inferior quality of the brand(s) associated with the website[2]; they make it difficult for customers to navigate with ease and hence make it difficult to process information on the website. There is sufficient evidence to suggest that when people find it difficult to process product information, their evaluation of the brand becomes negative.[3,4] According to an internet performance research firm, 95% of surveyed IT executives felt that poor websites resulted in loss of revenue,[5] hence it is important that web platforms are designed in ways that enable consumers to attain their wealth management goals with the least amount of friction. To that end, it is becoming common to involve end users in both development and testing of such websites.[6] However, the methods employed to elicit users' feedback remain subjective and rudimentary. Most web usability studies employ qualitative and/or quantitative surveys to collect users' assessment of a website, but self-reports and similar techniques are severely limited in their ability to bring forth users' insights objectively. Further, these measures fail to highlight specific areas that might need improvement, thus severely limiting a company's ability to make changes to the website to reflect user feedback.

Such shortcomings could be overcome if the fintech industry took advantage of the new developments in neuroergonomic research. Drawing on the findings from neuroergonomics, this research proposes a novel multimodal approach to assessing how websites could be tested by making end users perform different types of ecologically valid tasks in a systematic and repeated manner while behavioral activity and user physiological measures are logged. The approach used in this research combines the strengths of neurophysiological methods with traditional survey tools to help elicit rich and specific feedback that customary tools fail to provide. Recent developments in neuroergonomics have provided methodologies which can benefit web assessment studies. Neuroergonomics is defined as the study of the human brain in relation to performance while at work and in everyday settings.[7,8] Within this field, the portable and wearable brain imaging technology of functional near-infrared spectroscopy (fNIRS) has emerged as a practical sensor modality to capture cortical oxygenation changes in the brain. fNIRS has been used to assess variety of cognitive functions, such as working memory, attention, and decision-making.[9-11] These sensors have been used in real-world field settings to assess workload of air traffic controllers using a new generation of user interfaces, and in training unmanned vehicle operators with customized flight simulator interfaces.[12] More recently, miniaturized and battery-operated wireless versions of fNIRS have been developed[13] and used to monitor individuals in ambulatory and outdoor settings, such as when participants compare spatial navigation aid interfaces.[14] fNIRS as a noninvasive and portable technology can be used

Neuroergonomics. https://doi.org/10.1016/B978-0-12-811926-6.00030-0
Copyright © 2019 Elsevier Inc. All rights reserved.

with other data collection methods, and provides cognitive and affective-state-related brain activity data.[12,15–17] In this study we aimed to investigate usability of a new web-based investment tool via simultaneous fNIRS and eye-tracking measurements. The brain and body measures were used in addition to traditional self-reported survey measures and task performance measures on the website to provide a comprehensive assessment of modulated task difficulty and user engagement.

METHODS

This multimodal study combined four different metrics (survey measures, performance measures, neural correlates, and eye-tracking metrics) to assess how users perceived a new wealth management web platform.

The 37 participants in the study (46% female, mean age = 47. 54) were composed of three groups of users: first-time users of the website, the company's current clients, and its employees . Users interacted with the prototype of the new website and performed several investment-related tasks that they would normally undertake on a wealth management platform.

To assess the website an experimental protocol with two levels of task difficulty was devised for all task types: easy tasks were designed so that users could readily find the information needed to complete the task on the same webpage where they started, and more challenging, difficult tasks required users to search deeper for information necessary to complete the tasks. While easy tasks required little navigation and planning, difficult tasks required search, working memory, and planning, thus testing how easy it was to perform the given task on the website. Navigation included going to different webpages on the site, changing the format of the financial information presented, and using other functions of the webpage. Users went through 37 different tasks on the website, and all four metrics were collected for each task. Contrasting the four dimensions of metrics on task difficulty provided a rigorous test of the website and the approach.

RESULTS

Analyses of data across each of the four modalities were conducted using task difficulty (easy versus difficult) and group (first-time users versus clients versus employees) as fixed factors in a mixed-model effect analysis. Together these metrics revealed significant differences between easy and difficult tasks across all task domains on the website.

Survey Measures

The survey measures (Fig. 30.1) comprised a single item assessing participants' perception of the difficulty of all tasks on a nine-point Likert scale (1 = very difficult, 9 = very easy). The linear mixed-effect model revealed a significant effect of difficulty on the measure (F $(1, 351) = 131.05$, $P = .00$), indicating that the participants could differentiate between easy and difficult tasks. There was no main effect of group (F $(2, 32) = .22$, $P = .80$) or interaction between group and task difficulty (F $(2, 351) = .12$, $P = .88$). Participants found it easier to find information for easy tasks ($M_{easy} = 7.68$, S.D. = 1.79) compared to difficult tasks ($M_{difficult} = 6.20$, S.D. = 2.48).

Behavioral Measures

Accuracy on the task was used as the measure of user performance on the website (Fig. 30.2). The user obtained a score of 1 (in contrast to 0) on the task if he/she succeeded in accurately performing an investment-related task on the new platform. The linear mixed-effect model revealed a significant effect of difficulty on accuracy (F $(1, 1070) = 66.20$, $P = .00$), indicating that participants could differentiate between easy and difficult tasks. There was no main effect of group (F $(2, 34.2) = 1.78$, $P = .18$) or interaction between group and task difficulty (F $(2, 1070) = .51$, $P = .59$). Participants had higher accuracy in performing easy tasks (86%) compared to difficult tasks (66.2%).

Eye-Tracking Measures

Fixation count is the most commonly used eye-tracking metric to assess users' visual experience (Fig. 30.3). The number of fixations on screens with the tasks was used as the dependent variable in the mixed model. The linear mixed-effect model revealed a significant effect of difficulty on fixation count (F $(1, 1301.4) = 207.68$, $P = .00$). There was no main effect of group (F $(2, 31.9) = .30$, $P = .73$) or interaction between group and task difficulty (F $(2, 1301.4) = 2.19$, $P = .11$). Participants had fewer fixations for easy tasks (M = 25.27, S.D. = 21.33) compared to difficult tasks (M = 50.63, S.D. = 31.04).

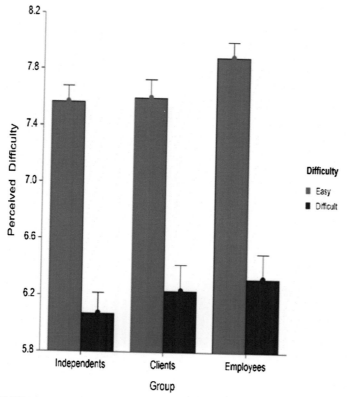

FIGURE 30.1   Survey measures (perceived difficulty) for easy versus difficult tasks.

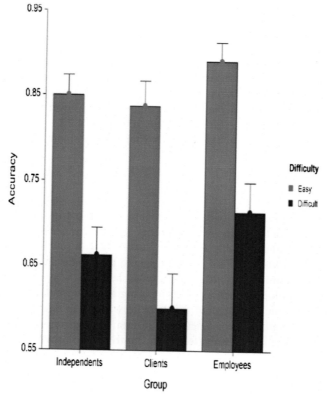

FIGURE 30.2   Performance measures (accuracy) for easy versus difficult tasks.

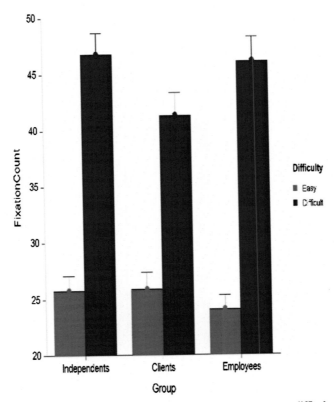

FIGURE 30.3   Eye-tracking measures (fixation count) for easy versus difficult tasks.

Neural Measures

While survey and performance measures helped distinguish between easy and difficult tasks, neural measures revealed additional insights that survey and performance measures could not uncover (Fig. 30.4). An analysis of the neural activity revealed patterns of brain activity that help explain why participants found the difficult tasks more challenging. Across all four quadrants, participants in all three groups displayed much higher mental effort (as measured by oxygenated hemoglobin changes) for difficult tasks compared to easy tasks (Q1: $F (1, 1009.2) = 14.07$, $P = .00$; Q2: $F (1, 1022.6) = 6.32$, $P = .01$; Q3: $F (1, 954.2) = 9.11$, $P = .00$; Q4: $F (1, 992) = 5.15$, $P = .02$). There was no main effect of group or interaction between group and task difficulty.

CONCLUSION

Given the large investment in web platforms, wealth management companies need better methods of testing their platforms before launching them. Traditionally, web usability studies have relied on survey measures as the only method to assess users' perceptions of a website. However, surveys are deficient in uncovering critical insights, and this concern has been echoed in the literature for a long time. In this research we proposed and tested a new paradigm in web usability which incorporates a multimodal approach. Using a systematic and repetitive block-design-based procedure coupled with technologies such as brain imaging and eye tracking, we gained objective measures of engagement and additional insights such as areas of the website which users found difficult to navigate, issues with symbolism, and effectiveness of visualization and delivery of data, among others. Put together, the four metrics lead to a more holistic and rigorous testing of the website which improves the provider's confidence in offering the new website to its clientele.

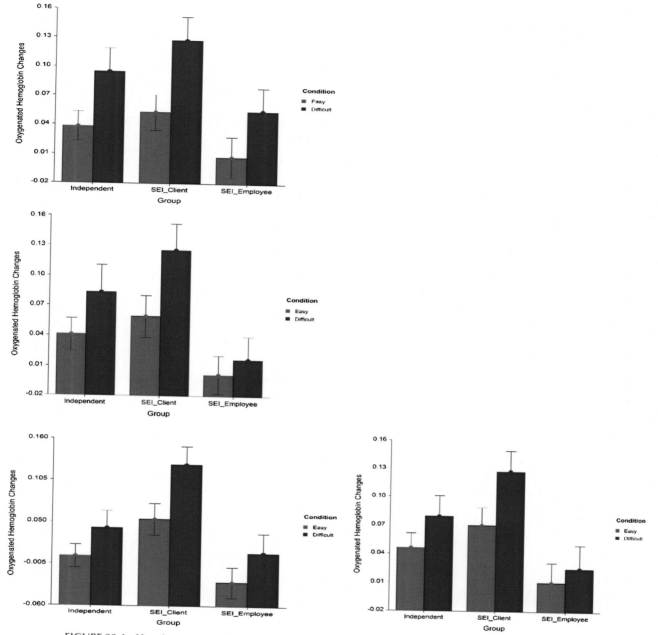

FIGURE 30.4   Neural measures (oxygenated hemoglobin changes in Q1, Q2, Q3, and Q4) for easy versus difficult tasks.

REFERENCES

1. KPMG International Cooperative. *The pulse of fintech Q4 2016.* 2017. Retrieved from: https://assets.kpmg.com/content/dam/kpmg/xx/pdf/2017/02/pulse-of-fintech-q4-2016.pdf.
2. Gregg DG, Walczak S. The relationship between website quality, trust and price premiums at online auctions. *Electronic Commerce Research* 2010;**10**(1):1–25.
3. Reber R, Schwarz N, Winkielman P. Processing fluency and aesthetic pleasure: is beauty in the perceiver's processing experience? *Personality and Social Psychology Review* 2004;**8**(4):364–82.
4. Song H, Schwarz N. If it's hard to read, it's hard to do: processing fluency affects effort prediction and motivation. *Psychological Science* 2008;**19**(10):986–8.

5. Dyn. *World's it executives Agree: Vital online Sales Hurt by poor website performance up to 50 percent of the time (web log post).* March 31, 2015. Retrieved October 22, 2017, from: http://dyn.com/blog/worlds-it-executives-agree-vital-online-sales-hurt-by-poor-website-performance-up-to-50-percent-of-the-time/.

6. Kujala S. User involvement: a review of the benefits and challenges. *Behaviour and Information Technology* 2003;**22**(1):1–16.

7. Parasuraman R. Neuroergonomics: research and practice. *Theoretical Issues in Ergonomics Science* 2003;**4**(1–2):5–20.

8. Parasuraman R, Rizzo M, editors. *Neuroergonomics: the brain at work.* Oxford University Press; 2008.

9. Ayaz H, Shewokis PA, Bunce S, Izzetoglu K, Willems B, Onaral B. Optical brain monitoring for operator training and mental workload assessment. *Neuroimage* 2012;**59**(1):36–47.

10. Di Domenico SI, Rodrigo AH, Ayaz H, Fournier MA, Ruocco AC. Decision-making conflict and the neural efficiency hypothesis of intelligence: a functional near-infrared spectroscopy investigation. *NeuroImage* 2015;**109**:307–17.

11. McKendrick R, Ayaz H, Olmstead R, Parasuraman R. Enhancing dual-task performance with verbal and spatial working memory training: continuous monitoring of cerebral hemodynamics with NIRS. *NeuroImage* 2014;**85**:1014–26.

12. Ayaz H, Onaral B, Izzetoglu K, Shewokis PA, McKendrick R, Parasuraman R. Continuous monitoring of brain dynamics with functional near infrared spectroscopy as a tool for neuroergonomic research: empirical examples and a technological development. *Frontiers in Human Neuroscience* 2013;**7**.

13. McKendrick R, Parasuraman R, Ayaz H. Wearable functional near infrared spectroscopy (fNIRS) and transcranial direct current stimulation (tDCS): expanding vistas for neurocognitive augmentation. *Frontiers in Systems Neuroscience* 2015;**9**.

14. McKendrick R, Parasuraman R, Murtza R, Formwalt A, Baccus W, Paczynski M, Ayaz H. Into the wild: neuroergonomic differentiation of hand-held and augmented reality wearable displays during outdoor navigation with functional near infrared spectroscopy. *Frontiers in Human Neuroscience* 2016;**10**.

15. Hill AP, Bohil CJ. Applications of optical neuroimaging in usability research. *Ergonomics in Design* 2016;**24**(2):4–9.

16. Mandrick K, Chua Z, Causse M, Perrey S, Dehais F. Why a comprehensive understanding of mental workload through the measurement of neurovascular coupling is a key issue for Neuroergonomics? *Frontiers in Human Neuroscience* 2016;**10**.

17. Mehta RK, Parasuraman R. Neuroergonomics: a review of applications to physical and cognitive work. *Frontiers in Human Neuroscience* 2013;**7**.

FURTHER READING

1. Dapp T, Slomka L, Hoffmann R. *Fintech—the digital (r) evolution in the financial sector.* Frankfurt am Main: Deutsche Bank Research; 2014.

2. Mackenzie A. The fintech revolution. *London Business School Review* 2015;**26**(3):50–3.

Chapter 31

Hybrid Collaborative Brain–Computer Interfaces to Augment Group Decision-Making

Davide Valeriani, Caterina Cinel, Riccardo Poli

University of Essex, Colchester, United Kingdom

INTRODUCTION

Traditionally, brain–computer interfaces (BCIs) convert neural activity into commands, allowing people with severe disabilities to control external devices or communicate using brain activity alone.[1] However, in recent years researchers have started to explore the use of neuroergonomic BCIs to augment human perception.[2,3] For example, in decision-making, BCIs have been used to detect and correct user errors.[4] Nevertheless, despite advances in neuroscience, the average individual assisted by a BCI still makes worse decisions than the average non-BCI user.

Years of research in decision-making have shown how groups are generally better than individuals in making decisions (wisdom of crowds),[5,6] thanks to groups' augmented perception and cognition achieved by integrating different views and percepts through the interaction of their members. However, there are circumstances in which groups can perform worse than individuals, such as in the presence of strong leadership[7,8] or time constraints.[5]

Previous research has investigated the possibility of using collaborative BCIs (cBCIs)[9] to improve single-user BCI performance through groups and also improve group decision-making through BCIs. This idea has been tested, for example in a simple discrimination task where groups of eight cBCI-assisted observers were more accurate and faster than single non-BCI users.[10] However, the performance of the cBCIs presented in that study was still worse than that achieved by non-BCI groups. Other studies have shown similar limitations.[11]

In previous research we have shown that hybrid cBCIs could be used to improve group performance in decision-making.[12] We adopted a hybrid approach, where neural signals and response times (RTs) were used to predict the confidence level of each observer in a decision. These confidence estimates were then used to weigh individual behavioral responses (actual mouse-button presses) and build group decisions, which were significantly better than both individual ones and, for the first time, decisions obtained by equal-sized non-BCI groups. This hybrid approach was successfully tested with *low-level* decision tasks involving *visual perception* only,[12,13] where very simple shapes were used.

This chapter investigates the possibility of using a hybrid cBCI to improve group performance in two tasks involving more realistic stimuli: *visual search* (Experiment 1), where users have to identify a polar bear in a scene including many distractors (penguins), and *complex speech perception* (Experiment 2), where participants have to decide whether certain target words are uttered in spoken sentences affected by noise. Group decisions are obtained by aggregating individual responses using either a simple majority or a weighted majority based on confidence estimated by the participants after each decision and by a cBCI using neural signals and RTs.

METHODS

Participants

Ten healthy volunteers (average age = 27.4 ± 5.5 years, five females) took part in Experiment 1. Ten healthy volunteers, all native English speakers (average age = 24.9 ± 4.9 years, two females) with normal hearing, did Experiment 2. All participants had normal or corrected-to-normal vision and gave written informed consent. This research received UK Ministry of Defence and University of Essex ethical approval in July 2014.

Neuroergonomics. https://doi.org/10.1016/B978-0-12-811926-6.00031-2
Copyright © 2019 Elsevier Inc. All rights reserved.

Experiments

Participants did a sequence of 8 blocks of 40 trials each, for a total of 320 trials for each experiment. Each trial started with the presentation of a fixation cross for 1 s, to allow participants to get ready for the stimulus and the electroencephalogram (EEG) signals to go back to the baseline. Then in Experiment 1 a display containing an image of an Arctic environment was presented, followed by a black and white 24×14 checkerboard mask shown for 250 ms. Participants had to decide whether or not a polar bear was present by pressing the left or the right mouse button, respectively. In Experiment 2 an audio recording was played and participants had to decide whether or not one of the target words ("route," "check," "grid," "lookout," "side," "trucks," and "village") was uttered. After responding, the participants in both experiments were asked to indicate their confidence in that decision (0–100% in steps of 10%) using the mouse wheel. This had to be done within 4 s of the response. The mouse was controlled using the preferred hand. RTs from the stimulus onset were recorded.

For Experiment 1, the set of stimuli was similar to those used by Valeriani et al.[14] and consisted of manually created realistic images representing an Arctic environment containing a variable number of penguins (distractors) and, possibly, a polar bear (target). The resulting dataset contained 68 stimuli with the target and 10 without it. In Experiment 2 we used audio recordings consisting of 41 sentences containing one target word and 42 sentences without any target word. Between 4 and 20 words (average length 9.3 ± 2.8 words) were uttered in each audio recording, which were spoken by a member of the army (male, native English speaker). The duration of the audio recordings was between 2.19 and 8.75 s (average duration 4.3 ± 1.4 s). For each recording we created five versions by superimposing multiple types of noise on the original audio files (white noise, environmental noise, volume changes, speed change, change of sampling rate, and audio dropouts), all of which are typical of real-world military communications. We called the resulting 415 stimuli the "standard" set. We then created an additional 415 stimuli in the same manner but with significantly more noise ("high-noise" set). Noise was added using the Pydub library (www.pydub.com).

In either experiment, the same sequence of stimuli was used for all participants to simulate, offline, concurrent group decisions. In Experiment 2, however, the difficulty of the audio tracks was dynamically varied by picking them from either the "standard" or the "high-noise" set, so as to keep the accuracy of all participants not too far from 80%. Stimuli containing the target were presented in 25% and 50% of the trials of Experiments 1 and 2, respectively. Volunteers were comfortably seated at about 80 cm from an LCD screen and participants in Experiment 2 were wearing in-ear earphones. In Experiment 2 volunteers undertook a memorization task before starting the actual experiment to memorize the set of target words. Participants were then familiarized with the task by undertaking two sessions of 10 trials each in both experiments. Preparation and task familiarization took approximately 40 min, while each experiment took about 35 min.

Data Recording and Group Decisions

A Biosemi ActiveTwo EEG system was used to record the neural signals from 64 electrode sites following the 10–20 international system. The EEG data was sampled at 2048 Hz, referenced to the mean of the electrodes placed on the earlobes and bandpass filtered between 0.15 and 40 Hz. Artifacts caused by eye movements were corrected by applying a method based on correlations to the average difference between channels Fp1-F1 and Fp2-F2.

For each trial, response-locked epochs starting 1 s before the user's response and lasting 1.5 s were extracted from the EEG data recorded at each channel for Experiment 1, and at locations FC5, C5, CP5, TP7, and T7 for Experiment 2, as in the auditory experiment we expected that key information could be found in the neural signals recorded in the left temporal lobe.[15] Stimulus-locked epochs starting on the onset of the stimulus and lasting 1.5 s were also extracted for Experiment 1, while this representation of the EEG data was not useful in Experiment 2 as the audio recordings had different lengths and the target word could be uttered at any time within them. The data of each epoch was passed through a filter with a pass band of 0–6 Hz, a stop band of 8–1024 Hz, and finally downsampled to 16 Hz. Similarly to Valeriani et al.,[13] the epochs were split into training and test sets using 10-fold cross-validation. Each epoch of the training set was assigned a label representing the correctness of the decision of the participant in that trial. We then used local temporal correlation common spatial patterns (LTCCSPs)[16] to extract two neural features from each epoch, representing the decision confidence of the user.[13] Hence in Experiment 1 the feature vector was composed of four LTCCSP features (two for each type of epochs) and the RT, while for Experiment 2 we only had two LTCCSP features and the RT (from which the length of the audio recording was subtracted to remove its dependency on the stimulus at hand). The feature vector associated with each decision was then fed into a least angle regressor[17] to obtain the confidence estimate.

All possible groups of sizes 2–10 were formed offline by combining the 10 participants. Each group decision was obtained by considering the sign of the weighted sum of its members' decisions (weighted majority), where the confidence values estimated by the hybrid cBCI were used as weights. We also simulated group decisions based on the reported confidence provided by participants after each decision by using these values (i.e., values in the set {0.0, 0.1, ..., 1.0}) to weigh individual decisions. The performance of these two weighted-majority-based systems was compared with that of non-BCI groups using a standard majority to make decisions.

RESULTS

Individual Performance

In Experiment 1 participants made correct decisions in 77.5±9.1% of the trials, while in Experiment 2 they correctly identified target words in 84.9±3.2% of the trials, showing the difficulty of both tasks for a single decision-maker. We should note that in Experiment 2 the average performance slightly deviated from the target (80%), as some participants were able to perform well in the task despite the use of high-noise stimuli.

Group Performance

Fig. 31.1 shows the percentage of correct decisions achieved by groups of increasing size using one of the three methods described above for the two experiments.

We used the Wilcoxon signed-rank test to compare statistically the group performance of the three methods for each experiment. Naturally, statistical significance could not be reached for groups of size 10, as we only have one such group. Hence we focus our analyses on groups of size 2–9.

In Experiment 1 groups assisted by our hybrid cBCI were significantly superior to equally sized groups using a standard majority for all group sizes ($p < 0.02$) and to groups using the weighted majority based on reported confidence values for even sizes ($p < 0.006$). The two weighted-majority methods performed on par for odd-sized groups. Moreover, the method based on the reported confidence was significantly better than the majority method for group sizes 2, 4, 5, 6, 7, and 8 ($p < 0.002$), while the two methods were on par for sizes 3 and 9.

In Experiment 2, however, groups making decisions using reported confidence were significantly superior to cBCI-assisted and majority-based groups for all group sizes 2–9 ($p < 0.003$). Groups assisted by the cBCI were better ($p < 0.006$) than majority-based groups for all sizes except 3, where the former was nearly statistically significantly better than the latter ($p = 0.076$).

CONCLUSIONS

This chapter shows that it is possible to improve group performance in realistic decision-making tasks involving visual or auditory stimuli with a hybrid collaborative BCI. The cBCI estimates the decision confidence of each participant from the neural signals and RTs, and uses these values to weigh individual responses and obtain group decisions. These results confirm that our previous findings obtained with low-level decision tasks extend to more complex tasks.

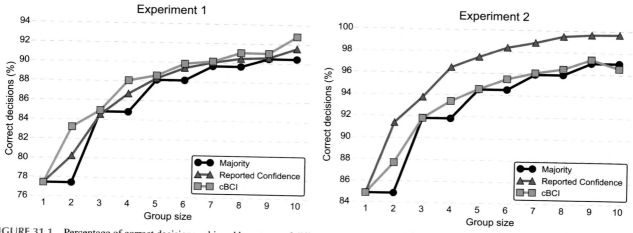

FIGURE 31.1 Percentage of correct decisions achieved by groups of different sizes using a standard majority (*black line*) or a weighted majority using either the confidence reported by the participants (*blue line*) or the confidence estimated by the cBCI (*orange line*) for Experiment 1 (left) and Experiment 2 (right).

Moreover, we have seen that in the visual search task the confidence estimated by the cBCI provided a better prediction of correctness than the reported confidence, while in the speech perception task the reported confidence was the best predictor of correctness. This suggests that the best confidence estimate varies across tasks. However, we should note that in the speech perception task the cBCI was only relying on EEG data from five electrode locations (to promote generalization) and response-locked epochs (due to the variability of the target word positions in the sentences) to estimate decision confidence. In future research we will focus on complementing the feature set used for confidence estimation with neural features extracted from the time, frequency, and time–frequency domains (e.g., wavelets), and features extracted from other physiological measures related to decision-making (e.g., skin conductance and pupil dilation).

ACKNOWLEDGMENTS

This work was supported by the Defence and Security National PhD Programme through DSTL.

REFERENCES

1. Wolpaw JR, McFarland DJ. Control of a two-dimensional movement signal by a noninvasive brain-computer interface in humans. *Proceedings of the National Academy of Sciences of the United States of America* December 2004;**101**(51):17849–54.
2. Parasuraman R, Rizzo M. *Neuroergonomics: the brain at work.* New York (USA): Oxford University Press; 2007.
3. Matran-Fernandez A, Poli R. BrainComputer interfaces for detection and localization of targets in aerial images. *IEEE Transactions on Biomedical Engineering* 2017;**64**(4):959–69.
4. Parra L, Spence C. Response error correction-a demonstration of improved human-machine performance using real-time EEG monitoring. *IEEE Transactions on Neural Systems and Rehabilitation Engineering* 2003;**11**(2):173–7.
5. Bahrami B, Olsen K, Latham PE, Roepstorff A, Rees G, Frith CD. Optimally interacting minds. *Science* 2010;**329**(5995):1081–5.
6. Kerr NL, Tindale RS. Group performance and decision making. *Annual Review of Psychology* 2004;**55**(1):623–55.
7. Branson L, Steele NL, Sung C-H. When two heads are worse than one: impact of group style and information type on performance evaluation. *Journal of Business and Behavioral Sciences* 2010;**22**(1):75–84.
8. Locke CC, Anderson C. The downside of looking like a leader: power, nonverbal confidence, and participative decision-making. *Journal of Experimental Social Psychology* 2015;**58**:42–7.
9. Wang Y, Jung T-P. A collaborative brain-computer interface for improving human performance. *PLoS One* 2011;**6**(5):e20422.
10. Eckstein MP, Das K, Pham BT, Peterson MF, Abbey CK. Neural decoding of collective wisdom with multi-brain computing. *NeuroImage* 2012;**59**(1):94108.
11. Cecotti H, Rivet B. Subject combination and electrode selection in cooperative brain-computer interface based on event related potentials. *Brain Sciences* 2014;**4**(2):335–55.
12. Poli R, Valeriani D, Cinel C. Collaborative brain-computer interface for aiding decision-making. *PLoS One* 2014;**9**(7):e102693.
13. Valeriani D, Poli R, Cinel C. Enhancement of group perception via a collaborative brain-computer interface. *IEEE Transactions on Biomedical Engineering* 2016;**99**:1.
14. Valeriani D, Poli R, Cinel C. A collaborative brain-computer interface for improving group detection of visual targets in complex natural environments. In: *In 7th international IEEE EMBS neural Engineering Conference.* 2015. p. 25–8.
15. Zatorre RJ. Neural specializations for tonal processing. *Annals of the New York Academy of Sciences* 2012;**930**(1):193–210.
16. Zhang R, Xu P, Liu T, Zhang Y, Guo L, Li P, Yao D. Local temporal correlation common spatial Patterns for single trial EEG classification during motor imagery. *Computational and Mathematical Methods in Medicine* 2013;**2013**:1–7.
17. Efron B, Hastie T. Least angle regression. *The Annals of Statistics* 2004;**32**(2):407–99.

Chapter 32

How to Recognize Emotions Without Signal Processing: An Application of Convolutional Neural Network to Physiological Signals

Nicolas Martin, Jean-Marc Diverrez, Sonia Em, Nico Pallamin, Martin Ragot
Usage and Acceptability Lab, Cesson-Sévigné, France

INTRODUCTION

Emotion recognition is currently a major challenge for affective computing.[1] Various modalities (e.g., speech, face) have been already explored,[2] and recognition from physiological data seems an interesting approach.[3] Nevertheless, the current solutions are generally based on laboratory sensors and/or a controlled situation. But the development of wearable devices with physiological sensors opens the possibility of recognizing emotions outside a laboratory context.

The chapter is organized as follows. First, an introduction to recognition of emotion based on physiological signals is presented. Next, the deep learning algorithms and their applications to physiological signals are outlined. Finally, an application of convolutional neural networks (CNNs) (a class of deep learning algorithms) to emotion recognition from raw physiological signals is described.

Emotion Recognition

To recognize emotion from physiological signals, the most common approach[4,5] is to use this processing chain: collecting physiological and subjective data during emotion generation; extracting statistical features from physiological data using signal processing; and using a training model to recognize specific patterns in data with machine learning algorithms (i.e., the ability to learn without being explicitly programmed).

Deep Learning Algorithms

Deep learning algorithms, a class of machine learning algorithms, have received a lot of interest in recent years.[6] Indeed, many studies showed unmatched performance of deep learning[7,8] in domains such as object recognition, speech recognition, etc. The growing interest in deep learning is mainly due to the availability for training of large datasets and the development of computing power, especially graphic processing units. Deep learning algorithms differ from traditional machine learning algorithms (e.g., a support vector machine) in the way they handle data. Indeed, with the traditional algorithms the training of classifiers are made from handcrafted features extracted from raw signals using signals processing (e.g., heart rate is estimated using peak detection on a photoplethysmography signal). With the deep learning algorithms, the feature extractor is trained on data at the same time as the classifier (i.e., an end-to-end trainable model). This solution offers the possibility to discover linear and nonlinear features from data and constructs an adapted internal representation of data. Technically, one or more layers of the model are dedicated to the extraction of statistical features from raw data and then suppress the signal-processing step.[9] Thus, deep learning can offer new possibilities in a mobility context and/or real time. Indeed, the deep learning models, after being trained, can provide classification relatively quickly and easily. For example, it is possible to train a model on a computer and transfer it into a smartphone to make real-time classification using the deep learning library Tensorflow.[10]

Neuroergonomics. https://doi.org/10.1016/B978-0-12-811926-6.00032-4
Copyright © 2019 Elsevier Inc. All rights reserved.

Deep Learning Applied to Physiological Signals

The performance of deep learning algorithms on physiological data has been tested in prior studies.[11–13] In these papers, the feature extractor inside the deep learning model is trained on data to avoid signal processing and feature selection. From these automatically extracted features, the classifier provides a probability of emotional state (e.g., positive or negative emotion) and shows a recognition accuracy similar to recognition based on handcrafted features.

CURRENT STUDY

In the current study, CNNs,[14] a class of deep learning algorithms, were used to recognize emotion from raw physiological data. The models which included the feature extractor and classifiers were trained end to end. In contrast to previous research, the classifiers predict a numerical value between 0 and 1. This approach seems to be more adapted to human functioning[15] and offers more precision. Thus the objective of this study is to test the performance of deep learning models for emotion recognition. If the models show satisfactory performance, it could be an interesting method for real-time emotion classification on mobile devices.

METHOD

The DEAP database[16] was used to train the CNNs and test their performance for emotion recognition based on physiological data. This database contains physiological data (electroencephalogram and peripheral physiological signals) from 32 participants. Each participant was exposed to 40 min of music videos as emotion inducers. For each inducer, evaluations on three dimensions (arousal, valence, and dominance) were provided. These evaluations are numerical values between 1 and 9 which are normalized between 0 and 1 to facilitate the training. As the objective is to build an emotion recognition system based on physiological signals, only four signals from the DEAP database were used: galvanic skin response, respiration belt, plethysmograph, and skin temperature. These signals can be extracted from actual consumer sensors (e.g., Microsoft Bandfn11[1]). Thus the deep learning models were trained to recognize emotional state. The models are composed of convolution and fully connected layers (see Fig. 32.1). The convolution layers comprise convolution, max-pooling (i.e., dimension reduction), and ReLU (i.e., activation function) layers. The fully connected layers are composed of neural layers, activation functions, and dropout. The convolution layers act as feature extractors by extracting the relevant important features and reducing the data dimensionality. To maximize the performance and reduce the data dimension (4 sensors × 7860 [60 s sampled at 128 Hz]), several convolution layers were stacked. Dropout was added to prevent overfitting on data.[17] Finally, the models were defined, trained, and tested using Tensorflow.[10]

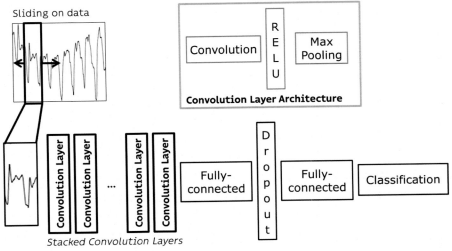

FIGURE 32.1 Model architecture.

---

1. https://www.microsoft.com/microsoft-band/en-us.

TABLE 32.1 Performance of Recognition

Recognition	Valence	Arousal	Dominance
Accuracy	0.79 (SD = 0.14)	0.75 (SD = 0.17)	0.72 (SD = 0.21)
Mean square error	0.06	0.09	0.12
Mean absolute error	0.20	0.25	0.27

RESULTS

To train and test the models, the DEAP dataset was split into two: a training dataset containing 90% of the data and a testing dataset with 10% of the data. Three models were trained: one for each emotional dimension (i.e., arousal, valence and dominance) evaluated on the dataset. To evaluate the model performance, accuracy, mean square error, and mean absolute error were computed. The results on the testing dataset show an accuracy of 0.79 for valence, 0.75 for arousal, and 0.72 for dominance (see Table 32.1).

DISCUSSION AND CONCLUSION

The original paper describing the DEAP dataset provides classification results for arousal and valence based on hand-crafted features. An accuracy of 0.57 for arousal and 0.62 for valence was established, and gives a baseline for comparison with our models. Our models offer better accuracy than the solution based on handcrafted features. Moreover, our models predict a continuous emotion level, whereas in the baseline paper the models provided a binary output (i.e., positive/negative). Thus deep learning algorithms with CNNs seem to be an interesting approach to recognize continuous emotion levels without feature extraction steps. Nevertheless, even if our solution provides better accuracy than a solution based on handcrafted features, CNNs tend to overfit on data. Indeed, the quantity of data is relatively limited compared to the number of parameters to train. This problem appears even if dropout layers are included in the models.

Futures Work

Future work may be carried out on physiological signals gathered with wearable sensors as Empatica E4.[2] Apart from providing a solution for real-time classification outside the laboratory, wearable sensors can offer an effective way to collect a large amount of data. Indeed, collecting enough data to prevent overfitting and increase the model accuracy seems essential. Lastly, implementation of trained models on mobile devices may be explored, especially given the possible problems due to limited computing power and memory size.

ACKNOWLEDGMENTS

We thank all those who participated in any way in this research. This work was carried out within the Institute of Research and Technology b<>com, dedicated to digital technologies. It received support from the French government under the program Future Investments bearing reference ANR-07-A0-AIRT.

REFERENCES

1. Petta P, Pelachaud C, Roddy C, editors. *Emotion-oriented systems: the humaine handbook.* Berlin: Springer; 2011.
2. Busso C, Deng Z, Yildirim S, Bulut M, Lee CM, Kazemzadeh A, Narayanan S. *Analysis of emotion recognition using facial expressions, speech and multimodal information.* ACM Press; 2004. p. 205. https://doi.org/10.1145/1027933.1027968.
3. Maaoui C, Pruski A. *Emotion recognition through physiological signals for human-machine communication.* 2010.
4. Haag A, Goronzy S, Schaich P, Williams J. Emotion recognition using bio-sensors: first steps towards an automatic system. In: André E, Dybkjær L, Minker W, Heisterkamp P, editors. *Affective dialogue systems.* Springer Berlin Heidelberg; 2004. p. 36–48. https://doi.org/10.1007/978-3-540-24842-2_4.

5. Martin N, Diverrez J-M. From physiological measures to an automatic recognition system of stress. In: Stephanidis C, editor. *HCI International 2016 – posters' extended abstracts*, vol. 618. Cham: Springer International Publishing; 2016. p. 172–6. https://doi.org/10.1007/978-3-319-40542-1_27.

6. Schmidhuber J. Deep learning in neural networks: an overview. *Neural Networks* 2015;**61**:85–117. https://doi.org/10.1016/j.neunet.2014.09.003.

7. Cireşan D, Meier U, Masci J, Schmidhuber J. Multi-column deep neural network for traffic sign classification. *Neural Networks* 2012;**32**:333–8. https://doi.org/10.1016/j.neunet.2012.02.023.

8. Silver D, Huang A, Maddison CJ, Guez A, Sifre L, van den Driessche G, Hassabis D. Mastering the game of Go with deep neural networks and tree search. *Nature* 2016;**529**(7587):484–9. https://doi.org/10.1038/nature16961.

9. Bărar A-P, Neagoe V-E, Sebe N. Image recognition with deep learning techniques. In: *Recent advances in image, audio and signal processing: Budapest, Hungary, December 10-12, 2013*. 2013.

10. Abadi M, Agarwal A, Barham P, Brevdo E, Chen Z, Citro C, Zheng X. *TensorFlow: large-Scale machine learning on heterogeneous systems*. 2015. Retrieved from: http://tensorflow.org/.

11. Martinez HP, Bengio Y, Yannakakis GN. Learning deep physiological models of affect. *IEEE Computational Intelligence Magazine* 2013;**8**(2): 20–33. https://doi.org/10.1109/MCI.2013.2247823.

12. Wang D, Shang Y. Modeling physiological data with deep belief networks. *International Journal of Information and Education Technology (IJIET)* 2013;**3**(5):505–11. 10.7763/IJIET.2013.V3.326.

13. Zheng WL, Zhu JY, Peng Y, Lu BL. EEG-based emotion classification using deep belief networks. In: *2014 IEEE International Conference on Multimedia and Expo (ICME)*. 2014. p. 1–6. https://doi.org/10.1109/ICME.2014.6890166.

14. LeCun Y, Bengio Y. Convolutional networks for images, speech, and time series. In: *The handbook of brain theory and neural Networks*. 1995. p. 3361 (10), 1995.

15. Barrett LF. Discrete emotions or Dimensions? The role of valence focus and arousal focus. *Cognition and Emotion* 1998;**12**(4):579–99. https://doi.org/10.1080/026999398379574.

16. Koelstra S, Muhl C, Soleymani M, Lee J-S, Yazdani A, Ebrahimi T, Patras I. DEAP: a database for emotion analysis using physiological signals. *IEEE Transactions on Affective Computing* 2012;**3**(1):18–31. https://doi.org/10.1109/T-AFFC.2011.15.

17. Srivastava N, Hinton G, Krizhevsky A, Sutskever I, Salakhutdinov R. Dropout: a simple way to prevent neural networks from overfitting. *Journal of Machine Learning Research* 2014;**15**:1929–58.

Entries From the Inaugural International Neuroergonomics Conference

Chapter 33

Technical Manifestations of the Everted Brain: The Impact and Legacy of Raja Parasuraman

Peter A. Hancock

University of Central Florida, Orlando, FL, United States

Technical prosthetics and orthotics are prior visions made material in the world. Predominantly, they replicate and magnify inherent human abilities. Such biomimetic propensities are applicable and evident in all everted human bodily systems. However, the brain is the most special of these, the elaboration for which has exerted and continues to exert the most profound impact upon the world around us. Codified as "neuroergonomics," Parasuraman's interdisciplinary conception sought, and continues to seek, to apply the revolution in the neurosciences to a more formal consideration of the brain at work. In my paper, I shall look to explicate the fundamental wellsprings of this latter marriage, focusing upon a number of classic, foundational contributions which serve to inspire Parasuraman's own development and theorizing throughout his career. In particular, I look at the genesis of adaptive human–machine systems to draw together these diverse strands and illustrate how Parasuraman's ideas of neuroergonomics derived directly from his earliest explorations of vigilance and the other varieties of attention. It is instructive to understand how the advances in sustained attention research (to which Parasuraman was such a major contributor) prefigured the ongoing automation revolution. Drawing up the work of luminaries such as Norbert Wiener, Earl Wiener, Tom Sheridan, John Senders, and Neville Moray, the progressive expulsion of the human operator from the inner loops of control naturally (and even necessarily) serves to invoke the vigilance imperative. Yet manifestly, human beings are poor at this pursuit, especially when the displays are ill designed and configured. Such considerations naturally lead to the issue of function allocation and the problems associated with the "remnant principle." This latter, de facto strategy means that the human retains each of the functions that current automation has yet to master. It has led to disaster. And now we see the expulsion of the human operator to even greater distance from the locus of control as ever-greater levels of autonomy are invoked. I shall examine this latter vector of evolution with an especial eye to the ideas, insights, and tenets that neuroergonomics can bring to this vital enterprise for future human progress and survival.

Neuroergonomics. https://doi.org/10.1016/B978-0-12-811926-6.00033-6
Copyright © 2019 Elsevier Inc. All rights reserved.

Chapter 34

Can We Trust Autonomous Systems?

Peter A. Hancock, Kimberly L. Stowers, Theresa T. Kessler
University of Central Florida, Orlando, FL, United States

AIM

The aim of this work is to evaluate the necessity of, and the propensity for, human trust in ever-more independent autonomous technical systems. Here, we focus upon our program of work with robotics systems. Robots are artificially embodied cognitions. They possess agency in the world through their physical presence and are the epitome of autobiomimesis. As such they provide a ready and facile testing arena in which to evaluate how humans should, and perhaps should not, trust these ever more powerful entities.

METHODS

To measure and assess how humans exercise trust, mistrust, distrust, and skepticism in nascent autonomy, we have conducted a series of experimental evaluations specifically concerning human–robot interaction (HRI). Under the overall zeitgeist of moving robots from "tools to team-mates," we have examined how factors such as proximity, reliability, physical appearance, and transparency affect implicit and explicit forms of trust. We have tested numerous variations of these differing robot characteristics and established the influence upon operator's and observer's responses. Objective performance assessments have been accomplished through reflections of speed and accuracy of individual operator response, individual robot response, and human–robot response in tandem. Subjective perceptions have been determined through a series of validated trust and workload scales and measures, some of which have been created in our own laboratory.

RESULTS

Our series of experiments have established that information rate, density, and complexity play vital roles in human–robot interaction. No difference in trust toward the robot is evident when such information is presented in text versus audio forms. However, information sharing is particularly important when the robot is malfunctioning. Users prefer additional information about the robot processes in cases of malfunction to sustain trust. Our experiments have further shown that variation in robot physicality affects trust. For example, both robot proximity and speed of approach affects trust in a manner reminiscence of the "time-to-contact" literature. Furthermore, our experiments have shown that physical characteristics of the robot itself (eg, shape, degree of anthropomorphism) affect human perceptions of the robot including trust.

CONCLUSIONS

Autonomous systems represent our future. Robots are the future of computational social orthotics. They are manifestations of neuroergonomics in the real-world. They may eventually prove to be our beneficial partner or the source of our imminent destruction. We conclude with a brief evaluation of these dichotomous eventualities.

Neuroergonomics. https://doi.org/10.1016/B978-0-12-811926-6.00034-8
Copyright © 2019 Elsevier Inc. All rights reserved.

Chapter 35

Learning and Modulating Spatial Probabilities in Virtual Environments

Amy L. Holloway[1], Peter Chapman[1], Alastair D. Smith[2]
[1]University of Nottingham, Nottingham, United Kingdom; [2]University of Plymouth, Plymouth, United Kingdom

BACKGROUND

Visual search behavior, in simple two-dimensional arrays, is sensitive to the spatial statistics of target distributions. When the target appears with greater likelihood in one-half of the array, participants become faster at locating it, in comparison to when targets are located on the other (low-probability) side. This phenomenon is known as probability cueing, and appears to operate below a level of conscious awareness (i.e., participants do not report awareness of the distribution when probed after the task). In contrast, studies that have employed large-scale search paradigms, in which full-body movements are required to inspect locations, report that probability cueing is more complex, requiring a combination of stable spatial cues and, when successful, appears to be explicit in nature.

AIMS

This research aimed to investigate whether probability cueing could be observed in a virtual environment—a three-dimensional medium between visual and large-scale search. The immersive qualities of the environment were manipulated by changing the display size and visual aspects across two experiments. In the final experiment, we investigated whether spatial cueing could be modulated by applying transcranial direct-current stimulation (tDCS) to the right posterior parietal cortex.

METHODS

Participants were required to search for a hidden target within a virtual arena, which was displayed on a computer monitor (**Study 1**) or within an environment simulator (**Study 2**). In Study 1, the environment consisted of textured walls and flooring. In Study 2, texture was replaced by solid colors (to reduce levels of simulator sickness in the more immersive environment). Participants were not informed that the target was more likely to be located on one side of the array than the other. In Study 3, participants firstly completed a set of baseline trials, followed by 15 min of 1 mA tDCS (anodal, cathodal, or sham) to the right parietal cortex, then two blocks of probability cueing trials on a computer monitor (the same as Study 1). In all experiments, participants were probed for awareness of the manipulation after testing.

RESULTS AND CONCLUSIONS

Study 1

Significant cueing effects were demonstrated, as exemplified by reduced search times to targets on the cued side of space. However, participants not express any conscious awareness of the probability cue. This is in line with the visual search literature, suggesting that physical effort may be necessary for explicit learning of the probability cue.

Study 2

No significant cueing effects were found. The removal of textures in the environment may have decreased optic flow, making it difficult for participants to accurately update their spatial location, and represent target statistics, as they explored the

Neuroergonomics. http://dx.doi.org/10.1016/B978-0-12-811926-6.00035-X
Copyright © 2019 Elsevier Inc. All rights reserved.

environment. Future experiments in immersive environments should, therefore, systematically explore whether probability cueing is related to the level of contextual information present in the environment.

Study 3

Significant cueing effects were demonstrated, as in Study 1. Anodal and cathodal tDCS had an effect on probability cueing, with both stimulation groups outperforming Sham tDCS as demonstrated by a reduction in the overall number of inspected locations on cued trials. This suggests that right parietal cortex may play a role in learning [HA1] the probability cue in a small-scale virtual environment, and may contribute to the planning of search-related movements. Future experiments should investigate the role of the right parietal cortex in probability cueing further, using other neurostimulatory or neuro-imaging techniques.

Chapter 36

Physiological Markers for UAV Operator Monitoring

Raphaëlle N. Roy, Thibault Gateau, Angela Bovo, Frédéric Dehais, Caroline P.C. Chanel
ISAE-SUPAERO, Université de Toulouse, Toulouse, France

Unmanned aerial vehicle (UAV) operating is a complex task performed in a dynamic and uncertain environment. During UAV operation, the human operator is often faced with difficult decisions that have to be made within a limited amount of time and that can result in dramatic consequences (in particular, linked with situational awareness). A solution to prevent or palliate the drops in operator performance that can result from such a task is to perform mental-state monitoring and to communicate the collected information about the human operator along with the UAV state to the decisional framework in charge of the mission progress, to properly update the control policy.

To design systems that consider the human operator, the mental states relevant for this type of task should be described and assessed in ecological settings. Examples of such mental states are the ones related to time-on-task increases—such as fatigue and mind wandering—and the ones related to mental workload—such as memory load and temporal pressure—as well as automation surprise. In particular, as regard time-on-task related mental states, as the systems grow more automated, the operators are requested to operate at irregular and interspaced intervals.[1] Hence, they can be waiting in a long monotonous phase. Although UAV operators' fatigue state has already been assessed at the behavioral and oculomotor levels, to our knowledge f literature is lacking regarding potential cardiac and cerebral markers for this particular application field.

AIMS OF THIS CHAPTER

The contributions of this chapter are twofold. First, a review of the potential mental states of interest for UAV operator monitoring is presented with a focus on mental fatigue and mind wandering as reflected by performance. Second, a preliminary study to investigate the physiological markers that could allow us to perform an online monitoring and estimation of a UAV operator's engagement is proposed.

METHODS

To investigate physiological markers of engagement—as reflected by performance—the preliminary study was performed on the data acquired from five volunteers who completed a 2 h UAV monitoring task. The task included an alarm-monitoring task and a target-identification task using a joystick. Only 10 alarms occurred during the session, among which only seven required identification by the operator. The investigated markers were of oculomotor (eye-tracking), cardiac [electrocardiography (ECG)], and cerebral [electroencephalography (EEG)] origin.

RESULTS

In addition to a significant modulation of the alpha power, the blink rate, and the number of fixations with time on task, the main results are a significant correlation of response times with both the cardiac low-frequency/high-frequency (LF/HF) ratio and the number of ocular fixations.

Neuroergonomics. http://dx.doi.org/10.1016/B978-0-12-811926-6.00036-1
Copyright © 2019 Elsevier Inc. All rights reserved.

CONCLUSIONS

This study paves the way toward neuroadaptive systems or biocybernetic systems.[2] The perspectives detail specific tools to achieve online mental-state monitoring called passive brain–computer interfaces,[3] as well as mixed initiative designs. Last, ongoing work on the potential use of supplementary markers such as eye fixation-related potentials (EFRPs) is mentioned.

REFERENCES

1. Cummings ML, Mastracchio C, Thornburg KM, Mkrtchyan A. Boredom and distraction in multiple unmanned vehicle supervisory control. *Interacting With Computers* 2013:iws011.
2. Fairclough SH. Fundamentals of physiological computing. *Interacting With Computers* 2009;**21**(1):133–45.
3. Zander TO, Kothe C. Towards passive brain-computer interfaces: applying brain-computer interface technology to human-machine systems in general. *Journal of Neural Engineering* 2011;**8**(2):025005.

Chapter 37

Estimating Cognitive Workload Levels While Driving Using Functional Near-Infrared Spectroscopy (fNIRS)

Anirudh Unni[1], Klas Ihme[2], Meike Jipp[2], Jochem W. Rieger[1]

[1]University of Oldenburg, Oldenburg, Germany; [2]DLR, Institute of Transportation Systems, Braunschweig, Germany

INTRODUCTION AND AIM

We envision that driver-assistive systems that adapt their functionality to the driver's cognitive state could be a promising approach to reduce road accidents due to human errors.[1] Workload is an important cognitive state because a cognitive overload or underload results in a decrease in human performance, which may result in fatal incidents while driving. Here, we investigate if it is possible to predict variations of cognitive workload levels (WL) from functional near-infrared spectroscopy (fNIRS) brain-activation measurements in a realistic driving scenario with multiple parallel tasks to inform adaptive assistive systems in future cars.

METHODS

In our study, we implemented the n-back working-memory task with several continuously changing load levels as a speed-regulation task into a realistic driving simulation to control and manipulate cognitive workload. We introduced five different workload levels (ie, 0-back to 4-back) and speed signs every 20 s while the participant was driving on a highway with concurrent traffic. Depending on the current n-back task, the participant was supposed to remember the previous "n" speed sequences and adjust his speed accordingly. A detailed explanation for the n-back experimental paradigm can be found in Ref. [2]. fNIRS data were recorded from the frontal and parietal cortices using a 32-channel NIRScout system (NIRx Medical Technologies LLC) from five participants during the course of the whole experiment which lasted for around 30 min.

RESULTS

We used the multivariate linear regression approach by combining fNIRS data from all channels to predict WL using the elastic net regularization model,[3] which combines the L1 and L2 penalties of the lasso and ridge regression techniques to get a continuous estimate of workload over time. Fig. 37.1 depicts a plot of WL induced by the n-back task (red curve) and the WL predicted by the model (blue curve) for an example participant. The correlation between the two curves is almost 0.8 for a 10-fold cross-validation. For the remaining four participants, we achieved correlations between 0.6 and 0.8, which were all statistically significant ($P < .05$).

DISCUSSION AND CONCLUSION

It can be seen from Fig. 37.1 that the predicted WL more or less follows the induced WL and could be used to predict if the WL is either increasing or decreasing. There are certain regions in the blue curve in which the model seems to over- or underestimate WL, which may be due to the incomplete model that currently neglects WL imposed by the concurrent driving task in the changing traffic situations. As a next step, we plan to introduce some nonlinearity into the workload model to get much better prediction rates.

Neuroergonomics. http://dx.doi.org/10.1016/B978-0-12-811926-6.00037-3
Copyright © 2019 Elsevier Inc. All rights reserved.

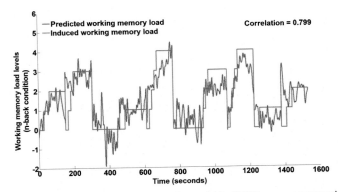

FIGURE 37.1 Ten-fold cross-validated prediction of workload from deoxyhemoglobin fNIRS measurements using multivariate regression analysis in an example participant.

ACKNOWLEDGMENTS

This work was funded by a grant of the Volkswagen Foundation and the Ministry of Science and Culture of Lower Saxony to the Research Centre on Critical Systems Engineering for Socio-Technical Systems. Anirudh Unni and Klas Ihme contributed equally.

REFERENCES

1. Parasuraman R. Human-computer monitoring. *Human Factors* 1987;**29**:695–706.
2. Unni A, Ihme K, Jipp M, Rieger JW. Assessing the driver's current level of working memory load with high density functional near-infrared spectroscopy: a realistic driving simulator study. *Front Hum Neurosci* 2017;**11**:167. doi:10.3389/fnhum.2017.00167.
3. Hastie T, Tibshirani R, Friedman JH. *The elements of statistical learning: data mining, inference, and prediction.* 2nd ed. New York: Springer; 2009.

Chapter 38

Auditory Neglect in the Cockpit: Using ERPs to Disentangle Early From Late Processes in the Inattentional Deafness Phenomenon

Sébastien Scannella, Raphaëlle N. Roy, Amine Laouar, Frédéric Dehais
ISAE-SUPAERO, Université de Toulouse, Toulouse, France

INTRODUCTION AND AIMS

Missing auditory alarms is a critical safety issue in many domains such as aviation.[1] Previous work has proposed that the underlying transitory cognitive impairments of auditory neglect could lie on early sensorial gating[2] or late attentional processes.[3] To investigate this phenomenon and its neural correlates, we analyzed auditory Event-Related Potentials (ERPs) in a double-task paradigm that comprised an active oddball task performed in a motion flight simulator, along with a flying scenario of low or high difficulty depending on the experiment block.

METHODS

Seven light-aircraft pilots took part in the experiment (mean age: 23.5 ± 2.1, mean flying hours: 98 ± 90). The control-flying scenario (low load) was a level-off scenario in nominal flying conditions, whereas a second flying scenario (high load) was an approach to Blagnac airport (France) in highly demanding and stressful conditions. High-load and low-load scenarios were randomly presented across pilots. Along with the flying tasks, pilots had to press the sidestick trigger whenever they heard a rare auditory sound (1250-Hz pure tone, 100-ms duration; 20% of sounds) labeled as a critical auditory alarm, interleaved with frequent nonrelevant auditory distractors (1000-Hz pure tones, 100-ms duration; 80% of sounds). A total of 80 auditory alarms were presented within each scenario. Reaction times and accuracy for alarm detection were recorded along with the ongoing EEG thanks to a 32-channel Biosemi system.

RESULTS

We found that pilots failed to respond to 56% (± 28) of the presented auditory alarms in the high-load scenario compared to only 1% (± 0.5) in the low-load one ($P < .001$). Regarding ERP results, permutation tests revealed an effect of the scenario load for both the frequent sound-related and the rare alarm-related ERPs ($P < .001$). These effects corresponded to a decrease of the frequent sound-related N100 and a large decrease of the alarm-related P300. Finally, analyses from alarm hit versus alarm miss showed significant effects on both the N100 and the P300 components ($P < .05$). In both cases, the high-load scenario induced a reduction of the components' amplitudes (see Fig. 38.1).

DISCUSSION AND CONCLUSION

The main objective of this paper was to investigate the inattentional deafness phenomenon under realistic settings. As expected, the behavioral results revealed that the rate of missed alarms increased between the two scenarios (1 and 56%, respectively). The related ERP analyses revealed that auditory neglect in an ecological situation, such as a highly demanding

Neuroergonomics. http://dx.doi.org/10.1016/B978-0-12-811926-6.00038-5
Copyright © 2019 Elsevier Inc. All rights reserved.

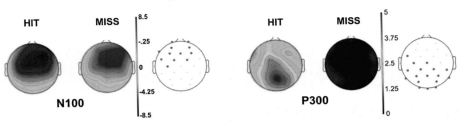

FIGURE 38.1 Alarm-related maps for the high-load scenario. Time windows: N100: 120–130 ms; P300: 340–400 ms. Red dots indicate significant differences between Hit and Miss conditions; $P_{perm} < 0.05$.

flying task, may rely on both perceptual stages—as shown by a reduction of the N100 component—and attentional processes, revealed by a P300 attenuation. We thus bring clues that conciliate diverging studies attributing the inattentional phenomenon to preattentive or attentional processes.

REFERENCES

1. Dehais F, Causse M, Vachon F, Régis N, Menant E, Tremblay S. Failure to detect critical auditory alerts in the cockpit evidence for inattentional deafness. *Human Factors: The Journal of the Human Factors and Ergonomics Society* 2014;**56**:631–44.
2. Scannella S, Causse M, Chauveau N, Pastor J, Dehais F. Effects of the audiovisual conflict on auditory early processes. *International Journal of Psychophysiology* 2013;**89**:115–22.
3. Giraudet L, St-Louis M-E, Scannella S, Causse M. P300 event-related potential as an indicator of inattentional deafness? *PLoS One* 2015;**10**:e0118556.

Chapter 39

Immediate Effects of Ankle Foot Orthosis During Gait Initiation: Impaired Balance Control and Change in Ankle Electromyographic Activity

Arnaud Delafontaine[1], Jean-Louis Honeine[1,3], Manh-Cuong Do[1], Olivier Gagey[1,2]

[1]University Paris-Sud, Orsay, France; [2]C.H.U Kremlin Bicêtre, Kremlin Bicêtre, France; [3]University of Pavia, Pavia, Italy

AIMS

The clinical objective of ankle foot orthosis (AFO) use is to reduce pain and improve postural equilibrium and gait kinematics. However, AFO impedes ankle joint mobility.[1] This should affect propulsive force generation and balance control. The present study examined biomechanical and electromyographic (EMG) activity changes in the anticipatory postural adjustments (APA) phase and step-execution phase of the gait initiation (GI) paradigm following wear of AFO.

METHODS

Nine healthy adult subjects participated in GI protocol.[2] They initiated gait at a self-paced speed and fixed step length with no ankle constraint (Control), or with AFO on stance, swing, or bilateral ankles.

RESULTS

AFO affects biomechanical and EMG parameters during APA and step-execution phase. The duration of APA increased, whereas bilateral Tibialis Anterior EMG and center of gravity (CG) instantaneous progression velocity at foot-off decreased. During execution phase, the braking of CG fall, EMG of stance Soleus activity, and anteroposterior progression velocity of CG at foot-contact were altered.

CONCLUSIONS

Similarly to experiment in which ankle hypomobility was induced by strapping,[3] AFO alters the generating propulsive force and balance control during GI more particularly on stance ankle. Moreover, unilateral AFO induces opposite motor changes that could be responsible for central neuromotor deconditioning at long term. On the neuroergonomic side, it will be interesting to create an AFO which stimulates the muscle that actively participates in the vertical braking of CG fall.

REFERENCES

1. Thoumie P, Sautreuil P, Faucher M. Evaluation des propriétés physiologiques des orthèses de cheville. Revue de littérature. *Annales de Réadaptation et de Médecine Physique* 2004;**47**:225–32.
2. Brenière Y, Do MC. Control of gait initiation. *Journal of Motor Behavior* 1991;**23**:235–40.
3. Delafontaine A, Honeine JL, Do MC, Gagey O, Chong RK. Comparative gait initiation kinematics between simulated unilateral and bilateral ankle hypomobility: does bilateral constraint improve speed performance? *Neuroscience Letters* 2015;**31**(603):55–9.

Neuroergonomics. http://dx.doi.org/10.1016/B978-0-12-811926-6.00039-7
Copyright © 2019 Elsevier Inc. All rights reserved.

Chapter 40

Toward a Better Understanding of Human Prioritization. A Dual-Task Study

Benoît Valéry[1,2], Nadine Matton[2], Sébastien Scannella[2], Frédéric Dehais[1]

[1]ISAE-SUPAERO, Université de Toulouse, Toulouse, France; [2]Ecole Nationale de l'Aviation Civile (ENAC), Université de Toulouse, France

OBJECTIVE

The present study aims to further investigate the prioritization process that sustains executive control in complex multitasking. It intends to provide a description of task-attributes processing, in which several of these attributes present opposite polarities. The general aim of this work is to propose a heuristic decision-making account[1] of prioritization.

BACKGROUND

Multitasking performance crucially depends on executive control processes[2,3] like prioritization,[4] which is the fundamental ability to orient resource allocation as well as task sequencing to favor the execution of the most-emphasized tasks in overloaded situations.[5,5a] Current models of resources allocation during multitasking[6,7] can be discussed regarding their plausibility toward human computational limitations.

METHODS

Participants had to handle a computerized dual n-back task, where each task was char-acterized by two attributes: a demand value (0, 1, or 2-back difficulty) and a payoff value (1 or 3 points). Participants were asked to maximize their score and were not biased toward any executive strategy. There were 27 experimental conditions, representing all the task demand/payoff combinations. We recorded eye gaze output as long as detection performance.

RESULTS

Preliminary results ($N=9$) showed that visual resource allocation was accounted by a demand×payoff interaction. More precisely, there was no impact of the payoff attribute over the executive strategy as long as the demand level was low enough to allow for maximum performance.

CONCLUSIONS

Because not all the task attributes shape continuously the allocation of visual resources, it is suggested that decision-making heuristics could support prioritization.

FUNDINGS

The present work was funded by the DGA-MRIS (France)

Neuroergonomics. http://dx.doi.org/10.1016/B978-0-12-811926-6.00040-3
Copyright © 2019 Elsevier Inc. All rights reserved.

REFERENCES

1. Gigerenzer G, Gaissmaier W. Heuristic decision making. *Annual Review of Psychology* 2011;**62**:451–82.

2. Burgess PW, Veitch E, de Lacy Costello A, Shallice T. The cognitive and neuroanatomical correlates of multitasking. *Neuropsychologia* 2000;**38**(6):848–63.

3. Meyer DE, Kieras DE. A computational theory of executive cognitive processes and multiple-task performance: Part 1. basic mechanisms. *Psychological Review* 1997;**104**(4):3–65.

4. Barabasi A-L. The origin of bursts and heavy tails in human dynamics. *Nature* 2005;**435**(7039):207–11.

5. Kurzban R, Duckworth A, Kable JW, Myers J. An opportunity cost model of subjective effort and task performance. *The Behavioral and Brain Sciences* 2013;**36**(6):661–79.

5a. Durantin G, Gagnon JF, Tremblay S, Dehais F. Using near infrared spectroscopy and heart rate variability to detect mental overload. *Behavioural brain research* 2014;**259**:16–23.

6. Salvucci D, Taatgen N. *The multitasking mind.* Oxford University Press; 2011.

7. Wickens CD, Gutzwiller RS, Santamaria A. Discrete task switching in overload: a meta-analyses and a model. *International Journal of Human-Computer Studies* 2015;**79**:79–84.

Chapter 41

Assessing Working Memory Load in Real Flight Condition With Wireless fNIRS

Frédéric Dehais[1], Hasan Ayaz[2,3,4], Thibault Gateau[1]

[1]ISAE-SUPAERO, Université de Toulouse, Toulouse, France; [2]Drexel University, Philadelphia, PA, United States; [3]University of Pennsylvania, Philadelphia, PA, United States; [4]Children's Hospital of Philadelphia, Philadelphia, PA, United States

INTRODUCTION AND AIMS

Many studies have emphasized that pilots' working memory (WM) is highly taxed when following air traffic control (ATC) instructions.[1] Several stressors such as message complexity may negatively impact pilots' ability to execute ATC clearances and jeopardize flight safety. To monitor such cognitive limitation, we adopted a neuroergonomics approach to measure neural correlates of pilots' WM performance in real flight conditions.

METHODS

Eleven visual flight rule (VFR) pilots participated in the experiment in the Higher Institute of Aeronautics and Space (ISAE-SUPAERO) DR400 light aircraft during an actual flight at 2500 ft altitude. We used the four-optode/12 channel mini-functional near-infrared spectroscopy (fNIRS) wireless portable device to record the pilots' hemodynamics of the prefrontal cortex.[2] Similar to Gateau et al. (2015),[3] pilots heard prerecorded ATC instructions and were instructed to read them back while flying the aircraft and maintaining altitude and heading. Two levels of difficulty were defined. In the low WM load, only one major digit per trial was used to set each flight parameter (eg,: 17 for "speed 170, heading 170, altitude 1700, vertical speed +1700"). In the high WM load, each flight parameter value was different from the previous one (eg,: "speed 179, heading 245, altitude 5800, vertical speed −1500"). The task consisted of 10 randomized repetitions of each difficulty for a total of 20 trials. Cognitive Optical Brain Imaging (COBI) Studio was used to collect raw fNIRS data that were moving average convergence/divergence (MACD) filtered.[1] The experiment was approved by the European Aviation Safety Agency (EASA60049235) and supported by the AXA Research Fund.

RESULTS

The participants committed on average 8.73 errors [standard deviation (SD) = 2.10] errors during the entire experiment, all occurring during the high-load trials. The analysis of variance (ANOVA) over the fNIRS data revealed a main effect of the oxygenation ($F (1, 10) = 9.39$; $P = .01$; partial $\eta^2 = 0.48$) with higher oxyhemoglobin ($\Delta[HbO_2]$) than deoxyhemoglobin ($\Delta[HHb]$) and a main effect of the load [$F(1,10) = 6.0$; $P = .03$; partial $\eta^2 = 0.37$] corresponding to higher peak response within the high-load condition. In addition, a significant interaction effect between load and oxygenation was found [$F(1,10) = 16.4$; $P < .01$; partial $\eta^2 = 0.62$] showing that the load effect was only present for $\Delta[HbO_2]$ ($P < .001$) (Fig. 41.1).

CONCLUSION

This study demonstrates the efficiency of fNIRS to monitor pilot's WM abilities under real flight settings and paves the way to the implementation of fNIRS-based brain–computer interface in the cockpit.

Neuroergonomics. http://dx.doi.org/10.1016/B978-0-12-811926-6.00041-5
Copyright © 2019 Elsevier Inc. All rights reserved.

FIGURE 41.1 Left: DR400 light aircraft—Right: Pilot, left-seated equipped with the mini-fNIRS; the safety pilot is right-seated.

REFERENCES

1. Durantin G, Scannella S, Gateau T, Delorme A, Dehais F. Processing functional near infrared spectroscopy signal with a Kalman filter to assess working memory during simulated flight. *Front Hum Neurosci* 2015;**9**:707.
2. Ayaz H, Onaral B, Izzetoglu K, Shewokis PA, McKendrick R, Parasuraman R. Continuous monitoring of brain dynamics with functional near infrared spectroscopy as a tool for neuroergonomic research: empirical examples and a technological development. *Front Hum Neurosci* 2015:e0121279.
3. Gateau T, Durantin G, Lancelot F, Scannella S, Dehais F. Real-time state estimation in a flight simulator using fnirs. *PLoS One* 2015;**10**(3):e0121279.

Chapter 42

Assessing Driver Frustration Using Functional Near-Infrared Spectroscopy (fNIRS)

Klas Ihme[1], Anirudh Unni[2], Jochem W. Rieger[2], Meike Jipp[1]

[1]DLR, Institute of Transportation Systems, Braunschweig, Germany; [2]University of Oldenburg, Oldenburg, Germany

INTRODUCTION AND AIM

Driving is a goal-directed task. During traffic, blocking obstacles occur, often eliciting driver frustration, and may result in more risky driving including speeding and aggressive behavior toward other traffic road participants. We envision using "adaptive automation"[1] to design driver-assistance systems that prevent those maladaptive driving behaviors by supporting the driver to reduce the current frustration level or increasing the level of automation. For this, it is essential that we can assess driver frustration already at an early stage. As affective states come along with a change in cognitive appraisal and a subjectively experienced feeling, brain activity seems to be a promising indicator of frustration. Thus, we aim to investigate if brain-activation fNIRS measurements could be used to assess realistically occurring driver frustration.

METHODS

Thirteen volunteers participated in a driving simulator study, in which frustration was induced through a combination of time pressure and blocking events. A cover story asked the participants to imagine that they had to deliver a parcel to a client within 6 min with the incentive of gaining 2 € per successful delivery. In the six frustrating scenarios (FRUST), the drives were blocked by heavy traffic, whereas in the six nonfrustrating scenarios (NOFRUST), participants could drive mostly unblocked. fNIRS data were recorded from the frontal and parietal cortices using a 32-channel NIRScout system (NIRx Medical Technologies LLC) during the course of the whole experiment.

RESULTS

We subjected the individual fNIRS data to a statistical parametric mapping (SPM)-based generalized linear model (GLM) analysis, which revealed relative concentration increases of oxyhemoglobin (HbO) and decreases of deoxyhemoglobin (HbR) during FRUST compared to NOFRUST drives. These changes were found bilaterally in ventrolateral frontal and temporo-occipital channels in the group analysis as well as at the individual level (see Fig. 42.1 for group level and an example of an individual analysis. Note the much higher t-values in the individual analysis).

DISCUSSION AND CONCLUSION

Our results indicate that brain-activity patterns of frustrated drivers are discernible from nonfrustrated drivers using fNIRS. The revealed brain areas displaying higher activity in the frustrating drives are in line with the relatively scarce literature on frustration-related neuroimaging results and seem related to cognitive appraisal, impulse control, and emotion-regulation processes. Because frustration is a multicomponent phenomenon, it could be helpful to measure behavioral and other physiological measures in combination with fNIRS data to acquire a robust assessment of frustration.

Neuroergonomics. http://dx.doi.org/10.1016/B978-0-12-811926-6.00042-7
Copyright © 2019 Elsevier Inc. All rights reserved.

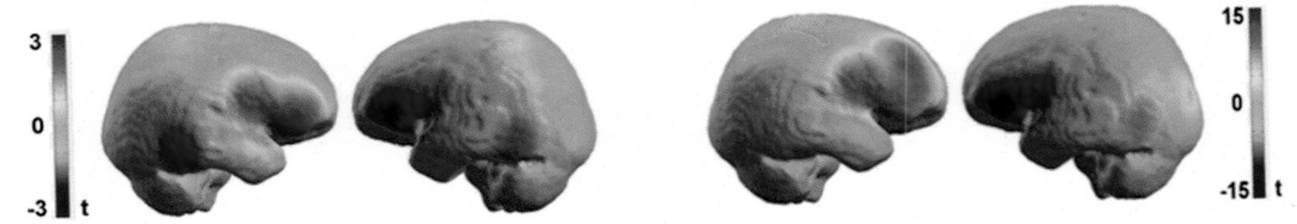

FIGURE 42.1   *T*-statistics maps of brain areas showing frustration-related deoxyhemoglobin (HbR) modulation across: (A) 13 participants from statistical parametric mapping (SPM) level 2 analysis; (B) example participant from SPM level 1 analysis.

ACKNOWLEDGMENTS

This work was funded by a grant of the Volkswagen Foundation and the Ministry of Science and Culture of Lower Saxony to the Research Centre on Critical Systems Engineering for Socio-Technical Systems. Klas Ihme and Anirudh Unni contributed equally.

REFERENCE

1. Parasuraman R. Human-computer monitoring. *Human Factors* 1987;**29**:695–706.

Chapter 43

The Spatial Release of Cognitive Load in Multi-Talker Situation

Guillaume Andéol[1], Clara Suied[1], Sébastien Scannella[2], Frédéric Dehais[2]

[1]Institut de Recherche Biomédicale des Armées, Brétigny sur Orge, France; [2]ISAE-SUPAERO, Université de Toulouse, Toulouse, France

Operators of complex systems often have to deal with multitalker situations (eg, pilots, air controllers). Multitalker situations can be facilitated by spatial separation between the channels thanks to virtual auditory displays.[1] However, it seems that the ability to take advantage of spatial separation could be determined by the cognitive resources of the listeners. However, the cognitive resources of listeners can be limited in many situations, either by age or fatigue or when task demands exceed listener mental capacity, for instance in a multitasking environment. Because of this limitation, we need to investigate the link between auditory cues in speech intelligibility and cognitive load. The present study investigated the role played by the levels of talkers in the spatial release of cognitive load in a multitalker situation. An experiment was designed in which participants had to report the speech emitted by a target talker in the presence of a concurrent masker talker. The spatial separation and the relative levels of the talkers were manipulated. The cognitive load was assessed with a prefrontal functional near-infrared spectroscopy. Data from 14 young normal-hearing listeners revealed that the target-to-masker ratio had a direct impact on the spatial release of cognitive load. Spatial separation significantly reduced the prefrontal activity only for the intermediate target-to-masker ratio and had no effect on prefrontal activity at adverse target-to-masker ratio. Therefore, the relative levels of the talkers might be a key point to determine the spatial release of cognitive load and, more specifically, the prefrontal activity induced by spatial cues in multitalker environments.[1a]

ACKNOWLEDGMENTS

This work was supported in part by the French Procurement Agency (Direction Générale de l'Armement, DGA). The authors warmly thank Jean Christophe Bouy for software development.

REFERENCES

1. Brungart DS, Simpson BD. *Improving multitalker speech communication with advanced audio displays.* RTO NATO; 2005. Available at: http://oai.dtic.mil/oai/oai?verb=getRecord&metadataPrefix=html&identifier=ADA454531.

1a. Andéol G, Suied C, Scannella S, Dehais F. The spatial release of cognitive load in cocktail party is determined by the relative levels of the talkers. *Journal of the Association for Research in Otolaryngology* 2017;**18**(3):457–464.

Neuroergonomics. http://dx.doi.org/10.1016/B978-0-12-811926-6.00043-9
Copyright © 2019 Elsevier Inc. All rights reserved.

Chapter 44

Decreased Intra-Hemispheric Prefrontal Connectivity and Impaired Performance After Induction of Cognitive Fatigue During a State of Sleep Deprivation. An Optical Imaging Study

Guillermo Borragán[1], Céline Guillaume[1], Hichem Slama[1,3], Carlos Guerrero-Mosquera[2], Philippe Peigneux[1]

[1]*Université Libre de Bruxelles (ULB), Brussels, Belgium; [2]Universitat Pompeu Fabra, Barcelona, Spain; [3]Erasme Hospital, Brussels, Belgium*

A situation of resource unavailability following sustained demands and a drop of performance are generally associated with the triggering of cognitive fatigue (CF).[1] There is not yet a clear agreement about the origin of this lack of resources.[2] To provide further insight in this issue, the present study investigated the neural dynamics of CF in a sleep-deprivation situation in which resources are naturally compromised. Using functional near-infrared spectroscopy (fNIRS), we recorded cortical brain activity in 16 participants at three different times during the night (8 p.m., 2 a.m., and 7 a.m.). The recordings were made while they performed with a demanding task susceptible to induce CF. Results showed that, although hemodynamic levels remain comparable at all sessions, lower performance levels were associated with a loss of connectivity at the end of the night in the left prefrontal cortex. These results support the dynamics models of stress and performance,[3,4] which posit a disruption in the access to the pool of resources as a cause of performance drop during sustained attention.

REFERENCES

1. Lim J, Wu W, Wang J, Detre JA, Dinges DF, Rao H. Imaging brain fatigue from sustained mental workload: an ASL perfusion study of the time-on-task effect. *NeuroImage* 2010;**49**(4):3426–35. http://dx.doi.org/10.1016/j.neuroimage.2009.11.020.
2. Hockey R. *The psychology of fatigue.* New York, United States of America: Cambridge University Press; 2013.
3. Hancock PA, Warm JS. A dynamic model of stress and sustained attention. *Human Factors* 1989;**31**(5):519–37.
4. Hockey. Compensatory control in the regulation of human performance under stress and high workload: a cognitive-energetical framework. *Biological Psychology* 1997;**45**(1–3):73–93. http://dx.doi.org/10.1016/S0301-0511(96)05223-4.

Neuroergonomics. http://dx.doi.org/10.1016/B978-0-12-811926-6.00044-0
Copyright © 2019 Elsevier Inc. All rights reserved.

Chapter 45

Bright Light Exposure Does Not Prevent the Deterioration of Alertness Induced by Sustained High Cognitive Load Demands

Guillermo Borragán, Gaétane Deliens, Philippe Peigneux, Rachel Leproult
Université Libre de Bruxelles (ULB), Brussels, Belgium

THEORETICAL FRAMEWORK AND STUDY GOAL

Several studies have reported beneficial effects of light exposure to counteract increased sleepiness and decreasing performance. However, little is known about the role of light exposure to prevent decreases of alertness due to high cognitive demands. In the current study, we investigated the effects of bright-light exposure to prevent increases in sleepiness and decreases in alertness induced by a dual working-memory task (TloadDback) in which high cognitive load (HCL) levels can be adapted to the individual's maximal capacity.

METHODS

In a randomized crossover study, 20 healthy participants were exposed to two sessions that included 20 min of light exposure (dim light or bright light). Subjective sleepiness (Karolinska Sleepiness Scale) and objective alertness (Psychomotor Vigilance Task) were assessed before light exposure and after light exposure before and after performing on the TloadDback task. Additionally, self-reported levels of affect, vigor, cognitive fatigue, and performance during the sustained attention task were assessed.

RESULTS

Bright light exposure did not prevent decreased alertness and increased sleepiness after the TloadDback task. Similarly, decreased on-task performance, increased feeling of cognitive fatigue, and changes in affect and vigor were not attenuated by bright-light exposure.

DISCUSSION AND CONCLUSION

This study used bright light to attempt preventing decreased alertness and improves the feeling of well-being following sustained cognitive demands. Our results suggest that bright light administered prior to the exposure to a cognitively demanding task is not beneficial to prevent impairments ensuing from high cognitive demands.

Neuroergonomics. http://dx.doi.org/10.1016/B978-0-12-811926-6.00045-2
Copyright © 2019 Elsevier Inc. All rights reserved.

Chapter 46

A Psychophysiology-Based Driver Model for the Design of Driving Assistance Systems

Franck Mars, Philippe Chevrel
CNRS, Central Nantes & IMT-Atlantique, Nantes, France

AIMS

Lane departures while driving a car may be caused directly by errors in steering behavior and indirectly because of driver distraction. Designing advanced driving assistance systems (ADAS) is a way to avoid these situations. One of the key problems is to monitor and predict driver behavior to make the ADAS intervene in a timely and efficient way. Our approach to achieve this goal is to incorporate a driver model in the design process. The model represents the visual and motor determinants of steering control. It has been used as the foundation of two types of ADAS: (1) a haptic-shared control system that exerts forces on the steering wheel in such a way that the automation blends into the driver's sensorimotor control loop, providing continuous support to lane keeping; and (2) a method of distraction estimation based on the comparison between steering behavior and the model prediction.

METHODS

Based on current knowledge of human sensorimotor functions, the model represents the visual anticipation of the road curvature fed by the angular deviation of a far point, the visual compensation of lateral position error fed by the angular deviation of a near point, and a neuromuscular system that transforms the output of the visual subsystem into a steering wheel torque.

For haptic-shared control, the approach consisted in designing a control law that optimizes performance and cooperation criteria, taking into account the predictions of the driver model. A comparison was performed between the driver behaviour when using this system and when using a shared control law that did not incorporate a driver model.

For distraction estimation, an identification of the driver model's parameters was conducted in five different distraction conditions: no distraction and cognitive, visual, motor and visuomotor distractions. The torque prediction error and the parameters values were analysed.

RESULTS

The results concerning haptic-shared control suggest that cooperation between the driver and the automation can be improved by the inclusion of the driver model in the control law. In particular, when the level of haptic authority was set at a high level, the driver acted 27% more in coherence with the system.

For the second application, the results show that the model prediction error is commensurate to a modification of steering behaviour caused by distraction. The parametric analysis suggests that an online identification of the model may be used to some extent to discriminate between different types of distraction.

CONCLUSIONS

Adopting a model-based approach to design ADAS is promising. The current model is limited to the representation of sensorimotor processes in humans. As such, it is adapted to be used in the context of real-time control embedded systems. In the future, it could be extended by incorporating a tactical analysis of the driving context.

Neuroergonomics. http://dx.doi.org/10.1016/B978-0-12-811926-6.00046-4
Copyright © 2019 Elsevier Inc. All rights reserved.

Chapter 47

Effect of Postural Chain Mobility on Body Balance and Motor Performance

Alain Hamaoui

Institut National Universitaire Champollion, Albi, France

AIMS

Although the analysis of motor performance in human beings generally refers to the ability of performing focal movements, it has been shown that postural adjustments occurring before, during, and after voluntary movements are also instrumental to preserve postural equilibrium.[1] The aim of this literature review was to specify the role of postural chain mobility in body balance, postural adjustments, and motor performance.

METHODS

Original articles investigating the effect of postural chain mobility on body balance and on simple motor tasks were analyzed using a systematic literature review. Special attention has been paid to biomechanical parameters assessing anticipatory postural adjustments and motor task performance.

RESULTS

In studies focusing on postural maintenance, it has been shown that restricted back mobility due to higher muscular tension impairs body balance,[2] especially when it is asymmetrical.[3] In experiments exploring simple motor tasks such as sitting ramp pushes,[4] it has been shown that a limited mobility of the postural chain was associated with smaller anticipatory postural adjustments and weaker performance.

CONCLUSIONS

In addition to the classical focus on focal movement, task analysis, and organization should also take into account postural adjustments and postural chain mobility, which are integral components of the motor pattern.

REFERENCES

1. Bouisset S, Do M-C. Posture, dynamic stability, and voluntary movement. *Neurophysiologie Clinique/Clinical Neurophysiology* 2008;**38**(6):345–62.
2. Hamaoui A, Friant Y, Le Bozec S. Does increased muscular tension along the torso impair postural equilibrium in a standing posture? *Gait Posture* 2011;**34**(4):457–61.
3. Hamaoui A, Hudson AL, Laviolette L, Nierat M-C, Do M-C, Similowski T. Postural disturbances resulting from unilateral and bilateral diaphragm contractions: a phrenic nerve stimulation study. *Journal of Applied Physiology* 2014;**117**(8):825–32.
4. Le Bozec S, Bouisset S. Does postural chain mobility influence muscular control in sitting ramp pushes? *Experimental Brain Research* 2004;**158**(4):427–37.

Neuroergonomics. http://dx.doi.org/10.1016/B978-0-12-811926-6.00047-6
Copyright © 2019 Elsevier Inc. All rights reserved.

Chapter 48

Effect of Seat and Backrest Sloping on the Biomechanical Strain Sustained by the Body

Nadège Tebbache, Alain Hamaoui
Institut National Universitaire Champollion, Albi, France

AIMS

The sitting posture is frequently adopted, both as a resting posture and as a transition toward the standing posture. This literature review aims to summarize the investigated effects of the seat pan and backrest inclination on the biomechanical strain sustained by the body while seated or performing the sit-to-stand task.

METHODS

A systematic search was performed for experimental studies on healthy volunteers, in which the effect of a sloping seat pan and/or backrest was assessed by means of biomechanical parameters. Papers focusing on static prolonged seated posture and dynamic sit-to-stand task were both investigated.

RESULTS

It is well documented that a forward-sloping seat pan favors the preservation of the lumbar lordosis[1] and an opening of the trunk–thigh angle.[2] However, this effect does not necessarily relieve the lumbar intervertebral discs.[3] More seriously, such a sloping creates a forward and downward sliding effect, which requires an overactivity of the quadriceps and soleus,[2] and might be harmful for the musculoskeletal system.

During prolonged sitting, compression of the soft tissues at specific spots in contact with the seat affects comfort (mainly by reducing blood flow). A reclined backrest offers an additional stabilizing support and modifies the distribution of the upper body weight on the more numerous contact points. However, this gain in static comfort affects the dynamic task of rising from the seat, the first phase for which implies trunk flexion.[4] Due to the lack of standardization around this complex movement, it remains unclear which angle of backrest inclination offers the better compromise between static and dynamic comfort.

CONCLUSIONS

Sloping seat pan and backrest both have various positive and negative influences on the biomechanical strain sustained by the body while seated or performing the sit-to-stand transition. Their use should vary as a function of task parameters.

REFERENCES

1. Bendix T, Biering-Sørensen F. Posture of the trunk when sitting on forward inclining seats. *Scandinavian Journal of Rehabilitation Medicine* 1983;**15**(4):197–203.
2. Hamaoui A, Hassaïne M, Zanone P-G. Sitting on a sloping seat does not reduce the strain sustained by the postural chain. *PLoS One* 2015;**10**(1).

Neuroergonomics. http://dx.doi.org/10.1016/B978-0-12-811926-6.00048-8
Copyright © 2019 Elsevier Inc. All rights reserved.

3. Claus A, Hides J, Moseley GL, Hodges P. Sitting versus standing: does the intradiscal pressure cause disc degeneration or low back pain? *Journal of Electromyography and Kinesiology* 2008;**18**(4):550–8.

4. Nuzik S, Lamb R, VanSant A, Hirt S. Sit-to-stand movement pattern. A kinematic study. *Physical Therapy* 1986;**66**(11):1708–13.

Chapter 49

Effect of Human Exposure to Whole-Body Vibration in Transport

Hiba Souissi[1], Alain Hamaoui[2]

[1]Université de Nice Sophia-Antipolis, Nice, France; [2]Institut National Universitaire Champollion, Albi, France

AIMS

Human beings are exposed to multiple sources of vibratory movements. Some activities of daily living favor vibration exposure, and can affect drivers, transport users, and industrial workers. However, whole-body vibration (WBV) frequencies ranging from 0.7 to 100 Hz have been reported to lead to adverse effects.[1] The aim of this study was to establish a database to assess the effect of different WBV frequencies occurring in transport on the human physiological functions.

METHODS

A systematic literature search was conducted on PubMed database. Studies that analyzed adverse health effects that may be engendered by the WBV exposure were included.

RESULTS

Findings showed that several years of exposure to WBV frequency beyond 50 Hz might contribute to lumbar syndrome.[2] Gastric diseases and vascular disorders have been observed in people submitted on a daily basis to low-frequency vibrations below 20 Hz.[3] Nervous system disorders are considered a characteristic pathology of vibration frequencies over 20 Hz.[3] The short-term exposure to WBV can cause a respiration modification at frequencies between 1 and 4 Hz,[4] and a vestibular function modification at very low frequencies or close to the whole body resonance.[5]

CONCLUSION

Low-frequency (<20 Hz) or very high-frequency (>70 Hz) vibrations can be considered most dangerous for the human body. These vibrations can be found in vehicles (<20 Hz), in air transportation (0.2–7 Hz), or in heavy machine equipment (>20 Hz).

REFERENCES

1. Malchaire J, Piette A, Cock N. In: *Vibrations mains-bras: stratégie d'évaluation et de prévention des risques*, vol. 23. Bruxelles: Ministère fédéral de l'emploi et du travail; 1998. p. 38–41.
2. Pointillart V. *Abord clinique des affections du rachis: Par le chirurgien*. 2009. p. 67.
3. Seidel H, Heide R. Long-term effects of whole-body vibration: a critical survey of the literature. *International Archives of Occupational and Environmental Health* 1986;**58**:1–26.
4. Magid EB, Coermann RR, Ziegenruecker GH. Human tolerance to whole body sinusoidal vibration. *Aerospace Medicine* 1960;**31**:915–24.
5. Sarembaud A. *Mal des Transports, Naupathie, Cinétose*. 2008. p. 157–9.

Neuroergonomics. http://dx.doi.org/10.1016/B978-0-12-811926-6.00049-X
Copyright © 2019 Elsevier Inc. All rights reserved.

Chapter 50

Electroencephalography (EEG) Activity Associated With Manual Lifting Tasks: A Neuroergonomics Study

Awad Aljuaid[1], Waldemar Karwowski[2], Petros Xanthopoulos[3], Peter A. Hancock[4]

[1]Mechanical Engineering Department, Taif University, Saudi Arabia; [2]Department of Industrial Engineering & Management Systems, University of Central Florida, United States; [3]Decision and Information Sciences, Stetson University, United States; [4]Department of Psychology, University of Central Florida, Orlando, FL, United States

Electroencephalography (EEG) has been shown to be a reliable tool in neuroergonomics studies due to the relatively low cost of brain data collection and limited body invasion. The use of EEG frequency bands (including theta, alpha, and beta waves) has a great potential for improving our understanding of brain signatures of physical work. The psychophysical approach has been used for decades to improve safe work practices by focusing on human abilities and limitations in manual material-handling tasks. The main objective of this research was to study the brain's EEG activity expressed by the power spectral density due to manual lifting tasks, related to the maximum acceptable weight of lift (MAWL).

The changes in EEG power spectral density were recorded during determination of MAWL under low, medium, and high lifting frequencies. A total of 20 healthy males participated in this experiment study. Subjects repeated the same experiment after 2 weeks. The main objective of experiment #1 was to assess the effect of lifting frequency (low vs. medium) and lifting task repetition on EEG signatures. The medium lifting frequency was set at one lift every 14 s (4.3 lifts/min), whereas the low lifting frequency was set at one lift every 60 s (1 lift/min). The main objective of experiment #2 was to assess the effect of lifting frequency (high vs. medium) and the effect of lifting task repetition on EEG signatures. The medium lifting frequency was defined as one lift every 14 s (4.3 lifts/min), whereas the high lifting frequency was set at one lift every 9 s (6.7 lifts/min). The subjects were provided with written informed consent prior to the experiment. All procedures were approved by The Institutional Review Board at the University of Central Florida. EEG was recorded using a (Cognionics Data Acquisition Software Suite) and a Cognionics High-Density 64-channel Dry Headset 64-channel EEG (COGNIONICS, Inc. San Diego, CA). Electrodes were attached to the scalp using a custom subset of the 10–5 configuration. The artifacts correction in experimental EEG data was done using the Artifact Subspace Reconstruction method (ASR). ASR uses an algorithm to remove nonstationary high-variance signals from EEG and rebuilds the missing data with a spatial mixing matrix (assuming volume conduction).

The brain's areas of interest included the frontal part (attention, judgment, and motor planning); the central part (sensorimotor control); and the parietal part (cognitive processing). Analysis of variance (ANOVA) showed significant differences in EEG power spectral density between different lifting frequencies at three main brain areas, ie, frontal, central and parietal. Specifically, in experiment #1 significant changes between low and medium lifting tasks in theta EEG activity power were found in most of the brain regions: frontal, central, and parietal. Significant changes in alpha EEG activity power were also found in the frontal and central regions. Furthermore, in experiment #2 significant changes between medium and high lifting tasks in alpha and beta EEG activity power were found in central and parietal brain regions. No significant changes in theta and gamma EEG activity power were found in most of the studied brain regions. This study is the first study of EEG activity during manual lifting tasks, including the assessment of MAWL by the psychophysical method. The results of this study are considered critical to our understanding of the neural signatures of human physical activities, and consequently, should have an impact on the success of workplace design that considers brain activity in relation to specific human abilities and limitations in manual lifting tasks. The results of this study has important implication for prevention of low-back injury due to manual lifting tasks.

Neuroergonomics. http://dx.doi.org/10.1016/B978-0-12-811926-6.00050-6
Copyright © 2019 Elsevier Inc. All rights reserved.

Chapter 51

Anticipatory Postural Control of Stability During Gait Initiation Over Obstacles of Different Height and Distance Under Reaction-Time and Self-Initiated Instructions

Eric Yiou[1,2], Romain Artico[1,2], Claudine Teyssedre[1,2], Ombeline Labaune[1,2], Paul Fourcade[1,2]

[1]Université Paris Sud, Université Paris-Saclay, CIAMS, Orsay, France; [2]CIAMS, Université d'Orléans, Orléans, France

AIMS

Despite the abundant literature on obstacle crossing in humans, the question how the central nervous system (CNS) controls postural stability during gait initiation (GI), with the goal to clear an obstacle, remains unclear. GI, which corresponds to the transient period between quiet standing and swing-foot contact with the ground, is a functional task classically used for studying balance-control mechanisms during complex whole-body movement.[1] GI-stabilizing features include anticipatory postural adjustments (APAs) and lateral swing-foot placement. Previous study has shown that APA features associated with rapid leg flexion depend on the level of temporal pressure.[2] This study tested the hypothesis that the GI-stabilizing features (1) are scaled according to the changes in the swing-phase duration that is associated with obstacle height and/or distance, and (2) are modified by the temporal pressure.

METHODS

Fourteen participants initiated gait at maximal velocity in three conditions of obstacle height (2.5, 5, and 10% of subject's height), three conditions of obstacle distance (10, 20, and 30% of subject's height), one obstacle-free control condition and two levels of temporal pressure: reaction-time (high-pressure) and self-initiated (low-pressure). Gait was initiated on a force plate located at the beginning of a 5-m track. A V8i VICON eight-camera motion capture system was used to record heel and toe movement of swing and stance leg. Classical biomechanical variables related to GI were quantified.[1] An adaptation of the "margin of stability" (MOS)[3] was used to quantify dynamic stability at heel contact along the mediolateral (ML) direction. A mechanical model of the body falling laterally under the influence of gravity and submitted to an elastic restoring force is proposed to assess the effect of initial (foot off) center-of-mass (CoM) position and velocity (or "initial center-of-mass set") on stability at foot contact.

RESULTS

Results showed that the duration of the swing phase of GI, the anticipatory peak of ML center-of-pressure shift (CoP), and the initial ML CoM velocity increased with obstacle height, but not with obstacle distance. In contrast, postural stability remained unchanged across conditions. Mechanical model revealed how postural stability would be degraded in case the initial CoM set would not be scaled in function of swing-phase duration. The anteroposterior component of APAs varied also according to obstacle height and distance, but in an opposite way to the ML component. Indeed, the anticipatory peak of backward CoP shift and the initial forward CoM set decreased with obstacle height, probably to limit the risk of tripping over the obstacle, whereas the forward CoM velocity at foot off increased with obstacle distance, allowing a further step to be taken. These effects of obstacle height and distance were globally similar under low- and high-temporal pressure.

Neuroergonomics. https://doi.org/10.1016/B978-0-12-811926-6.00051-8
Copyright © 2019 Elsevier Inc. All rights reserved.

CONCLUSIONS

The present findings show that the CNS is able to precisely predict the potential instability elicited by obstacle clearance and that it scales the spatiotemporal parameters of APAs according to this prediction. The results offer a better understanding of how the body adapts to environmental constraints to ensure safe and efficient whole-body progression.

REFERENCES

1. Brenière Y, Cuong do M, Bouisset S. Are dynamic phenomena prior to stepping essential to walking? *Journal of Motor Behavior* 1987;**19**(1):62–76.
2. Hussein T, Yiou E, Larue J. Age-related differences in motor coordination during simultaneous leg flexion and finger extension: influence of temporal pressure. *PLoS One* 2013;**8**(12):e83064.
3. Hof AL, Gazendam MG, Sinke WE. The condition for dynamic stability. *Journal of Biomechanics* 2005;**38**(1):1–8.

Chapter 52

Effect of Age on Behavioral Performance and Metabolic Brain Activity During Dual-Task

Nounagnon F. Agbangla[1], Michel Audiffren[1], Jean Pylouster[1], Cédric T. Albinet[1,2]
[1]Université de Poitiers, Poitiers, France; [2]Université de Toulouse, INU Champollion, Albi, France

AIM AND METHODS

The objective of this study was to examine age-related effects on behavioral performance and metabolic activity of the prefrontal cortex (PFC) in a dual-task paradigm involving fine-motor control and executive-function control. Thirty-one adults (20 ± 1.1 years) and 33 older adults (70.7 ± 5.1 years) performed, separately and then concurrently, a modified Fitts task on targets of different sizes (Index of Difficulty: ID = 3; 4; 5 bits) and the random number generation task (RNG). Movement time (MT) and count score (CS) were the dependent variables for the modified Fitts task and the RNG task, respectively. Relative changes in concentrations of oxyhemoglobin [HbO_2] and deoxyhemoglobin [HHb] were recorded continuously on the left and right PFCs, with near-infrared spectroscopy (Oxymon MkIII-Artinis) with an acquisition rate of 10 Hz.

RESULTS

Single Task

At the sensorimotor level, older participants' MT was consistently longer than younger participants' MT ($P < .05$; $\eta^2 = 0.06$). At the cognitive level, older participants only tended to show greater CS (lower cognitive performance) than the young participants ($P = .08$).

Dual-Task

At the sensorimotor level, the significant group × condition interaction ($P < .01$; $\eta^2 = 0.35$) on MT indicated that, compared to the single-task condition, the young participants were faster ($\approx -5\%$) in the dual-task condition, whereas the older participants increased their MT ($\approx +15\%$). At the cognitive level, the main effect of group on CS ($P = .01$; $\eta^2 = 0.10$) indicated that regardless of sensorimotor complexity, the older participants showed higher CS than the young participants.

The analyses of the hemodynamic parameters showed a main effect of group on [HbO_2] ($P = .04$; $\eta^2 = 0.06$) only during the dual-task condition. In addition, the significant condition × complexity × group interaction ($P = .002$; $\eta^2 = 0.09$) indicated that during the dual-task condition, the young participants showed higher increases in [HbO_2] than the older ones and that this activation difference was prominent during the most complex condition (ie, performing the RNG task and aiming movements at ID = 5 bits).

DISCUSSION AND CONCLUSION

This study showed large motor dual-task cost in seniors compared to young adults, but no clear dual-task cost in the cognitive domain, neither for the young nor for the older adults. These results contradict somewhat with previous results using a similar procedure.[1] They may reflect that the seniors in the present study prioritized the cognitive performance,

Neuroergonomics. http://dx.doi.org/10.1016/B978-0-12-811926-6.00052-X
Copyright © 2019 Elsevier Inc. All rights reserved.

but consequently they did not have enough resources to perform the sensorimotor task correctly. At the hemodynamic level, our results showed a prominent activation of the PFC in the young adults compared to the seniors during the dual-task condition. This result partly agrees with those of Beurskens et al.,[2] who showed that gross-motor dual-task (walking + visual checking task) was accompanied by lower PFC activity in seniors, and extend them to fine-motor dual-task. In sum, our results showed the involvement of PFC in the management of two tasks and depicted age-related effects on this involvement.

REFERENCES

1. Albinet C, Tomporowski PD, Beasman K. Aging and concurrent task performance: cognitive demand and motor control. *Educational Gerontology* 2006;**32**:1–18.
2. Beurskens R, Helmich I, Rein R, Bock O. Age-related changes in prefrontal activity during walking in dual-task situations: a fNIRS study. *International Journal of Psychophysiology* 2014;**92**(3):122–8.

Chapter 53

Using Machine Learning Algorithms to Develop Adaptive Man–Machine Interfaces

Dargent Lauren[1], Branthomme Arnaud[2], Kou Paul[2], Girod Hervé[2], Morellec Olivier[2]

[1]ISAE-SUPAERO, Toulouse, France; [2]Dassault Aviation, Saint-Cloud, France

CONTEXT

Automation has been introduced in aircraft cockpits to reduce pilot workload and increase safety. However, a number of reports mention "human factors" issues and misunderstandings of an automated-system behavior or its displays by the crew as major contributors leading to flight incidents. The interfaces should play though a crucial role in improving the man–machine cooperation,[1] by displaying "the right information at the right time." This need of adapted displays and interfaces is more important than ever as the missions are becoming more and more complex, especially in the military domain. Moreover, many factors in workload mitigation are identified as crew or mission dependent and are highly variable from one flight to another, such as the cognitive demands of the current phase of flight or mission situation, the pilot's experience or "airmanship," or individual physiological parameters. Thus, we can think of an adaptive intelligent interface that would monitor the automated system–pilot team as well as the mission operational context, to provide the correct display and controls to the user and enable better cooperation between the human operator and the machine to match the current demands of the operational situation.[2]

AIM

We aim to investigate the potential of different machine-learning algorithms to develop adaptive interfaces. This intelligent agent should take into account the mission situation, the aircraft situation, and the user's cognitive status to choose the most-adapted set of displays and controls for the user. We also aim to identify the limitations of such algorithms in achieving actual implementation of such adaptive interfaces in cockpits.

METHOD

As we aim to train a supervised algorithm, we need to collect data to train the machine. Thus, we designed an experiment where we created a simple arcade-like "2D - game" putting the participant in position of a military combat-system manager, in which he or she has to achieve the best operational performance while maintaining his/her system alive. To do so, the participant can choose between four display and control configurations designed to help him/her achieve their goals. In addition to game parameters, participant's eye gaze and parameters are recorded using a Tobii X1-Light eye-tracker.[3] Then different classification algorithms are fed with these data to learn the best display and control configuration to provide to the player at any point in the game. Thirty-five people, all employed at Dassault Aviation, participated in the data collection process.

EXPECTED RESULTS

We expect that the final version of the game, integrating the trained machine, will be able to provide the participant with the "right" display and control configuration designed to help him "at the right time":

- helping him/her to increase significantly their performance at the game
- adapting to his/her "playing strategy"
- reducing his/her workload[4] [evaluation with a National Aeronautics and Space Administration Task Load Index (NASA-TLX) test at game over]

Neuroergonomics. http://dx.doi.org/10.1016/B978-0-12-811926-6.00053-1
Copyright © 2019 Elsevier Inc. All rights reserved.

CONCLUSION

At this stage of the experiment, five machine-learning algorithms [support vector machine, random forest, deep neural network, k-nearest neighbors (kNN), and adaptive boosting (adaboost)] are being evaluated for their ability to implement an efficient adaptive man–machine interface concept based on performance and ocular metrics. Preclusterization based on experimenter's observations or performance metrics is also considered to improve accuracy of the machine-learning algorithms' predictions.

REFERENCES

1. Sheridan TB, Parasuraman R. Human-automation interaction. *Reviews of Human Factors and Ergonomics* 2005;**1**(1).
2. Bonner M, Taylor R, Fletcher K, Muller C. Adaptive automation and decision aiding in the military fast jet domain. In: *Proceedings of human performance, situation awereness, and automation*. 2000. p. 154–9.
3. Tobii Website. http://www.tobii.com.
4. Hou M, Kobiersky R, Brown M. Intelligent adaptive interfaces for the control of multiple UAVs. *Journal of Cognitive Engineering and Decision Making* 2007:327–62.

Chapter 54

Link Between Out-of-the-Loop Performance Problem and Mind Wandering: How to Keep the Operator in the Loop

Jonas Gouraud, Bruno Berberian, Arnaud Delorme
Office National d'Etudes et de Recherche Aérospatiales, Salon-de-Provence, France

Increasing safety in the aerospace industry is paramount. To achieve this goal, engineers design increasingly reliable and robust systems. However, interaction between these systems and pilots are regularly at fault. Specifically, the out-of-the-loop (OOL) performance problem arises when operators suffer from complacency and vigilance decrement: they are able to neither understand what the system is doing, nor take back manual control. If the OOL performance problem represents a key challenge for the human factor community, it remains difficult to characterize and quantify after decades of research.[1]

The phenomenon of mind wandering (MW) could be closely linked to the OOL problem and offer new possibilities to study the issue. MW is the human-mind propensity to drift away from the task at hand toward inner unrelated thoughts. This occurs most of the time without intention or even awareness of the subject. Similar to OOL, MW increases reaction time, lowers accuracy, and decreases performance. Moreover, MW seems to increase when interacting with highly automated systems.[2] This could be due to the sensory attenuation created during low demanding and/or boring tasks, which decouples the subject from its environment.[3]

We aim at understanding the link between the two phenomena. Our hypothesis is that MW is one cause of OOL: if the system changes its mode or makes an error when the operator is MW, then it would likely result in OOL. If we understand this link, OOL researchers might be able to use all identified MW markers—brain dynamics, eye movements, perspiration, heart rate, reaction time, and accuracy. It would allow detecting and fixing OOL issues dynamically in operational environments.

Our first experiment will investigate the influence of automation over MW, because this question has only been addressed by Casner and Schooler[3] in a flight simulator. We will use the Sustained Attention to Response Task (SART)—an extensively used paradigm in MW studies—and modify the level of automation between two states, active and passive. We hope to observe changes in brain dynamics—alpha wave and event-related potential P3—and reported characteristics of the MW episodes—number, length, and depth. Depending on the results, we will address more directly the link between OOL and MW and see how they correlate within highly automated environments.

REFERENCES

1. Baxter G, Rooksby J, Wang Y, Khajeh-Hosseini A. The ironies of automation … still going strong at 30?. In: *Proceedings of ECCE 2012 conference. Edinburgh, North Britain.* 2012.
2. Casner SM, Schooler JW. Thoughts in flight automation use and pilots' task-related and task-unrelated thought. *Human Factors: The Journal of the Human Factors and Ergonomics Society* 2013. http://dx.doi.org/10.1177/0018720813501550.
3. Benedek M, Schickel RJ, Jauk E, Fink A, Neubauer AC. Alpha power increases in right parietal cortex reflects focused internal attention. *Neuropsychologia* 2014;**56**:393–400. http://dx.doi.org/10.1016/j.neuropsychologia.2014.02.010.

Neuroergonomics. http://dx.doi.org/10.1016/B978-0-12-811926-6.00054-3
Copyright © 2019 Elsevier Inc. All rights reserved.

Chapter 55

Applied Neuroergonomic: Recent Updates From Automotive Industry Case Studies

Ivan Macuzic, Evanthia Giagloglou, Ivana Živanovic-Macuzic, Branislav Jeremic
University of Kragujevac, Kragujevac, Serbia

AIMS

The main aim of performed and undergoing studies is to explore possibilities for implementation of various psychophysiology measurements [electroencephalography (EEG), electrodermal activity (EDA), heart-rate variability (HRV), and pulse oximeter (SpO2)] in ergonomic studies oriented to identification of operators' reaction to various mental and physical inputs from real workplace environments. Additionally, a number of characteristic operational, workplace task-related parameters were included in measurement procedures integrated in a joint multimodal data-acquisition and signal-processing system. The first automotive-industry case study was dedicated to the workplace, which represents the typical example of manual assembly tasks with monotonous and repetitive operations. The second one is oriented toward manual-handling operations of trolleys pushing and pulling.

METHODS

For both case studies, faithfully replicated industrial-workplace infrastructures were created in laboratory conditions. Wireless EEG and EDA measurement devices were integrated with job-related parameter measurements through a Lab Streaming Layer (LSL) platform to achieve full synchronization of such a multimodal system. Data analysis was performed offline using EEGLAB and MATLAB. For the first repetitive-operations industrial case study, several investigations were performed: (1) propagation of the P300 event-related potential (ERP) component's amplitude during the course of simulated operation together with propagation of the reaction times; (2) introduction of frequent micro-breaks and their impact on the operators' attention level; (3) potential hand-alteration influence on maintaining higher level of operators' attention. For the second pushing-and-pulling workplace case study, initial investigations of twisting-and-turning influence on EDA operators' response were performed, while additional measurements with introduction of mobile EEG were undergoing to ensure results of mental operators' response on demand full manual-handling tasks.

RESULTS AND CONCLUSIONS

Initial results and findings show significant potential in implementation of wireless, mobile EEG devices (and other psychophysiology measurements) in ergonomics research, as they enable identification of operators' mental response, combined with other subjective and objective workplace parameters.

ACKNOWLEDGMENTS

This research is financed under EU FP7 Marie Curie Actions FP7-PEOPLE-2011-ITN: Project "Innovation through Human Factors in risk analysis and management" - InnHF.

FURTHER READING

1. Mijović P, Ković V, De Vos M, Mačužić I, Jeremić B, Gligorijević I. Benefits of instructed responding in manual assembly tasks: an ERP approach. *Frontiers in Human Neuroscience* 2016;**10**:171.
2. Mijović P, Ković V, De Vos M, Mačužić I, et al. Towards continuous and real-time attention monitoring at work: reaction time versus brain response. *Ergonomics* 2016. p. 1–14. http://dx.doi.org/10.1080/00140139.2016.1142121. Published online 08. 03. 2016.

Copyright © 2019 Elsevier Inc. All rights reserved.

Chapter 56

Multi-Brain Computing: BCI Monitoring and Real-Time Decision Making

Anton Nijholt

University of Twente, Enschede, The Netherlands

In this chapter, we survey recent research on multibrain applications. That is, applications in which synchronized brain activity of multiple users is measured and integrated to use their joint brain activity to make real-time decisions about communication with and control of devices in smart environments.[1] Interestingly, we can go back to early brain–computer interface research of the 1970s to see many ideas and sometimes implementations of synchronized multibrain "computing." Usually they can be found in the artistic domain. In this decade (2010–20), we see growing attention in this research area, partly because of the availability of affordable electroencephalographic (EEG) devices and partly because of the interest of human–computer interaction researchers in affective computing.[2]

This additional interest is now responsible for a focus on brain–computer interface (BCI) research that has changed from clinical applications to applications that are of interest to industry, specific groups of professionals, or to the general population.[3] Traditional BCI researchers are not always open to these developments,[3,4] in which, rather than focusing on Amyotrophic Lateral Sclerosis (ALS) patients, this new research focuses on entertainment, games, art, and playful applications in the domestic and public domains.

Among the many applications of multibrain computing as adapted and extended from Stoica (2012)[5] are: (1) Joint decision-making in environments requiring high accuracy, and/or rapid reactions, or feedback; (2) joint/shared control and movement planning of vehicles or robots; (3) assessment of team performance, stress-aware task allocation, and rearrangement of tasks; (4) characterization of group emotions, preferences, and appreciations; (5) social interaction research (two or more people); (6) arts, entertainment, and games.

We discuss some examples from multibrain computing and focus on possible ways of joint decision making (or otherwise using the measured brain activity of multiple users). We will also emphasize the possibilities that are offered by the multimodal context, that is, considering brain–computer interfacing as one of the many possible modalities to obtain information about a user's or a group of users' affective states, preferences, and decisions. How to fuse information coming from different modalities and from different users needs to be discussed.[2] For that, we can learn from multimodal interaction research in human–computer interaction, including observations on sequential and parallel multimodality.

REFERENCES

1. Nijholt A. Competing and collaborating brains: multi-brain computer interfacing. In: Hassanieu AE, Azar AT, editors. *Brain-computer interfaces: current trends and applications. Intelligent systems reference library series*, vol. 74. Springer; 2015. p. 313–35.
2. Nijholt A. Multimodal and multi-brain computer interfaces. A review. In: *Proceedings 10th International Conference on Information, Communications and Signal Processing (ICICS 2015), IEEE Xplore*. December 2–4, 2015. p. 1–5. Singapore.
3. Nijholt A. The future of brain-computer interfacing. In: *5th International Conference on Informatics, Electronics & Vision (ICIEV)*. 13–14 May 2016. Dhaka, Bangladesh. IEEE Xplore, 156–61.
4. Brunner C, Birbaumer N, Blankertz B, Guger C, Kübler A, Mattia D, Millán J del R, Miralles F, Nijholt A, Opisso E, Ramsey N, Salomon P, Müller-Putz GR. BNCI horizon 2020: towards a roadmap for the BCI community. *Brain-Computer Interfaces* 2015;2(1):1–10.
5. Stoica A. MultiMind: multi-brain signal fusion to exceed the power of a single brain. In: *Proceedings of the 3rd Conference on Emerging Security Technologies*. 2012. p. 94–8.

Copyright © 2019 Elsevier Inc. All rights reserved.

Chapter 57

Toward Mental Workload Measurement Using Multimodal EEG–fNIRS Monitoring

Hubert Banville[1], Mark Parent[2], Sébastien Tremblay[2], Tiago H. Falk[1]

[1]Université du Québec, Montreal, Canada; [2]Université Laval, Quebec City, Canada

AIMS

A key goal of neuroergonomics is to develop tools to measure mental workload during task execution, potentially enabling safer and more efficient human–machine interaction in diverse working environments.[1] A promising approach to developing these high-performance and robust cognitive monitoring systems is to combine neurophysiological modalities, such as electroencephalography (EEG) and functional near-infrared spectroscopy (fNIRS), that can offer complementary information on brain activity.[2,3] Here, innovative temporal dynamics-related EEG and fNIRS features are described that will enable development of highly effective multimodal mental-workload measures.

METHODS

We recorded full-head EEG and partial-head NIRS data from nine participants performing seven different mental tasks that elicit various levels of mental workload. Several EEG and fNIRS temporal dynamics features were extracted from the recorded data, including recently proposed frequency-modulation features, measuring the interaction of different oscillatory processes in the brain.[4] These features were then correlated with subjective National Aeronautics and Space Administration Task Load Index (NASA TLX) workload ratings, and the most highly correlated features were selected for further analysis.

RESULTS

Although existing benchmark EEG and fNIRS features exhibited little to no relationship with the studied subjective dimensions, several of the suggested EEG and fNIRS features were found to be significantly correlated, particularly with the temporal demand and performance dimensions (18 and 15 features, respectively). Moreover, cortical regions of interest were identified based on these features.

CONCLUSION

This work paves the way to cognitive-monitoring systems based on multimodal neurophysiological techniques, and shows that innovative features can outperform conventional ones. Further investigations on the complementarity of such multimodal features will reveal if higher performance can be attained when used in combination.

ACKNOWLEDGMENTS

The authors wish to acknowledge funding from Fonds de Recherche du Québec - Nature et Technologies (FRQNT) and the Natural Sciences and Engineering Research Council of Canada (NSERC). The authors thank R. Gupta for help with data acquisition.

Neuroergonomics. http://dx.doi.org/10.1016/B978-0-12-811926-6.00057-9
Copyright © 2019 Elsevier Inc. All rights reserved.

REFERENCES

1. Parasuraman R. Neuroergonomics: Research and practice. *Theoretical Issues in Ergonomics Science* 2003;**4**(1–2):5–20.
2. Coffey EBJ, Brouwer AM, van Erp JBF. Measuring workload using a combination of electroencephalography and near infrared spectroscopy. In: *Proceedings of the human factors and ergonomics society annual meeting, SAGE Publications*; 2012;**56**(1).
3. Ahn S, et al. Exploring neuro-physiological correlates of drivers' mental fatigue caused by sleep deprivation using simultaneous EEG, ECG, and fNIRS data. *Frontiers in Human Neuroscience* 2016;**10**.
4. Fraga FJ, et al. Characterizing Alzheimer's disease severity via resting-awake EEG amplitude modulation analysis. *PLoS One* 2013;**8**(8):e72240.

Chapter 58

The Effects of Transcranial Direct Current Stimulation (tDCS) on Adapting to Temporal Lag in Virtual Environments

Hayley Thair, Roger Newport
University of Nottingham, Nottingham, United Kingdom

AIMS

With the rapid increase in Virtual Reality (VR) systems, more research is needed to understand how we process sensory information in these environments. Previous research has shown that movement times and error rates increase with greater temporal VR lag.[1,2] This study investigated whether transcranial direct-current stimulation (tDCS) can modulate sensory integration of a virtual hand, which may, in turn, affect movement times and reach accuracy.

METHODS

Participants placed their right hand inside a Mirrors And Genius (MIRAGE)-mediated reality device, which allows them to see live video of their virtual hand in the same physical location as their real hand. MIRAGE allows the incremental online manipulation of different properties of the virtual hand including appearance, spatial location, and temporal lag. Participants completed simple reach-to-point movements to targets appearing in pseudorandom locations within the work-space. They pointed to 70 targets with 0 or 150 ms additional lag (blocked), while movement kinematics were recorded using a motion tracker attached to the index finger. tDCS to the Posterior Parietal Cortex (PPC) was applied during the task with participants receiving all three stimulation types across three separate testing sessions.

RESULTS

Preliminary analysis suggests that a 150-ms lag condition results in more performance errors than 0-ms lag, and that active stimulation modulates performance compared to sham stimulation.

CONCLUSIONS

Differences in performance between active and sham conditions suggests that tDCS could be used as a tool for manipulating sensory integration in a virtual environment, particularly if errors caused by lag can be reduced.

REFERENCES

1. Mackenzie I, Ware C. Lag as a determinant of human performance in interactive systems. In: *INTERCHI*. 1993. p. 488–93.
2. Lee I, Choi S. Discrimination of virtual environments under visual and haptic rendering delays. In: *IEEE*. 2007. p. 554–9.

Neuroergonomics. https://doi.org/10.1016/B978-0-12-811926-6.00058-0
Copyright © 2019 Elsevier Inc. All rights reserved.

Chapter 59

Why Do Auditory Warnings During Steering Allow for Faster Visual Target Recognition?

Christiane Glatz[1,2], Heinrich H. Bülthoff[1], Lewis L. Chuang[1]

[1]Max-Planck Institute for Biological Cybernetics, Tübingen, Germany; [2]International Max Planck Research School, Tübingen, Germany

INTRODUCTION AND AIM

Auditory cues are often used to capture and direct attention away from an ongoing task to a critical situation. In the context of driving, previous research has shown that looming sounds, which convey time-to-contact information through their rising-intensity profiles, promote faster braking times to potential front collisions.[1] The current experiment investigates the role of auditory warnings in facilitating the identification of visual objects in the visual periphery during steering. This approximates the use of auditory warnings for cueing possible candidates for side collisions. We expected faster response times for visual targets cued by a looming sound compared to a constant sound. Electroencephalography (EEG) was recorded to determine whether faster response times were due to either earlier or stronger neural responses to the visual target. We hypothesize: (1) earlier event-related potentials (ERPs) for cued compared to noncued visual targets, and (2) larger amplitudes for visual targets that were cued by looming versus constant sounds.

METHODS

While performing a primary steering task, participants ($N=20$) had to identify visual stimuli (ie, Gabor patches) presented in the periphery and to discriminate them for their tilt orientation. In 50% of the trials, visual stimuli were preceded by a 400-Hz sound with either constant or looming profiles. The looming sounds' intensity profile increased exponentially over time, whereas the intensity profile for the constant sound did not change across the 500 ms.

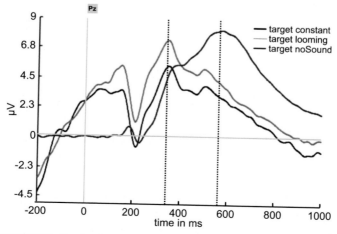

This figure shows the grand-average ERP to the visual target onset preceded by a constant (*black*), looming (*red*), or no sound (*blue*) curve.

Neuroergonomics. http://dx.doi.org/10.1016/B978-0-12-811926-6.00059-2
Copyright © 2019 Elsevier Inc. All rights reserved.

RESULTS

Our results show that participants responded faster to cued targets than to trials without a warning [$t(19) = -9.054$, $P < .001$]. The maximum peaks in the ERPs to visual targets were earlier for those that were cued with an auditory warning (black and red curves) compared to those without warning cues (blue curve). Next, looming warnings resulted in faster visual discrimination than constant sounds [$F(1,19) = 6.934$, $P = .016$]. The maximum peaks of ERPs to cued visual targets were larger for those that were cued by looming sounds compared to those that were cued by constant sounds.

DISCUSSION AND CONCLUSION

The maximal ERP peak that we report is likely the P3 component that is related to visual object-recognition performance. This response occurs later in the absence of a warning cue. Looming auditory warnings might have induced a larger P3 component to visual targets, relative to constant warnings, by being more effective attentional cues. Our EEG data corresponds to behavioral benefit of looming auditory cues observed in faster reaction times. Interestingly, warning signals can prepare the brain to respond earlier to visual events, even with a predictability of only 50%. These findings can directly be applied to the design of auditory warnings in which fast but also accurate reactions are preferable.

REFERENCE

1. Gray R. Looming auditory collision warnings for driving. *Human Factors: The Journal of the Human Factors and Ergonomics Society* 2011;**53**(1):63–74.

Chapter 60

Attending to the Auditory Scene Improves Situational Awareness

Menja Scheer, Heinrich H. Bülthoff, Lewis L. Chuang
Max-Planck Institute for Biological Cybernetics, Tübingen, Germany

AIM

Early studies suggested that auditory stimuli are only (cognitively) processed if they are relevant for the task.[1] However, studies that are more recent show that this is not necessarily true. Rather, it depends on the nature of the auditory stimuli. Environmental sounds are processed even when they are irrelevant for the task and even when participants are engaged in demanding visual tasks.[2] These latter results can be interpreted within the framework of situational awareness. To maintain situational awareness, it is essential to continuously scan the environment for unexpected sounds that might not be of immediate task relevance but could inform us about important changes in the environment. In the current study, we investigate whether and how the scanning for, and processing of, environmental sounds—in other words, situational awareness of the auditory scene—is influenced by auditory attention manipulations. Here, auditory attention was manipulated by requiring participants to perform an auditory oddball-detection task or not, while task-irrelevant environmental sounds were occasionally presented in the background. The current study answers the following questions: (1) Is the processing of environmental sounds influenced by manipulations of auditory attention or is it an automatic process? (2) Is the processing of irrelevant environmental sounds attenuated or enhanced by auditory attention? On the one hand, the processing of the irrelevant environmental sounds could be attenuated, because the additional task increases the demand for auditory attentional resources.[2] On the other hand, the processing of the environmental sounds could be enhanced because more attention is directed toward the auditory channel to optimize performance in the auditory oddball-detection task.

METHODS

Two groups of participants ($N=48$) were involved in a manual-steering task, while being probed with environmental sounds as well as beep tones. The first group of participants was instructed to only perform a steering task ("auditory irrelevant"). The second group was instructed to additionally perform an oddball-detection task by detecting the oddball beep tones (auditory relevant). The environmental sounds were task irrelevant for both groups of participants. The event-related potentials (ERPs) to the environmental sounds are illustrated in Fig. 60.1.

RESULTS

We found that the ERPs to the irrelevant environmental sounds were selectively enhanced when the auditory channel was task relevant. This was, however, specific to one component of the ERP, the lP3a (blue in Fig. 60.1). No influence of our manipulation on the other measured ERP components, MMN and eP3a, was observed.

CONCLUSION

We found that task relevance in the auditory channel influences the processing of environmental sounds. Therefore, we conclude that this process is not fully automatic but depends whether the auditory channel is task relevant. Directing attention toward the auditory scene led to an enhanced processing of the environmental sounds. Thus, situational awareness toward environmental sounds can be improved by adding an additional task in the auditory domain

Neuroergonomics. https://doi.org/10.1016/B978-0-12-811926-6.00060-9
Copyright © 2019 Elsevier Inc. All rights reserved.

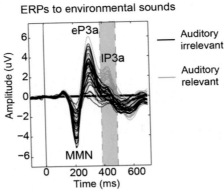

FIGURE 60.1 Grand-average waveform of the ERPs to the environmental sound for the condition when auditory attention is directed toward sounds (*gray*) or not (*black*). Each line represents the electrical potential at one electrode.

to direct the attention toward the auditory scene. Surprisingly, this improvement does not concern early processes like the detection of the unexpected event (reflected in MMN). Thus, our results show that directing the attention toward the auditory scene does not improve the detection and orienting to irrelevant environmental sounds per se, but specifically improves the later processing step of extracting the meaning of the environmental sound that is reflected by the IP3a.[3]

REFERENCES

1. Wickens C, Kramer L, Vanasse L, Donchin E. Performance of Concurrent tasks: a psychophysiological analysis of the reciprocity of information-processing resources. *Science* 1984;**221**(4615):1080–2.
2. Escera C, Alho K, Winkler I, Näätänen R. Neural mechanisms of involuntary attention to acoustic novelty and change. *Journal of Cognitive Neuroscience* 1998;**10**(5):590–604.
3. Escera C, Corral M. The distraction potential (DP), an electrophysiological tracer of involuntary attention control and its dysfunction. In: *The cognitive neuroscience of individual differences*. 2003. p. 63–76.

Chapter 61

Individual Differences and Detection Response Task Reaction Times

Antonia S. Conti*, Moritz Späth*, Klaus Bengler

Technical University of Munich, Garching, Germany

AIMS

The Detection Response Task (DRT) is an applied tool used to measure the "attentional effects of cognitive load."[1] Through a simple reaction-time paradigm[2] performed in parallel with other tasks of interest, the DRT provides reaction times (RTs) and hit rate (HR) as performance metrics. These metrics are interpreted as the degree to which the tasks of interest occupied a test persons' resources, altering the test person's ability to react to a presented signal (ie, the DRT). As RTs can change due to person-based factors,[3] the aim of this experiment was to investigate whether scores associated with different individual-based characteristics were correlated with DRT RTs. The following characteristics were investigated: age, gender, concentration ability, intelligence (speed component), fatigue, and divided attention.

METHODS

Thirty-six persons participated in this experiment. Two age groups were tested: 21–29 years of age ($n = 18$; $M = 23.4$, $SD = 1.72$; nine males) and 64–79 years of age ($n = 18$; $M = 70.7$, $SD = 4.63$; nine males). All participants were volunteers. The Mini-Mental-Status Test (MMST) and Multiple-Choice Vocabulary Intelligence Test (Mehrfachwahl–Wortschatz Intelligenztest; MWT-B) were used to screen participants prior to the experiment. After this phase, a demographic questionnaire was administered before the experiment began. Two experimental blocks were implemented: a psychological test block and a DRT block, presented in a counterbalanced order. In the psychological test block, participants performed to the best of their ability the following tests: Test of Attentional Performance (relevant for divided attention), d2-R Test (relevant for concentration ability), and Number Connection Test (Zahlen–Verbindungs Test; relevant for intelligence), Stanford Sleepiness Scale, and Epworth Sleepiness Scale (both relevant for fatigue). Prior to the DRT block, a vibration motor was placed on the shoulder area of participants and a button was affixed to their left index finger, which was pressed against a steering wheel to respond to the DRT vibration signal. Participants were instructed to respond to the signal as fast and accurately as possible. After participants felt comfortable with the DRT, the DRT block began. In the DRT block, the DRT was tested under four conditions: alone as a baseline, together with a simple or difficult cognitive task (n-back task), or with the difficult n-back task, plus a continuous-tracking task.

RESULTS

Through median splits, participant data were divided into two groups for each aforementioned psychological category and seven separate 4 (condition) × 2 (median split groups) mixed analyses of variance (ANOVAs) were conducted to evaluate the DRT RTs. Additionally, six bivariate Pearson correlations between the DRT RTs and the psychological test data were conducted. Alpha levels were adapted due to multiple comparisons. Age, intelligence, and concentration ability revealed significant interaction effects, $P < .001$ each, which stemmed from group-specific RT slopes across conditions. No interaction effects were found for gender, divided attention, or fatigue. For all characteristics and split groups, condition had a significant effect on DRT RTs, each $P < .001$. Correlations between RTs and age, intelligence, divided attention, and concentration ability, respectively, were significant with at least $P = .003$, except one condition each for both divided attention and concentration ability, which were nonsignificant.

---

* These authors contributed equally to this work.

Neuroergonomics. http://dx.doi.org/10.1016/B978-0-12-811926-6.00061-0
Copyright © 2019 Elsevier Inc. All rights reserved.

CONCLUSIONS

Generally, younger persons and test persons with higher cognitive abilities reacted faster. Those able to divide their attention better reacted faster on the DRT despite task difficulty. Concentration ability was increasingly relevant for difficult task conditions.

REFERENCES

1. ISO/DIS 17488. *Road vehicles – transport information and control systems – detection-response task (DRT) for Assessing attentional effects of cognitive load in Driving*. Geneva: International Organization for Standardization; 2015. http://www.iso.org/iso/catalogue_detail.htm?csnumber=59887.
2. Luce RD. *Response times: their role in inferring elementary mental organization*. New York: Oxford University Press; 1986.
3. Strayer DL, Cooper JM, Turrill J, Coleman JR, Hopman RJ. *Measuring cognitive distraction in the automobile III: a comparison of ten 2015 in-vehicle information systems*. Washington, DC: AAA Foundation for Traffic Safety; 2015.

Chapter 62

Cognitive Components of Path Integration: Implications for Simulator Studies of Human Navigation

Alastair D. Smith[1], Lydia Dyer[2]

[1]University of Plymouth, Plymouth, United Kingdom; [2]University of Nottingham, Nottingham, United Kingdom

AIMS

Efficient daily navigation is underpinned by path integration, the mechanism by which we use self-movement information to update our location in space. Although this process is well understood at a motor level, it is unclear whether there is a cognitive component to the behavior. I will present data from a triangle-completion paradigm. During the task, participants were required to perform a concurrent verbal task—we were interested in whether the nature of the task affected their path-integration accuracy (for both distance and heading calculations).

METHOD

Participants were blindfolded and led along two legs of a right-angled triangle. They were then asked to walk along a final trajectory back to the starting position, completing the triangle. In a within-subjects design, we manipulated concurrent verbal task: in one condition participants sequentially counted their steps aloud and in another they generated random numbers.

RESULTS

Analysis of distance error revealed an effect of concurrent task, with significantly poorer performance on random-number trials. In contrast, analysis of heading error revealed no difference between conditions.

CONCLUSIONS

These data illustrate a dissociation between distance and heading calculations, with distance being more susceptible to interference from a concurrent task. This suggests a cognitive component to path integration, which has been argued to be a uniquely human solution. The findings also carry implications for simulator studies of human navigation that require participants to perform cognitive tasks while learning virtual environments.

Neuroergonomics. http://dx.doi.org/10.1016/B978-0-12-811926-6.00062-2
Copyright © 2019 Elsevier Inc. All rights reserved.

Chapter 63

Vector-Based Phase Analysis Approach for Initial Dip Detection Using HbO and HbR

Amad Zafar, Keum-Shik Hong, Muhammad J. Khan
Pusan National University, Busan, Republic of Korea

INTRODUCTION

The initial dip, also referred to as the fast response or the early deoxygenation, is the small decrease in the concentration change in oxyhemoglobin (ΔHbO) or increase in the concentration of deoxyhemoglobin (ΔHbR) at the locus of neural activity.[1] This early deoxygenation has been reported to occur prior to the increase of blood oxygenation. However, due to its small amplitude and short duration, it has been difficult to detect. The initial work related to its detection results in vector-based phase analysis.[2]

AIMS

A new threshold circle to minimize possible misclassification of initial dips in the functional near-infrared spectroscopy (fNIRS) signals using the vector-based phase analysis is investigated. In contrast to the work in Hong and Naseer[3] (i.e., the square root of the sum of the squares of oxy- and deoxyhemoglobins), the peak value of oxy- or deoxyhemoglobin during the resting state is used.

METHODS

Mental arithmetic was used as an activity for five healthy subjects in the experiment. The paradigm consists of 120-s rest for the initial baseline and 40 s for the trial. The trial was further divided into 10 s and 30 s for the activity task and rest, respectively. The brain signals were acquired using a frequency domain fNIRS system (ISS Imagent, ISS Inc.) at a sampling rate of 31.25 Hz. Eight sources were used to investigate each side of the prefrontal cortex. The total eight channels of a source–detector pair were formed. To detect initial dips, vector-based phase analysis method based on an orthogonal-vector coordinate plane defined by ΔHbO and ΔHbR signals were used.[2] To minimize any misclassification of initial dips, we incorporated a threshold circle as a decision criterion in the vector-based phase analysis. The radius of the threshold circle was set to the value of maximum peak detected in ΔHbO and ΔHbR signals in the final 60 s of the initial baseline (during resting state) of 120 s for each channel.

RESULTS

In comparison to a previous study,[3] in which the maximum value of magnitude during the resting period was used as a decision criterion, the peak value of HbO or HbR [i.e., max(HbO) or max(HbR)] during the resting state is used to draw the threshold circle. With the new criterion, the radius of the circle becomes smaller than that in Hong and Naseer,[3] and earlier detection of initial dips are possible. For the given arithmetic task, the channels of detecting the initial dips were not the same over the subjects, which reflect that the activated brain region in association with the performed task spreads in the prefrontal cortex. In addition, the new method can further reduce the misinterpretation of large variations in the resting state and even during the task period. This new criterion is suitable for the online applications. In future, further studies will be done on the classification of the initial dips detected in the same or different brain regions.

Neuroergonomics. http://dx.doi.org/10.1016/B978-0-12-811926-6.00063-4
Copyright © 2019 Elsevier Inc. All rights reserved.

CONCLUSIONS

An initial dip detection strategy using a threshold circle as a decision criterion incorporated in the vector-based phase analysis method was proposed. The radius of threshold circle was set to the peak value detected in the ΔHbO and ΔHbR signals during the resting period. The results showed that this decision criterion works well to minimize false detection of initial dips from the fNIRS signals.

ACKNOWLEDGMENTS

This work was supported by the National Research Foundation of Korea under the Ministry of Science, ICT and Future Planning, Korea (grant no. NRF-2014 –R1A2A1A10049727).

REFERENCES

1. Frostig RD, Lieke EE, Tso DY, Grinvald A. Cortical functional architecture and local coupling between neuronal activity and the microcirculation revealed by in vivo high-resolution optical imaging of intrinsic signals. *Proceedings of the National Academy of Sciences of the United States of America* 1990;**87**(16):6082–6.
2. Yoshino K, Kato T. Vector-based phase classification of initial dips during word listening using near-infrared spectroscopy. *Neuroreport* 2012;**23**(16):947–51.
3. Hong K-S, Naseer N. Reduction of delay in detecting initial dips from functional near-infrared spectroscopy signals using vector-based phase analysis. *International Journal of Neural Systems* 2016;**226**(3):1650012.

Chapter 64

Subjective and Objective Methods to Continuously Monitor Workload

Horia A Maior, Sarah Sharples, Max L. Wilson
University of Nottingham, Nottingham, United Kingdom

As technology pervades our everyday life, tasks are increasingly *"dominated by mental rather than physical task components."*[1] Moreover, with the recent advancements in technology, the human role has moved toward a supervisory and decision-making one, this further increasing the demands on our limited mental resources. It is, therefore, crucially important to consider, measure, and evaluate human limitations in terms of their cognition.

In Human–Computer Interaction (HCI), we are concerned with understanding and assessing users' Mental Workload (MWL) to evaluate the demands placed upon them while interacting with computer-based systems. There are a variety of subjective and objective methods used for measuring MWL including primary and secondary task analysis, physiological or psychophysiological techniques, as well as user opinions using subjective techniques.[1] However, to be useful for HCI, the methods of collecting useful data about users should allow normal interaction and relatively unrestricted users. Functional Near-InfraRed Spectroscopy (fNIRS) has recently been shown more suitable for allowing insights into MWL during HCI user studies compared to other brain-sensing techniques.[2,3]

In this paper we present a comparison between two methods for continuously assessing workload:

- The instantaneous self-assessment of workload technique (ISA), using subjective ratings of workload based on 30-s interval periods during the task, and
- Measures of oxygenation in the prefrontal cortex (PFC) using fNIRS.

We found strong correlations between task demand and both ISA continuous subjective scale (see Fig. 64.1A) and fNIRS oxygenation in the PFC (see Fig. 64.1B), with ISA requiring user's ability to self-report, and fNIRS being an independent-of-user, consistent, objective reference to MWL.

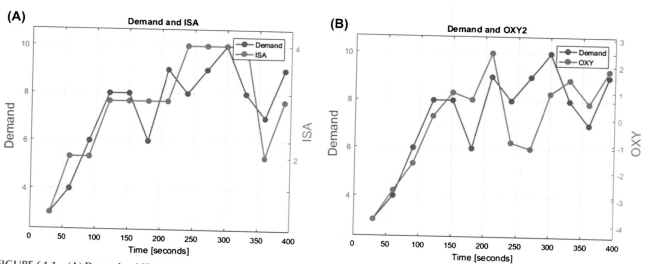

FIGURE 64.1　(A) Demand and ISA. (B) Demand and OXY (Filtered fNIRS data)

Neuroergonomics. https://doi.org/10.1016/B978-0-12-811926-6.00064-6
Copyright © 2019 Elsevier Inc. All rights reserved.

REFERENCES

1. Sharples S, Ted M. Definition and mesurement of human workload. In: Wilson John R, Sharples S, editors. *Evaluation of human work*. CRC Press; 2015.
2. Lukanov K, Maior HA, Wilson ML. Using fNIRS in usability testing: understanding the effect of web form layout on mental workload. In: *Proceedings of the 2016 CHI conference on human factors in computing systems*. ACM; May 2016. p. 4011–6.
3. Maior HA, Pike M, Sharples S, Wilson ML. Examining the reliability of using fNIRS in realistic HCI settings for spatial and verbal tasks. In: *Proceedings of the 33rd annual ACM conference on human factors in computing systems*. ACM; April 2015. p. 3039–42.

FURTHER READING

1. Jordan CS, Brennen SD. *Instantaneous self-assessment of workload technique (ISA)*. [4] using subjective ratings of workload based on 30-s interval periods during the task, and, 1992. Retrieved from: http://www.skybrary.aero/bookshelf/books/1963.pdf.

Chapter 65

Decision Making and Executive Functioning in Aortic Valve Resection: About the Design of an Ergonomic Aortic Valve Resection Tool

René Patesson[1], Eric Brangier[1,2]

[1]*Université Libre de Bruxelles, Bruxelles, Belgique;* [2]*Université de Lorraine, Metz, France*

Aortic stenosis is the most important acquired heart disease, with a prevalence of 4.8% in patients aged more than 75 years and represent more than 60% of the indications for cardiac surgery in elderly patients. Regardless of the good results reported with conventional aortic valve replacements (AVRs) in elderly patients, still many patients are denied surgery by the high operative risk. In this frail population, Transcatheter Aortic Valve Implantation (TAVI) can be a good compromise to achieve good results and minimize morbidity and mortality. This technique allows implanting a balloon-expandable bioprosthesis, without resecting the native aortic valve. Several complications are described as consequence of the residual, highly calcified valve squeezed between the aortic wall and the stent of the implant. In this perspective, the concept of endovascular resection of the aortic valve, before TAVI, was developed and enhanced by Astarci, Glineur, Elkhoury, and Raucent.[1]

Our paper aims to present an accurate ergonomic analysis of the design of an aortic valve resection tool.

Ergonomics also studies the impact of the design on the safe and effective use of devices, especially in the medical field. Errors in the use of these devices are often linked, at least, to the design of user devices that have a lack of compatibility between physical dimensions of technologies, cognitive aspects of operators, work organization, and tasks that the operators have to achieve. It is, therefore, important that medical devices are designed by adding expertise in ergonomics in addition to those from biomedical engineering, to ensure an efficient, usable, and safe device. However, requirements to include ergonomics in the design process are often implicit. Our project makes these requirements explicit, and applies to the design process of a specific product or a device.

For this project, it becomes necessary to design a new device for a procedure that does not yet exist. The future device will integrate new functions, the future users of which cannot imagine today, such as cutting a valve in few seconds, removing the native leaflets in a short time, and positioning an aortic valve in a click, while the patient is under rapid pacing. From an ergonomic point of view, this project is particularly complex and delicate. The complex task must be conducted in a very short time (60s, because the patient is under rapid pacing, and his/her heart has been stopped). In fact, the method habitually used for TAVI requires rapid ventricular pacing at 180–220 beats per minute, which lowers the systolic blood pressure to ≤60mmHg. The heart then beats so fast that it "freezes" and the blood no longer circulates, which also blocks blood circulation in the brain. As a result, such blockage can only be very limited in time—to 1 min—and the surgeon is working within a highly restricted time frame. In this task, decision-making and executive functioning are impaired with potentially serious consequences, particularly when humans interact with a dynamic, uncertain, or stressing environment. Consequently, to have the maximum of relevant information to design an ergonomic device, it is necessary to accumulate knowledge about ergonomics in operating rooms by doing real observations (filmed and analyzed with Actogram Kronos software) of valve operations through transapical access.

Neuroergonomics. http://dx.doi.org/10.1016/B978-0-12-811926-6.00065-8
Copyright © 2019 Elsevier Inc. All rights reserved.

Our main goal is that surgeons and assistants will be able to understand the principles of each function of our tool to fit better in their work environments, their decision-making processes, executive processes, and psychosocial practices to ensure a high level of performances of the device. From this perspective, our approach has to:

- Build a cognitive model of the task analysis of valve operations through transapical access;
- Model the cognitive task by means of a hierarchical model of the task that was then created using K- Mad-e software;
- Provide an ergonomic perspective on the future design cycle by incorporating a user-centered approach;
- Define the new tasks related to people in operating rooms;
- Characterize the operating modes by a cognitive analysis;
- Model the overall task and especially model the "minute" corresponding to the resection and placement of the valve (when the heart is under rapid pacing);
- Provide a time estimation of each subtask;
- Find a simple representation of the procedure, easy to understand, with a first approach to possible errors, to reduce difficulties;
- Design an efficient tool in collaboration with engineers involved in the project;

This led to several models, mockups, three-dimensional (3D) printed parts, and finally a demonstrator based on the ergonomics principles and task analysis. Ergonomic specifications have been concretely translated by the development of four mockups and finally a usable demonstrator.

REFERENCE

1. Astarci P, Glineur D, Elkhoury G, Raucent B. A novel device for endovascular native aortic valve resection for transapical transcatheter aortic valve implantation. *Interactive Cardiovascular and Thoracic Surgery* 2012;**14**:378–80.

FURTHER READING

1. Patesson R, Brangier E. The ten characteristics of the critical task. Ergonomic analysis of vitality requirements in aortic valve surgery. In: Duffy V, editor. *Digital human modeling: applications in health, safety, ergonomics and risk management*. 2016. p. 42–53. http://dx.doi.org/10.1007/978-3-319-40247-5_5.
2. Patesson R, Brangier E. Analyse d'une tâche critique provoquée en chirurgie cardiaque : perspectives ergonomiques pour la pose d'endovalve. In: *Actes de la conférence Ergo'IA 2016. Bidart Biarritz.6, 7, 8 juillet 2016*. 2016.

Chapter 66

Short-Duration Affective States Induced by Emotional Words Improve Response Inhibition: An Event-Related Potential Study

Magdalena Senderecka, Michal Ociepka, Magdalena Matyjek, Bartlomiej Kroczek
Jagiellonian University, Krakow, Poland

AIM

The impact of emotions on higher level cognitive processes, such as memory, attention allocation, planning, and decision-making, has been reflected in the results of a wide range of studies. However, relatively less is known about emotion's influence on response inhibition. This function refers to the ability to supress thoughts and actions that are inappropriate in a given context, and has crucial role in determining human behavior.[1] The aim of the present study was to examine the influence of short-duration affective states induced by emotional words on response inhibition using event-related potentials.

METHOD

Thirty-one participants (20 females, mean age 23.4 years), performed an emotional stop-signal task that required response inhibition to negative (e.g., anger, death, punishment), positive (e.g., love, miracle, promotion) or neutral words (e.g., feature, product, document), which acted as the stop-signal. Emotional stop-signals, consisting of 27 negatively- and 27 positively-valenced, high arousing nouns, were retrieved from the Nencki Affective Word List (NAWL[2]). Neutral stop-signals consisted of 27 neutrally- valenced, low arousing nouns from the same database. Nouns from three categories were matched for imageability ratings and several objective psycholinguistic features as frequency and number of letters.

RESULTS

The behavioral data revealed that negative words facilitated inhibitory processing by decreasing the stop-signal reaction time and increasing the inhibitory rate relative to neutral ones. The P3 amplitude was more pronounced in successfully inhibited trials than in unsuccessfully inhibited trials for all categories of words, however, the size of this effect was largest for negative stop-signals.

Additionally, the LPP component, reflecting the continued allocation of attention to stimuli, was significantly enhanced in the negative trials compared to the positive and neutral trials.

CONCLUSIONS

These results support the hypothesis that negative, arousing words improve cognitive control operations and suggest a positive influence of short-duration affective states, induced by verbal stimuli, on response inhibition.

Neuroergonomics. http://dx.doi.org/10.1016/B978-0-12-811926-6.00066-X
Copyright © 2019 Elsevier Inc. All rights reserved.

ACKNOWLEDGMENTS

This work was supported by an Opus 10 grant 2015/19/B/HS6/00341 from the National Science Centre of Poland awarded to Magdalena Senderecka.

REFERENCES

1. Mostofsky SH, Simmonds DJ. Response inhibition and response selection: two sides of the same coin. *Journal of Cognitive Neuroscience* 2008;**20**:751–61.
2. Riegel M, Wierzba M, Wypych M, Zurawski L, Jednorog K, Grabowska A, Marchewka A. Nencki Affective Word List (NAWL): the cultural adaptation of the Berlin Affective Word List – Reloaded (BAWL-R). *Behavior Research Methods* 2015;**47**:1222–36.

Chapter 67

Predicting Audience Preferences for Television Advertisements Using Functional Brain Imaging

Atahan Agrali[1], Siddharth Bhatt[1], Rajnesh Suri[1], Kurtulus Izzetoglu[1], Banu Onaral[1], Hasan Ayaz[1,2,3]

[1]Drexel University, Philadelphia, PA, United States; [2]University of Pennsylvania, Philadelphia, PA, United States; [3]Children's Hospital of Philadelphia, Philadelphia, PA, United States

Consistent with the Neuroergonomics approach, brain activities measured in response to viewing of natural stimuli in everyday settings could be used to assess audience preferences.[1,2] Neuroimaging tools, such as functional magnetic resonance imaging (fMRI) and Electroencephalogram (EEG), have been utilized in proof-of-concept marketing studies; however, various challenges such as high operating cost, restrictions on the user during data acquisition, and the efficiency of sensor setup, limit further use of these in large-scale deployment as well as in actual field conditions. Functional near-infrared spectroscopy (fNIRS) is the youngest and still an emerging neuroimaging technique that utilizes near-infrared light to measure oxygenation changes in the outer cortex. Latest generation of optical brain imaging utilizes wearable and wireless sensor pads to enable measurement of brain activity in nontethered and ambulatory settings.[3]

In this study, we record brain activity from a group of naive individuals while viewing popular, previously broadcast television content, Super Bowl advertisements. The Super Bowl is the largest sporting event in the United States with more than 100 million viewers each year with well-documented preferences from large audience. We have utilized fNIRS to monitor anterior prefrontal cortex of 11 volunteers. Participants viewed 30 Super Bowl advertisements (30 s videos) of which 15 were ranked highest (high-rated) and the other 15 ranked lowest (low-rated) based on USA Today's Ad Meter scores compiled from thousands of online viewers' self-reported measures. In our protocol, each advertisement followed by a set of self-reported measures to further capture the likeability, participant's familiarity with the advertisement, intend to purchase and recommend for the products/services of the advertised brand and the level of excitement of the advertisement.

Preliminary results indicate that brain activation predicts the self-reported rating by a broad audience. Consistently high cortical oxygenated-hemoglobin concentration changes occurred during the viewing episode of low-rated advertisement videos. Data acquisition for the study is still ongoing and further analysis will compare if ratings of the larger audience are more related to the accuracy or brain activity than those of the individuals from whom the brain data were obtained (Fig. 67.1).

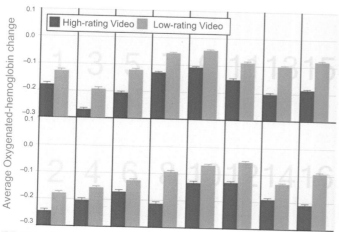
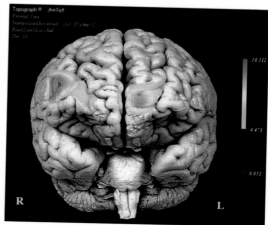

FIGURE 67.1 Average oxygenated-hemoglobin changes in prefrontal cortex while watching the advertisements indicate consistent pattern across high- and low-rated videos (left). F-statistics map of significant difference (right).

Neuroergonomics. https://doi.org/10.1016/B978-0-12-811926-6.00067-1
Copyright © 2019 Elsevier Inc. All rights reserved.

265

REFERENCES

1. Ariely D, et al. Neuromarketing: the hope and hype of neuroimaging in business. *Nature Reviews Neuroscience* 2010;**11**(4):284–293.
2. Plassmann H, et al. Consumer neuroscience: applications, challenges, and possible solutions. *Journal of Marketing Research (JMR)* 2015;**52**(4):427–435.
3. Ayaz H, et al. Continuous monitoring of brain dynamics with functional near infrared spectroscopy as a tool for neuroergonomic research: empirical examples and a technological development. *Frontiers in Human Neuroscience* 2013;**7**:1–13. https://doi.org/10.3389/fnhum.2013.00871.

Chapter 68

When Does the Brain Respond to Information During Visual Scanning?

Nina Flad[1,2], Heinrich H. Bülthoff[1], Lewis L. Chuang[1]

[1]Max-Planck Institute for Biological Cybernetics, Tübingen, Germany; [2]IMPRS for Cognitive and Systems Neuroscience, Tübingen, Germany

AIMS

High-stress work environments such as a flight deck (or surveillance systems) present operators with multiple instruments that have to be constantly monitored with eye movements. Eye tracking allows us to infer when an operator's overt attention has been assigned to an instrument, namely when fixation begins. However, the brain could already be processing information prior to its fixation. When does the brain respond to information when the operator is free to scan the environment freely? Is it the emergence of a target stimulus? Or rather the fixation on a target stimulus?

METHODS

In our study, participants were required to continuously monitor four regions-of-interest (ROIs) that presented three-letter-strings and to respond with a key press to the appearance of a target string. This is comparable to tasks such as instrument monitoring. Allowing for self-paced visual scanning gave rise to two different conditions in our study. These two conditions differ with respect to the sequence of target appearance and target fixation: (1) There was a fixation on the target position before the target appeared. (2) The target appeared before there was a fixation on the target. In the latter condition, the event evoking a change in the electroencephalogram (EEG) signal could either be the target onset or the fixation onset (ie, start of target fixation). All event-related potentials (ERPs) from the naturalistic visual scanning scenario were compared to the event-related potential (ERP) of a baseline condition that only had one ROI and prohibited eye movements.

RESULTS

Our results show that ERPs that were attributed to the target onset were similar, regardless of eye movements (see Fig. 68.1, black, red, blue lines). In contrast, the ERP that was attributed to the fixation onset was atypical (pink line).

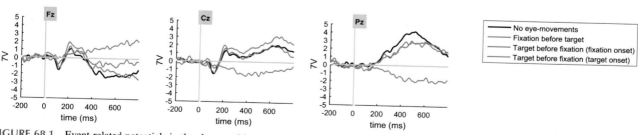

FIGURE 68.1 Event-related potentials in the absence (*black line*) or presence of eye movements. Fixation on the target could occur before the target appeared (*red line*) or afterward. In the latter case, attributing the potential to fixation onset (*magenta line*) leads to a response that differs from all other responses, including the potential attributed to target onset (*blue line*).

Neuroergonomics. http://dx.doi.org/10.1016/B978-0-12-811926-6.00068-3
Copyright © 2019 Elsevier Inc. All rights reserved.

CONCLUSION

It is commonly assumed that a visual stimulus is processed when it is fixated. However, we demonstrate that visual perception can take place even prior to fixation. Experimental studies have shown how far peripheral vision is sufficient for recognizing animals in complex scenes.[1] Here, we show that target onsets gave rise to brain responses even before the targets were fixated. This poses a challenge for the use of EEG/ERP in visual scanning environments. If fixation onset is not necessarily the onset of perception, they cannot always be used for attributing ERP data as has been the assumption of previous research[2,3] in natural scene viewing and reading. Future studies that seek to employ EEG/ERP measures in visual scanning work environments should take into account that the brain can respond to to-be-fixated information prior to fixation and to exercise caution in determining when this is.

REFERENCES

1. Thorpe SJ, Gegenfurtner KR, Fabre-Thorpe M, Buelthoff HH. Detection of animals in natural images using far peripheral vision. *European Journal of Neuroscience* 2001;**14**(5):869–76.
2. Dimigen O, Sommer W, Hohlfeld A, Jacobs AM, Kliegl R. Coregistration of eye movements and EEG in natural reading: analyses and review. *Journal of Experimental Psychology: General* 2011;**140**(4):552–72.
3. Rämä P, Baccino T. Eye fixation-related potentials (EFRPs) during object identification. *Visual Neuroscience* 2010;**27**(5–6):187–92.

Chapter 69

Brain–Computer Interface: Analysis of Different Virtual Keyboards for Improving Usability

Liliana Garcia[1], Véronique Lespinet-Najib[1], Mathilde Menoret[1], Bernard Claverie[1], Jean M. André[1], Ricardo Ron-Angevin[2]

[1]ENSC-Bordeaux INP, Bordeaux, France; [2]University of Málaga, Málaga, Spain

Brain–computer interface (BCI) technology translates voluntary choices in active command using brain activity. In fact, brain electrical signals, particularly Evoked Related Potentials (ERPs), produced some ms after cognitive tasks, are frequently used to activate commands, being the visual-P300 based BCI system the main interface used for communication and control purpose. In the P300-speller BCI, the user's task consists in visualizing a matrix of 6×6 rows and columns and then focusing attention on a desired character (target, rare event). Since its implementation, some authors have carried out different research for optimizing the P300-visual speller BCI to increase accuracy performance or improve usability (ie,[1,2]). Until now, no studies have been conducted for manipulating speller sizes to ensure best conditions to user experience.

The aim of this study was to analyze the effect of different spellers sizes on participants' usability while using a visual P300 BCI system based on the one developed by Farwell and Donchin.[3] The main objective was to better understand how easily a user can carry out the task comfortably and efficiently by analyzing the same speller system type while using different BCI speller sizes to identify problems that users could have interacting with its complex system.

A group of healthy subjects participated in this study. Three different speller sizes were evaluated. After a calibration phase, participants had to spell some words. Subject's interaction with the system was tested by assessing the effect of the system on performance, workload [based on National Aeronautics and Space Administration Task Load Index (NASA-TLX) test], fatigue, motivation, and other subjective measures. P300 parameters were analyzed to find robust and reproducible physiological markers allowing us objective assessment during the task.

The obtained results show significant difference in performance between the three different speller sizes, with the largest speller size being the one with the highest error percentage. Regarding subjective measures, the obtained results suggest the largest speller size as the favorite due to its bigger symbol size and then its ease of use. Combining all the results, the proposed medium-size speller seems to be the most suitable to increase usability.

The study carried out in this experiment shows how important it is to consider the ergonomics of the interface to obtain higher usability. In this particular study, three different speller sizes were evaluated. It is important to note that subjective measures did not correlate with the best performance. The obtain results have important implications for the development of future speller configuration-based BCI technology.

ACKNOWLEDGMENTS

This work was partially supported by the University of Málaga, by the Spanish Ministry of Economy and Competitiveness (MINECO) through the project LICOM [DPI2015-67064-R (MINECO/ European Fund of Regional Development FEDER)] and by the European Regional Development Fund (ERDF).

Neuroergonomics. http://dx.doi.org/10.1016/B978-0-12-811926-6.00069-5
Copyright © 2019 Elsevier Inc. All rights reserved.

REFERENCES

1. Sellers EW, Donchin E. A P300-based brain–computer interface: initial tests by ALS patients. *Clinical Neurophysiology* 2006;**117**(3):538–48.
2. Thulasidas M, Guan C, Wu J. Robust classification of EEG signal for brain-computer interface. *IEEE Transactions on Neural Systems and Rehabilitation Engineering* 2006;**14**(1):24–9.
3. Farwell L, Donchin E. Talking off the top of your head: toward a mental prosthesis utilizing event related brain potentials. *Electroencephalography and Clinical Neurophysiology* 1988;**70**(6):510–23.

Chapter 70

A Method for Prediction of Behavioral Errors From Single-Trial Electrophysiological Data

Hiroki Ora, Yoshihiro Miyake

Tokyo Institute of Technology, Yokohama, Japan

AIMS

Behavioral errors may cause serious results. For example, while driving a car, if the intention is to stop and the accelerator pedal is mistaken for the brake pedal, a serious traffic accident may result. If electrophysiological phenomena that involve behavioral errors are identified, we may be able to avoid behavioral errors by detecting the error precursor. We have already revealed a neural sign of behavioral errors in our previous study.[1] In the previous study, we applied a spatiotemporal analysis to high-density electroencephalogram (EEG) signals recorded during a visual discrimination task, a d2 test of attention. We demonstrated that, during trials with error outcomes, positive deviation of scalp amplitude with latency of ~ 30 ms was observed in frontal regions, positive deviation with latency of ~ 125 ms was observed in parietal regions, and then a positive deviation with latency of ~ 160 ms was observed in the occipital region. In this study, we propose a single-trial prediction method of behavioral errors by detecting the electrophysiological sign of behavioral errors.

METHODS

We used nonlinear support vector machine (SVM) to predict behavioral errors from electrophysiological signals just before reactions. To evaluate the method, we applied the method to the data reported in our previous study. In the previous study, participants ($n = 10$) performed a d2 test of attention during EEG recording (128 channels).

RESULTS

The nonlinear SVM classifier was able to detect trials with error outcomes (mean AUC$=0.72$) from single-trial electrophysiological data.

CONCLUSION

The results suggest that the nonlinear SVM classifier was able to predict trials with error outcomes during d2 test of attention. With progress of the line of this study, we may be able to avoid behavioral errors by recognizing the error precursor.

REFERENCE

1. Ora H, Sekiguchi T, Miyake Y. Dynamic scalp topography reveals neural signs just before performance errors. *Sci Rep* 2015;**5**:12503. doi:10.1038/srep12503.

Neuroergonomics. http://dx.doi.org/10.1016/B978-0-12-811926-6.00070-1
Copyright © 2019 Elsevier Inc. All rights reserved.

Chapter 71

Neuro-Functional Correlates of the Out-of-the-Loop Performance Problem: Impact on Performance Monitoring

Bertille Somon[1,2], Aurélie Campagne[2], Arnaud Delorme[3,4], Bruno Berberian[1]

[1]Office National d'Etudes et de Recherche Aérospatiales, Salon-de-Provence, France; [2]Université Grenoble Alpes, CNRS, LPNC UMR 5105, F-38000, Grenoble, France; [3]Université Toulouse III – Paul Sabatier, CNRS, CerCo, Toulouse, France; [4]University of California, San Diego, CA, United States

The Out-Of-the-Loop (OOL) phenomenon represents a key challenge in neuroergonomics, and understanding its impact on human operators is crucial for successful design of new automated systems. At the behavioral level, one of the direct consequences of the OOL performance problem, is an insufficient monitoring and checking of the state of the system.[1] This monitoring failure is difficult to characterize. Recent works suggest that event-related potentials[2–4] as well as neural sources of self-performance monitoring[5,6] are similar to the ones observed when we monitor the actions of others. However, in the aeronautics field, the operator most often monitors an automated system and not his own actions nor the ones of another human agent. Our electroencephalogram (EEG) study aims (1) at determining whether the neural correlates of performance monitoring are similar when we supervise our actions, the ones of another human agent or of an artificial system; and (2) at characterizing the physiological markers of performance monitoring during the OOL performance problem.

In laboratory conditions, participants performed a modified Erisken flanker task (which is error-prone) according to three types of supervision: self-monitoring, monitoring of another human agent, and monitoring of an automated system. The OOL performance problem was induced by manipulating the difficulty and duration of the task. A better characterization and quantification of the OOL performance problem could help in monitoring mental state of operators and pilots in real time to prevent this phenomenon based on their cerebral activity.

ACKNOWLEDGMENTS

This project is supported by a grant of the Région Provence-Alpes-Côte d'Azur and the French Aerospace Lab (ONERA) Salon de Provence.

REFERENCES

1. Kaber DB, Endsley MR. Out-of-the-loop performance problems and the use of intermediate levels of automation for improved control system functioning and safety. *Process Safety Progress* 1997;**16**(3):126–31. http://dx.doi.org/10.1002/prs.680160304.
2. van Schie HT, Mars RB, Coles MGH, Bekkering H. Modulation of activity in medial frontal and motor cortices during error observation. *Nature Neuroscience* 2004;**7**(5):549–54. http://dx.doi.org/10.1038/nn1239.
3. Koban L, Pourtois G, Vocat R, Vuilleumier P. When your errors make me lose or win: event-related potentials to observed errors of cooperators and competitors. *Social Neuroscience* 2010;**5**(4):360–74. http://dx.doi.org/10.1080/17470911003651547.
4. de Bruijn ERA, von Rhein DT. Is your error my concern? An event-related potential study on own and observed error detection in cooperation and competition. *Frontiers in Neuroscience* 2012;**6**. http://dx.doi.org/10.3389/fnins.2012.00008.
5. Bonini F, Burle B, Liegeois-Chauvel C, Regis J, Chauvel P, Vidal F. Action monitoring and medial frontal cortex: leading role of supplementary motor area. *Science* 2014;**343**(6173):888–91. http://dx.doi.org/10.1126/science.1247412.
6. Cracco E, Desmet C, Brass M. When your error becomes my error: anterior insula activation in response to observed errors is modulated by agency. *Social Cognitive and Affective Neuroscience* September 2015. http://dx.doi.org/10.1093/scan/nsv120.

Neuroergonomics. http://dx.doi.org/10.1016/B978-0-12-811926-6.00071-3
Copyright © 2019 Elsevier Inc. All rights reserved.

Chapter 72

Recent Advances in EEG-Based Neuroergonomics for Human–Computer Interaction

Jérémy Frey[1,2], Martin Hachet[2], Fabien Lotte[2]

[1]University of Bordeaux, Bordeaux, France; [2]INRIA Bordeaux Sud-Ouest/LaBRI, Talence, France

Human–Computer Interfaces (HCIs) are increasingly ubiquitous in multiple applications including industrial design, education, art, or entertainment. As such, HCIs could be used by very different users, with very different skills and needs. This thus requires user-centered design approaches and appropriate evaluation methods to maximize User eXperience (UX). Existing evaluation methods include behavioral studies, test beds, questionnaires, and inquiries, among others. Although useful, such methods suffer from several limitations as they can be either ambiguous, lack real-time recordings, or disrupt the interaction. Neuroergonomics can be an adequate tool to complement traditional evaluation methods. Notably, Electroencephalography (EEG)-based evaluation of UX has the potential to address these limitations, by providing objective, real-time, and nondisruptive metrics of the ergonomics quality of a given HCI.[1] We present here an overview of our recent works in that direction. In particular, we show how we can process EEG signals to derive metrics characterizing (1) how the user perceives the HCI display (HCI output) and (2) how the user interacts with the HCI (HCI input).

First, at the output level, we conducted some experiments to study the visual comfort experienced by users of stereoscopic displays (eg, 3D TV). We showed them several objects presented at different stereoscopic depths in front or behind the screen while recording their EEG signals. Some of these depths were voluntarily uncomfortable (either too close or too far from the eyes). By analyzing Event-Related Potentials (ERPs) following a stereoscopic object appearance using advanced EEG signal-processing techniques, we showed we could discriminate, in a single trial (ie, in 1s of signal), ERPs corresponding to comfortable displays versus ERPs corresponding to uncomfortable ones.[2]

Second, we asked our users to perform different cognitive tasks involving different levels of cognitive workload (ie, mental efforts), while recording their EEG signals. Here as well, by using signal processing and suitable machine learning, we could discriminate low mental workload from high mental workload in short windows (2s of signal) of EEG.[3] By refining these algorithms we showed we could then discriminate mental efforts in complex interaction tasks, notably during 3D object manipulation tasks,[4] as well as during navigation tasks in a video game.[5] We also showed we could use such methods to compare the mental workload induced by different interaction devices or techniques.

Overall, these recent results suggest that EEG can be used as an objective and complementary evaluation technique to characterize and assess different HCIs. This thus opens the door to a new generation of HCIs, designed by exploiting EEG-based neuroergonomics.

REFERENCES

1. Frey J, Mühl C, Lotte F, Hachet M. Review of the use of electroencephalography as an evaluation method for human-computer interaction. In: *Proc PhyCS*. 2014.
2. Frey J, Appriou A, Lotte F, Hachet M. Classifying EEG signals during stereoscopic visualization to estimate visual comfort. *Computational Intelligence and Neuroscience* 2016a.
3. Mühl C, Jeunet C, Lotte F. EEG-based workload estimation across affective contexts. *Frontiers in Neuroscience* 2014;**8**:114.
4. Wobrock D, Frey J, Graef D, de la Riviére J-B, Castet J, Lotte F. Continuous mental effort evaluation during 3D object manipulation tasks based on brain and physiological signals. In: *Proc Interact*. 2015.
5. Frey J, Daniel M, Castet J, Hachet M, Lotte F. A framework for electroencephalography-based evaluation of user experience. In: *Proc CHI*. 2016b.

Neuroergonomics. http://dx.doi.org/10.1016/B978-0-12-811926-6.00072-5
Copyright © 2019 Elsevier Inc. All rights reserved.

Chapter 73

Attention and Driving Performance Modulations Due to Anger State: Contribution of Electroencephalographical Data

Franck Techer[1,2], Christophe Jallais[1], Yves Corson[2], Alexandra Fort[1]
[1]Université de Lyon, IFSTTAR, TS2, LESCOT, Lyon, France; [2]University of Nantes, LPPL, Nantes, France

Experiencing negative emotions such as anger in driving can promote driving errors.[1–3] Anger, defined by a negative and highly arousing state, may impact attention and therefore the driving activity. In one hand, anger could have a positive impact on the alerting network efficiency[4] and so may become useful when driving with advanced driving assistance systems (ADAS) providing alerting cues. In another hand, negative emotions[5] are commonly associated with attentional disruptions and mind-wandering so that they may interfere with the use and the efficiency of ADAS. However, the relationship between emotions and attention while using ADAS is not well developed. The aim of this study was to assess the influence of anger on attention and driving performance while using an alerting system. For this purpose, 33 participants completed a simulated driving scenario once in an anger state and once in a natural mood. This scenario consisted in following a motorcycle on a simulated straight rural road. A warning system informed the participants about imminent motorcycle braking. Event-related potentials (ERP) were recorded so as to reflect attentional modulations that may be undetectable with behavioral data. Results indicated that anger impacted driving performance and attention, provoking an increase in lateral variations while reducing the amplitude of the visual N1 peak. To our knowledge, those results are the first to reveal such impact of an anger state on ERPs. However, further research is needed to corroborate this finding as well as assessing the impact of other emotions on ERPs in a driving environment. This kind of physiological data may be used to monitor driver state and provide specific help corresponding to their current needs.

REFERENCES

1. Stephens A, Groeger JA. Driven by anger: the causes and consequences of anger during virtual journeys. In: *Advances in traffic psychology*. Ashgate; 2012. p. 3–15.
2. Abdu R, Shinar D, Meiran N. Situational (state) anger and driving. *Transportation Research Part F: Traffic Psychology and Behaviour* 2012;**15**(5):575–80.
3. Roidl E, Frehse B, Hoeger R. Emotional states of drivers and the impact on speed, acceleration and traffic violations-A simulator study. *Accident Analysis and Prevention* 2014;**70**:282–92. https://doi.org/10.1016/j.aap.2014.04.010.
4. Techer F, Jallais C, Fort A, Corson Y. Assessing the impact of anger state on the three attentional networks with the ANT-I. *Emotion* 2015;**15**(3):276–80. https://doi.org/10.1037/emo0000028.
5. Smallwood J, Fitzgerald A, Miles LK, Phillips LH. Shifting moods, wandering minds: negative moods lead the mind to wander. *Emotion* 2009;**9**(2):271–6. https://doi.org/10.1037/a0014855.

FURTHER READING

1. Deffenbacher JL, Deffenbacher DM, Lynch RS, Richards TL. Anger, aggression and risky behavior: a comparison of high and low anger drivers. *Behaviour Research and Therapy* 2003;**41**(6):701–18.
2. Sullman MJM. The expression of anger on the road. *Safety Science* 2015;**72**:153–9. https://doi.org/10.1016/j.ssci.2014.08.013.

Neuroergonomics. https://doi.org/10.1016/B978-0-12-811926-6.00073-7
Copyright © 2019 Elsevier Inc. All rights reserved.

Chapter 74

Development of Intelligent Early Warning System for Hypoglycemia Attacks

Ali Berkol[1], Emre O. Tartan[1], Gozde Kara[2]

[1]Baskent University, Ankara, Turkey; [2]Steinbeis Advanced Risk Technologies GmbH, Stuttgart, Germany

Hypoglycemia is the clinical syndrome that describes the condition of blood sugar (blood glucose) level less than 70 mg/dL. Its symptoms and signs develop quickly and include nervousness or anxiety, sweating, chills, anger, stubbornness, confusion, blurred and/or impaired vision, lack of coordination, seizures, and unconsciousness. Late treatment gives rise to more severe symptoms, including coma, and may result in death.[1] Most common reason why hypoglycemia happens is the side effect of medications used for diabetes mellitus treatment, such as insulin, sulfonylureas, and biguanides. People without diabetes may suffer hypoglycemia caused by: accidentally taken diabetes drugs, excessive alcohol consumption, some critical illnesses, insulin overproduction due to pancreas tumor and/or hormone deficiencies, severe infections, inborn error of metabolism, hypothyroidism, starvation, etc. Surely, the risk is higher on diabetes with the above factors.

The brain communicates by sending nerve impulses from one cell to another in the form of electrical charges thru neurons. When considering the fact that hypoglycemic seizures start in the brain (because decreased glucose supply causes impairment and dysfunction of neurons), electroencephalography (EEG) is an adequate testing method as it picks up this neural electrical activity. This study develops an early warning diagnostic system against hypoglycemic attacks by analyzing and processing related EEG outputs with metaheuristic soft-computing techniques. Proposed methodology collects the EEG data taken from both healthy and hypoglycemic subjects. Then acquired data will be classified and trained to determine the state of hypoglycemia with due accuracy. Classifications will be made by utilizing some universal classifications, namely Support Vector Machines (SVM), Artificial Neural Networks (ANN), and Adaptive Neuro Fuzzy Inference System (ANFIS). Designed system will ensure the collection of real-time EEG inputs from patient continuously and operate as to distinguish the neural abnormality relative to hypoglycemia by referencing the refined EEG data. Hence, the system is able to detect changes in neural activities typical to hypoglycemic symptoms at the earliest possible instance. Upon verification of a case, the system will send immediate warnings before the patient suffers imminent attack and drive the patient to urgent treatment. Hence, the possible devastating effect of hypoglycemic attack is avoided. Similar studies exist for epilepsy[2] and diabetes but not for this rare case of hypoglycemia. This study has its novelty by designing a predictive health management system particular to hypoglycemia prognostics and further by involving advanced computational techniques as fundamental enablers that leverage the implementation.

ACKNOWLEDGMENTS

We would like to thank Associate Prof. Dr. Hamit Erdem and Mr. Tansel Kasar for their valuable comments.

REFERENCES

1. Haumont D, Dorchy H, Pelc S. EEG abnormalities in diabetic children: influence of hypoglycemia and vascular complications. *US National Library of Medicine National Institutes of Health* December 1979;**18**(12):750–3.
2. Costa RP, Oliveira P, Rodrigues G, Leitão B, Dourado A. "Epileptic seizure classification using neural Networks with 14 Features" Knowledge-based intelligent information and engineering systems volume 5178 of the series lecture notes in computer science. p. 281–288, ISBN: 978-3-540-85564-4.

Copyright © 2019 Elsevier Inc. All rights reserved.

Chapter 75

Out-of-the-loop (OOL) Performance Problem: Characterization and Compensation

Bruno Berberian

Office National d'Etudes et de Recherche Aérospatiales, Salon-de-Provence, France

The world surrounding us has become increasingly technological. Nowadays, the influence of automation is perceived in each aspect of everyday life and not only in the world of industry. Automation certainly makes some aspects of life easier, faster, and safer. Nonetheless, empirical data suggests that traditional automation has many negative performance and safety consequences. Particularly, in cases of automatic equipment failure, human supervisors seemed effectively helpless to diagnose the situation, determine the appropriate solution, and retake control, a set of difficulties called the "out-of-the-loop" (OOL) performance problem. Because automation is not powerful enough to handle all abnormalities, this difficulty in "takeover" is a central problem in automation design.

The OOL performance problem represents a key challenge for both systems designers and human factor society. After decades of research, this phenomenon remains difficult to grasp and treat. Recent tragic accidents remind us of the difficulty for human operators to interact with highly automated systems. The general objective of our research project is to improve our comprehension of the OOL performance problem. To address this issue, we aim (1) to identify the neurofunctional correlates of the OOL performance problem, (2) to propose design recommendations to optimize human–automation interaction and decrease OOL performance problem occurrence. Behavioral data and brain-imaging studies will be used to provide a better understanding of this phenomenon at both physiological and psychological levels.

Regarding the first objective, we propose to use the recent insights in neuroscience about the characterization of the processes of vigilance, attention, and performance monitoring to (1) characterize the degradation of the human activity during OOL episodes and (2) understand the dynamics toward such degraded state. We will focus on (1) the identification of the physiological markers of the performance monitoring functions when engaged in supervisory tasks and the characterization of the degradation of this monitoring function during OOL episodes, (2) the change in the attention and vigilance mechanisms during OOL episodes, with a particular interest for the mind-wandering phenomenon. Taking together, these results will help us to identify both physiological markers and physiological precursors of the OOL phenomenon and help in the monitoring of the operator state regarding this phenomenon.

Regarding the second objective, we propose to use the recent insight in psychology about the mechanism underlying the control of cooperative action to design more collaborative automation technology. Particularly, we assume that using the tools proposed by the framework of agency, ergonomists could design automation interfaces that are more predictable, and therefore more acceptable and more controllable. In this sense, we will try to understand how to design predictable system.

Taking together, this work aims to go further the current knowledge about OOL phenomenon and to clearly modelize the impact of automation on cognitive mechanisms.

ACKNOWLEDGMENTS

This project is supported by an ANR grant (Young researcher program).

Copyright © 2019 Elsevier Inc. All rights reserved.

Chapter 76

The Impact of Visual Scan Strategies on Active Surveillance Performance: An Eye-Tracking Study

Jean-Denis Thériault[1], Benoit Roberge-Vallières[1], Daniel Lafond[2], Sébastien Tremblay[1], François Vachon[1]

[1]Université Laval, Québec, QC, Canada; [2]Thales Research & Technology, Québec, QC, Canada

AIMS

There has been substantial investment in the field of closed-circuit television (CCTV) technology to increase the level of surveillance efficiency and system performance for the security of citizens and the protection of public infrastructures. In control rooms, operators have to actively monitor a number of camera feeds that generally surpass the number of displays available. Operators tend to prioritize some visual scenes to cope with the sheer amount of visual content to monitor.[1]

However, it is still not clear how content prioritization impacts operators' performance as incidents can also occur on cameras that are judged less important. The objective of the present study is to assess the impact on surveillance performance of different monitoring strategies adopted by (nonexpert) operators using a highly realistic CCTV operation simulation. The measurement of gaze location was taken as an indicator of attention allocation over the visual scenes and served to determine the surveillance strategy of participants.

METHODS

Placed in a context of crowd surveillance during an outdoor festival – a major event that poses potential threats to urban security – 65 participants (36 women, 29 men; age: $M = 25$; SD $= 6.7$) were instructed to actively monitor eight camera feeds to detect critical incidents (eg, physical injuries, lost children). Specifically, participants were asked to detect events for which they detained information about what and where to look for, as well as any suspicious or dangerous events that could appear on any of the eight camera feeds. The visual scan strategy used by the participants was assessed objectively by a mathematical index based on the total duration of gazes on each camera feed, and subjectively via self-reported measures.

RESULTS

Results allowed us to identify two distinct surveillance strategies: One that we refer to as "equally distributed" by which participants tend to look at every camera feed equally (under 25% divergence from an equal allocation pattern), and a strategy called "prioritized" in which gaze duration on some camera feeds were longer compared to other feeds. As shown by the behavioral index, more than half of the subjects (58.5%) tended to prioritize some camera feeds. Furthermore, data revealed a 66.2% correspondence between the strategy reported in the questionnaire and scan behavior extracted from the eye-tracking data. This correspondence was more important for prioritizing participants (ratio of 2.58 in favor of the prioritized strategy).

Performance was measured by detection rate and detection speed, and did not significantly differ according to the visual scan strategy employed by the participants, $F(2, 126) = 1.73$, $P = .18$, $\eta_p^2 = 0.03$ and $F(2, 60) = 0.38$, $P = .69$, $\eta_p^2 = 0.01$, respectively.

Neuroergonomics. http://dx.doi.org/10.1016/B978-0-12-811926-6.00076-2
Copyright © 2019 Elsevier Inc. All rights reserved.

CONCLUSIONS

The present study suggests that measurement of eye movements during CCTV monitoring provides an invaluable source of data that is helpful to better understand how CCTV monitoring is performed. It seems that spending more time looking at specific scenes than others does not necessarily increase surveillance performance even if operators have access to information about the type and location of incidents that might occur. Thus, operators should be allowed to choose which strategy to use rather than being imposed by authorities or through training. Before concluding that operators should be free to choose which scanning strategy to use (rather than one strategy being imposed by authorities or through training), further research into the potential effects of the nature of the events to surveil and also of the mental model and intent of the operator is needed.

REFERENCE

1. Wallace E, Diffley C. CCTV: making it work. In: *CCTV control room ergonomics*. Home Office Police Scientific Development Branch; 1998. 14/98.

Chapter 77

Using Neural Correlates for Enhancing Customer Experience Through Effective Visual Price Placement

Rajneesh Suri[1], Nancy M. Puccinelli[2], Dhruv Grewal[3], Anne L. Roggeveen[3], Atahan Agrali[1], Hasan Ayaz[1,4,5], Kurtulus Izzetoglu[1], Banu Onaral[1]

[1]Drexel University, Philadelphia, PA, United States; [2]University of Bath, Claverton Down, Bath, United Kingdom; [3]Babson College, Boston, MA, United States; [4]University of Pennsylvania, Philadelphia, PA, United States; [5]Children's Hospital of Philadelphia, Philadelphia, PA, United States

This research seeks to understand if consumers expect price information accompanying a product in a price promotion to appear in specific spatial location next to the product and how neural correlates could pinpoint the location for such an expectation. Our survey (study 1) of 727 prices for six different retailers suggested that electronics retailers (eg, Best Buy) are likely to present their prices more towards the right side of the product (68 vs. 50%, $z = 9.64$, $P < .001$), whereas discount retailers (eg, Target) present such prices more to the left-hand side of the product (77 vs. 50%, $z = 14.46$, $P < .001$). Thus, although a retail standard does not appear to exist, these results suggest that the location of a price is relevant and price perception is a complex process often impacted by factors beyond economics.[1] Building from research in visual field effects, we find that under low engagement, price information shown on the left will be more salient, causing that price information to dominate in product evaluations (see Fig. 77.1).

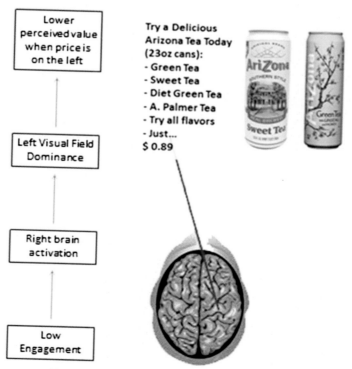

FIGURE 77.1   Price perception under low engagement.

Neuroergonomics. https://doi.org/10.1016/B978-0-12-811926-6.00077-4
Copyright © 2019 Elsevier Inc. All rights reserved.

Conversely, under high engagement, price information shown on the right of the product will be more salient. This will cause that price to be more dominant in evaluations and lead to lower value perception.

Study 2 tested the sales of a product for an actual retailer as a function of the left versus right visual field location of price information for the product. The study was a single three-level factor (visual field location: left, right, control) between designs. Sales data were collected over a 3 wk period and we featured a promotion for a product (ie, Arizona iced tea; see Fig. 77.1 for the signage) by posting a large sign on the front of a product cooler in a location of national convenience store. A price featured in the left visual field led to a nonsignificant increase in sales ($M_{left} = 2.12$, $M_{control} = 1.75$, $t(166) = 1.22$, $P > .22$), whereas a price featured in the right visual field led to significantly more sales ($M_{right} = 2.52$, $M_{control} = 1.75$, $t(166) = 2.19$, $P < .05$). These findings suggest that prices to the right increase buying behavior.

The aim of study 3 was to measure neural correlates as preliminary evidence of left-hemisphere activation under high engagement. Ten right-handed people from the community ($M_{age} = 26$ years, 63% male) participated in the study for a stipend. The study adopted a 2 (engagement: low, high) $\times$ 2 (visual field location: left, right) mixed design. Participants were shown six prices in a row on the screen in front of them. Participants were then shown one of the prices alone and asked to click the mouse on the location on the screen in front of them where that price had appeared previously. To assess the attentional focus of participants we measured the distance between the actual location and the participant's recall of the price's location. Further, to assess the relative level of activation in the left versus right hemisphere during the task, we measured the blood flow in the prefrontal cortex. To examine asymmetric activation in the prefrontal cortex, we used functional near infrared spectroscopy (fNIRS), a noninvasive neuroimaging tool for measuring activity in the prefrontal cortex.[2] As expected participants showed a leftward bias in recall of the price locations under low engagement but not high engagement ($P < .05$). Further, greater blood flow was observed in the anterior left dorsolateral prefrontal cortex (DLPFC) ($F(1, 5) = 22.5$, $P < .01$).

REFERENCES

1. Frederick S, Novemsky N, Wang J, Dhar R, Nowlis S. Opportunity cost neglect. *Journal of Consumer Research* December 2009;**36**:553–61.
2. Ayaz H, Shewokis PA, Bunce S, Izzetoglu K, Willems B, Onaral B. Optical brain monitoring for operator training and mental workload assessment. *Neuroimage* 2012;**59**(1):36–47. https://doi.org/10.1016/j.neuroimage.2011.06.023.

Chapter 78

Human Decision-Making During Crowd Evacuations: The Role of Stress, Conflicting Information, and Social Interactions

Nikolai W.F. Bode

University of Bristol, Bristol, United Kingdom

Emergencies or false alarms make it necessary to quickly and safely evacuate large numbers of people from confined spaces, such as buildings or vehicles. In this context, individuals have to make many decisions, such as when to leave or which exit route to take in often stressful situations and under the influence of a lot of, and potentially conflicting, information. For example, emergency exit signs might indicate exit routes that are different from routes individuals are used to using entering buildings. Additionally, route choices of early evacuees may influence subsequent evacuees. To ensure the safety of evacuees, it is necessary to understand how individuals make these decisions and what affects them. To this end, we have developed experiments in virtual environments.[1-4]

In these virtual environments we tested what affects the decisions of individuals in a safe and highly controlled way. Participants were presented with a top-down view of a building with multiple rooms and they controlled the movement of one pedestrian with mouse clicks, which allowed them to move through the building in the presence of a computer-simulated crowd of pedestrians. In the experiments participants had to escape from the building and choose between different exit routes. By putting participants under time pressure, by changing the behavior of the simulated crowd and by adjusting additional information participants receive (eg, via exit signs or messages), we tested what affects route choices in humans. Across the four experiments,[1-4] we tested the behavior of over 1700 participants.

We found that under a stress-inducing treatment, participants were less able or willing to adjust their original exit choice in the course of the evacuation, even when this was detrimental to their evacuation time.[1,3] For example, participants were less likely to avoid a congested exit by changing their original decision to move toward it. We also tested the effect of combining multiple sources of information, such as signs, crowd movement, and memorized information.[2] In isolation, some of these sources of information did not affect route choices significantly. However, when we combined such unconvincing information sources with additional directly conflicting sources of directional information, we found that the signals participants observed more closely in isolation did not simply overrule alternative sources of directional information.

Our experiments are not a direct test of behavior in real evacuations and care should be taken when extrapolating from our findings. However, they provide useful insights into the role different types of information and stress play in real human decision-making. This is likely to be important for identifying topics for future study on real human crowd movements and for developing more realistic agent-based simulations. At present, this research still focuses on behavioral responses and has not yet progressed to consider brain mechanisms. Therefore, I believe this field could be a useful application area for neuroergonomics that is ripe for investigation.

REFERENCES

1. Bode NWF, Codling EA. Human exit route choice in virtual crowd evacuations. *Animal Behaviour* 2013;**86**(2):347–58.
2. Bode NWF, Kemloh Wagoum AU, Codling EA. Human responses to multiple sources of directional information in virtual crowd evacuations. *Journal of the Royal Society, Interface* 2014;**11**(91):20130904.
3. Bode NWF, Kemloh Wagoum AU, Codling EA. Information use by humans during dynamic route choice in virtual crowd evacuations. *Royal Society Open Science* 2015;**2**:140410.
4. Bode NWF, Miller J, O'Gorman R, Codling EA. Increased costs reduce reciprocal helping behaviour of humans in a virtual evacuation experiment. *Scientific Reports* 2015;**5**:15896.

Neuroergonomics. http://dx.doi.org/10.1016/B978-0-12-811926-6.00078-6
Copyright © 2019 Elsevier Inc. All rights reserved.

Chapter 79

Toward an Online Index of the Attentional Response to Auditory Alarms in the Cockpit: Is Pupillary Response Robust Enough?

Alexandre Marois, Johnathan Crépeau, Sébastien Tremblay, François Vachon
Université Laval, Québec City, QC, Canada

AIMS

Auditory alarms are widely used in aviation to provide information or elicit attention (re)orienting toward emergency situations. However, a high number of incidents result from an absence of reaction to theses alarms.[1] This failure to detect critical auditory warnings has been shown to arise from limitations of sustained perceptual and attentional processes.[2] In this context, the fast and timely detection of the pilot's failure to notice an auditory alarm could help preventing accidents through the automated application of appropriate countermeasures. Given that eye-tracking systems can be successfully embarked in the cockpit,[3] pupillometry may offer the possibility to index in real time the attentional response to alarms. Indeed, there is evidence that sounds that deviate from the auditory environment, such as alarms, can elicit a pupil dilation response[4] (PDR). However, the pupil diameter can be affected by various external influences present in the cockpit such as eye movements and variations in luminance. To test the potential validity of the PDR as an index of auditory alarm's detection in the cockpit, the present study sought to determine whether reliable PDR could be triggered by deviant sounds in contexts in which luminance level and gaze position were varied systematically.

METHODS

We used the irrelevant sound paradigm in which deviant sounds were occasionally embedded within a to-be-ignored steady-state auditory sequence composed of the repetition of the same sound. To promote gaze displacement, participants in Experiment 1 performed a text comprehension task in which the irrelevant sound was presented during the reading phase. In Experiment 2, participants performed a serial recall task in which they had to recall the order of a visual sequence of digits presented concurrently to the to-be-ignored auditory stream. The successive appearance and disappearance of the visual items composing the to-be-remembered sequence created local changes in the level of luminance.

RESULTS

In both experiments, analyses showed that performance on the visual task was disrupted by the presentation of deviant sounds, as shown by the poorer performance in deviant trials compared to steady-state trials. Moreover, the analysis of pupil size showed that the presentation of deviant sounds triggered significantly larger PDRs than the presentation of steady-state standard sounds.

CONCLUSIONS

Our results showed that noticeable PDRs were elicited by deviant sounds despite concurrent changes in gaze position and luminance level. These findings are promising as they suggest that this physiological index of attention orienting to sound is robust to some of the external influences typically found in the cockpit. However, further research into the

Neuroergonomics. http://dx.doi.org/10.1016/B978-0-12-811926-6.00079-8
Copyright © 2019 Elsevier Inc. All rights reserved.

robustness of the PDR is required, and within an incremental approach from the lab to the field, it could eventually be possible to reliably use the online monitoring of the pupil diameter to determine whether or not an auditory alarm has been detected by the pilot. Such monitoring could then be coupled with an automated system responsible, for example, for triggering countermeasures designed to promote the conscious detection of the alarm or executing automatically an appropriate alarm-related action.

REFERENCES

1. Bliss JP. Investigation of alarm-related accidents and incidents in aviation. *The International Journal of Aviation Psychology* 2003;**13**:249–68. http://dx.doi.org/10.1207/S15327108IJAP1303_04.
2. Dehais F, Causse M, Vachon F, Régis N, Menant E, Tremblay S. Failure to detect critical auditory alerts in the cockpit: evidence for inattentional deafness. *Human Factors* 2014;**56**:631–44. http://dx.doi.org/10.1177/0018720813510735.
3. Imbert JP, Hurter C, Peysakhovich V, Blättler C, Dehais F, Camachon C. Design requirements to integrate eye trackers in simulation environments: aeronautical use case. In: *KES-IDT*. 2015.
4. Wetzel N, Buttelmann A, Schieler A, Widmann A. Infant and adult pupil dilation in response to unexpected sounds. *Developmental Psychobiology* 2015;**58**:382–92. http://dx.doi.org/10.1002/dev.21377.

Chapter 80

Functional Neuroimaging of Prefrontal Cortex Activity During a Problem Solving Versus Motor Task in Children With and Without Autism

Nancy Getchell[1], Ling-Yin Liang[2]
[1]University of Delaware, Newark, DE, United States; [2]University of Evansville, Evansville, IN, United States

AIMS

Autism spectrum disorder (ASD) is a common neurodevelopmental disorder characterized with deficits in social communication, repetitive behavioral patterns, and frequently, movement dysfunction. Researchers have linked some of these issues to atypical prefrontal cortex (PFC) activity. However, it is unclear if what role (if any) the PFC plays in motor dysfunction seen in this population. In the current study, we used functional near infrared spectroscopy (fNIRS) to compare prefrontal cortex activity in boys with and without ASD as they performed two tasks with similar movement requirements but varying levels of executive function. The executive function task was a computer-based Tower of Hanoi (TOH) disk-transfer task for which the PFC should be an active neural structure related to task-dependent problem solving. In contrast, the movement task was self-paced tapping using the same surface areas as the TOH task; this task should show far less PFC activity due to low problem-solving demands. Our aim was to narrow the potential points of impairment on the perception–cognition–action continuum in ASD by comparing PFC activation between tasks and groups.

METHOD

We compared changes in cerebral oxygenation in the PFC during the Tower of Hanoi and tapping (TAP) in two groups of children: Ten typically developing (TD) male children (12.3 ± 2.6 years old) and six male children with ASD (11.8 ± 3.5 years old). Both tasks were performed for 2 min. In the TOH, participants began with a three-disk puzzle. If they succeeded within the time frame, they were presented with a four-disk puzzle. TOH behavioral performance was measured by number of moves to complete each level and the highest level of successful performance (three or four disks). fNIRS data were taken throughout each condition, and relative changes in concentration of oxygenation (Δoxy-Hb) and deoxygenation (Δdeoxy-Hb) were measured.

RESULTS

A two-way repeated measures Analysis Of Variance (ANOVA) was used to examine the effects of group (TD vs. ASD) and condition (tapping vs. Tower of Hanoi) on Δoxy-Hb and Δdeoxy-Hb. Participants with ASD used significantly more moves than TD participants on the three-disk level ($P = .012$). Seventy percent of TD and 33% of ASD participants completed the four-disk level and did not differ in number of moves (TD: 22.4 ± 3.7, ASD: 22.5 ± 4.9). Number of taps did not differ between groups (TD: 216.7 ± 73.4; ASD: 206.3 ± 69.0). A significant interaction between group and task was found in Δdeoxy-Hb ($P = .005$). There was no interaction between group and task in Δoxy-Hb. There were regional activation differences between TD and ASD groups during the TOH task, with ASD on average showing more medial and TD, more lateral activation of the PFC (see Fig. 80.1).

Neuroergonomics. http://dx.doi.org/10.1016/B978-0-12-811926-6.00080-4
Copyright © 2019 Elsevier Inc. All rights reserved.

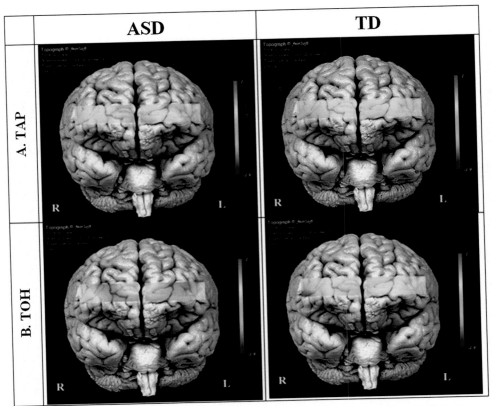

FIGURE 80.1   Average activation patterns in the Tower of Hanoi (TOH) and tapping (TAP) task per group.

CONCLUSIONS

Autism spectrum disorder (ASD) is a common neurodevelopmental disorder characterized with deficits in social interaction, atypical behavioral patterns, and difficulties in movement. Our results indicate that the ASD group used a different strategy to solve the TOH task. Furthermore, the lower performance scores in the cognitive task combined with different oxygenation patterns suggests they had difficulty organizing and planning during the TOH task. Although our findings are preliminary, our results support the notion that movement dysfunction in this population may result from differences in cognition.

Chapter 81

Differential Within and Between Effects on Prefrontal Hemodynamics of fNIRS Guided HD-tDCS

Ryan McKendrick[1,2], Melissa Scheldrup[2], Raja Parasuraman[2], Hasan Ayaz[3,4,5]

[1]Northrop Grumman, Redondo Beach, CA, United States; [2]George Mason University, Fairfax, VA, United States; [3]Drexel University, Philadelphia, PA, United States; [4]University of Pennsylvania, Philadelphia, PA, United States; [5]Children's Hospital of Philadelphia, Philadelphia, PA, United States

Noninvasive brain stimulation via transcranial direct current stimulation (tDCS) has seen a surge in research popularity. This technology has shown promise in terms of enhancing cognition, knowledge acquisition, and psychomotor abilities in both clinical and nonclinical populations. However, only a handful of studies have attempted to measure the effects that tDCS have on brain activity during said enhancement. To add to this growing literature, we monitored the effects of high-density transcranial direct current stimulation (hd-tDCS) while individuals performed a spatial working-memory task. Prior to stimulation, a montage was selected based on brain activity related to increased spatial working-memory performance as observed via functional near-infrared spectroscopy (fNIRS) of the prefrontal cortex and a finite-element analysis of current flow to this brain region. Stimulation was implemented as both within and between subjects. The control group received two 15 min sessions of sham stimulation (ie, ramp up to 1 mA and immediate ramp down to 0 mA), whereas the experimental group received an initial session of the same sham stimulation followed by a 15 min session of actual stimulation at 1 mA. After the second session, both groups continued to perform the task for an additional 15 min. Both groups' prefrontal cortexes were continuously monitored with fNIRS from the start to the end of the testing session. The experimental group that received stimulation showed an increase in performance across the testing session. Interestingly, different between and within effects of stimulation were observed for changes in oxygenation (oxygenated—deoxygenated hemoglobin). Within subjects receiving stimulation, the level of cerebral oxygenation increased over the test session prior to the onset of the hemodynamic response; this was observed throughout the right prefrontal cortex. Between subjects the rate of oxygenation change during the hemodynamic response in the second testing session (active stimulation) was significantly reduced in the stimulation group. However, this was due to the rate of oxygenation change in the sham group during this session being significantly positive and the rate being flat in the stimulation group. Overall, the tDCS to a region of the right prefrontal cortex enhanced spatial working-memory performance, and this was related to an increase in overall cerebral oxygenation along with a relative decrease magnitude of the hemodynamic response. Noninvasive brain stimulation may have multiple effects on cerebral metabolism, and going forward these differential effects need to be considered as more mature models of the physiological effects of noninvasive stimulation are developed.

Neuroergonomics. http://dx.doi.org/10.1016/B978-0-12-811926-6.00081-6
Copyright © 2019 Elsevier Inc. All rights reserved.

Chapter 82

Neuroergonomics In Situ: Differentiation Between Naviation Displays

Ryan McKendrick[1,2], Raja Parasuraman[2], Rabia Murtza[2], Alice Formwalt[2], Wendy Baccus[2], Martin Paczynski[2], Hasan Ayaz[3,4,5]

[1]Northrop Grumman, Redondo Beach, CA, United States; [2]George Mason University, Fairfax, VA, United States; [3]Drexel University, Philadelphia, PA, United States; [4]University of Pennsylvania, Philadelphia, PA, United States; [5]Children's Hospital of Philadelphia, Philadelphia, PA, United States

Highly mobile computing devices promise to improve quality of life, productivity, and performance. Increased situational awareness and reduced mental workload are two potential means by which this can be accomplished. However, it is difficult to measure these concepts in the "wild." We employed ultraportable battery-operated and wireless functional near-infrared spectroscopy (fNIRS) to noninvasively measure hemodynamic changes in the brain's prefrontal cortex. Measurements were taken during navigation of a college campus with either a handheld display, or an augmented-reality wearable display. Hemodynamic measures were also paired with secondary tasks of visual perception and auditory working memory to provide behavioral assessment of situational awareness and mental workload. Navigating with an augmented-reality wearable display produced the least workload during the auditory working-memory task, and a trend for improved situational awareness in our measures of prefrontal hemodynamics. The hemodynamics associated with errors were also different between the two devices. Errors with an augmented-reality wearable display were associated with increased prefrontal activity and the opposite was observed for the handheld display. This suggests that the cognitive mechanisms underlying errors between the two devices differ. These findings show fNIRS is a valuable tool for assessing new technology in ecologically valid settings and that augmented-reality wearable displays offer benefits with regard to mental workload while navigating, and potentially superior situational awareness with improved display design.

Neuroergonomics. http://dx.doi.org/10.1016/B978-0-12-811926-6.00082-8
Copyright © 2019 Elsevier Inc. All rights reserved.

Chapter 83

Embodied and Situated Cognitive Neuroscience

Ryan McKendrick[1,2], Ranjana Mehta[2], Hasan Ayaz[3,4,5], Melissa Scheldrup[2], Raja Parasuraman[2]

[1]Northrop Grumman, Redondo Beach, CA, United States; [2]George Mason University, Fairfax, VA, United States; [3]Drexel University, Philadelphia, PA, United States; [4]University of Pennsylvania, Philadelphia, PA, United States; [5]Children's Hospital of Philadelphia, Philadelphia, PA, United States

We assessed performance and brain activity while cognitive work coincided with physical work in simple and complex environmental settings. Previous studies examining embodied cognition have found performance decrements in working memory with concurrent physical activity. Additionally, neuroimaging has revealed increases and decreases in prefrontal oxygenated hemoglobin. However, the effect of environment on cognitive–physical dual tasking has not been previously considered. Participants were monitored with wireless functional near-infrared spectroscopy (fNIRS) as they performed an auditory 1-back task while sitting, walking indoors, and walking outdoors. Relative to sitting and walking indoors, auditory working memory performance only declined when participants were walking outdoors. Sitting during the auditory 1-back elicited an increase in oxygenated hemoglobin along with a decrease in deoxygenated hemoglobin in the bilateral prefrontal cortex. Walking reduced the total hemoglobin available to bilateral prefrontal cortex. Finally, walking outdoors reduced oxygenated hemoglobin and increased deoxygenated hemoglobin in the bilateral prefrontal cortex. Overall, we observed that during executive processing loading of selective attention and physical work resulted in deactivation of the bilateral prefrontal cortex and degraded working-memory performance. This suggests that selective attention and physical work in certain situations compete with executive processing and may supersede executive processing in the distribution of mental resources. Further, research is needed to determine if in situations in which executive functioning is paramount precautions should be taken to eliminate competition from physical work and selective attention.

Neuroergonomics. http://dx.doi.org/10.1016/B978-0-12-811926-6.00083-X
Copyright © 2019 Elsevier Inc. All rights reserved.

Chapter 84

Workload Transition, Cognitive Load States, and Adaptive Autonomous Transportation

Ryan McKendrick[1,2]
[1]Northrop Grumman, Redondo Beach, CA, United States; [2]George Mason University, Fairfax, VA, United States

Using a curvilinear function previously observed in left lateral orbitofrontal cortex (LLOFC) as an index for different cognitive-load states, we tested the effects of adapted-workload transitions to different cognitive-load states. In an initial session, we replicated the effects observed in the previous study (i.e., the presence of a cubic function with increasing cognitive load) and used these effects to identify cognitive-load states in individual performers. We then tested the effects of transitioning to different cognitive-load states. Cognitive load-state transitions caused a deviation between behavioral measures and induced a significant change in the cubic function relating LLOFC oxyhemoglobin (HbO) and working-memory load. We concluded that a change in cognitive load could not sufficiently account for the observed effects of workload transitions on behavior and prefrontal brain activity. Instead, to account for our effects and their deviation from previously observed effects, we present a preliminary hypothesis associating workload transitions with disruption of cognitive-process integration and an increase in cognitive dissatisfaction. This work directly relates to autonomous transportation in which levels of autonomy within the system are expected to change abruptly, inducing changes in the cognitive-load state of the human operator. Designers of such systems should be sensitive to transitions in cognitive-load state-caused system-autonomy transitions as transitions between cognitive-load states could have detrimental effects on cognitive-process integration and operator performance.

Neuroergonomics. http://dx.doi.org/10.1016/B978-0-12-811926-6.00084-1
Copyright © 2019 Elsevier Inc. All rights reserved.

Chapter 85

Towards Neuroadaptive Technology: Implicitly Controlling a Cursor Though a Passive Brain–Computer Interface

Thorsten O. Zander[1], Lauens R. Krol[1], Klaus Gramann[2]

[1]Technische Universitaet Berlin, Berlin, Germany; [2]Center for Advanced Neurological Engineering, University of California San Diego, San Diego, CA, United States

Today's interaction with technology is asymmetrical in the sense that:

- the operator has access to any and all details concerning the machine's internal state, whereas the machine only has access to the few commands explicitly communicated to it by the human, and
- although the human user is capable of dealing with and working around errors and inconsistencies in the communication, the machine is not.

With increasingly powerful machines, this asymmetry has grown, but our interaction techniques have remained the same, presenting a clear communication bottleneck: users must still translate their high-level concepts into machine-mandated sequences of explicit commands, and only then does a machine act. During such asymmetrical interaction, the human brain is continuously and automatically processing information concerning its internal and external context, including the environment of the human and the events happening there. We investigate how this information could be made available in real time and how it could be interpreted automatically by the machine to generate a model of its operator's cognition. This model then can serve as a predictor to estimate the operator's intentions, situational interpretations, and emotions, enabling the machine to adapt to them. Such adaptations can even replace standard input, without any form of explicit communication from the operator.

The above-mentioned cognitive model can be refined continuously by giving agency to the technological system to probe its operator's mind for additional information. It could deliberately and iteratively elicit, and subsequently detect and decode cognitive responses to selected stimuli in a goal-directed fashion. Effectively, the machine can pose a question directly to a person's brain and immediately receive an answer, potentially even without the person being aware of this happening. This cognitive probing allows for the generation of a more fine-grained user model. It can be used to fully replace any direct input to the machine, establishing effective, goal-oriented implicit control of a computer system. We investigated that in an experiment.

A cursor could move discretely over the nodes of a grid in two dimensions (up to eight possible directions: horizontal, vertical, and diagonal). One of the corners of the grid was designated the target. The cognitive probe was each cursor movement itself: the cursor would autonomously move into one direction, the system would evaluate the subsequent neuronal response of the observing participant by a passive brain–computer interface (BCI), and would then adjust the user model underlying the cursor's movement. In this case, the user model consisted of the participant's preferences for certain directions, and the better preferred a direction was, the higher was the chance that the cursor would move into that direction. Each cursor movement elicited an automatic, specific neuronal response depending on whether or not the participant felt this was an appropriate move or not. For example, if the cursor moved southwest, whereas the target was toward the northeast, the system would detect this as an inappropriate move based on the elicited electroencephalographic (EEG) activity. For example, having identified southwest as inappropriate in this single trial, the probability of this direction for subsequent movements was reduced. After a few such implicit interaction cycles, the model strongly weighted the correct direction toward the target. This scenario describes the online application of the paradigm—before that, the classifier was

Neuroergonomics. https://doi.org/10.1016/B978-0-12-811926-6.00085-3
Copyright © 2019 Elsevier Inc. All rights reserved.

calibrated on data gathered from a simply randomly moving cursor. During online application, the cursor reached its target significantly faster than the randomly moving cursor, and bridged the gap by over 80% between a mathematically perfect reinforcement and no reinforcement.

This approach fuses human and machine information processing, introduces fundamentally new notions of "interaction", and allows completely new neuroadaptive technology to be developed. This technology bears specific relevance to autoadaptive experimental designs, but opens up paradigm-shifting possibilities for technology in general, addressing the issue of asymmetry and widening the above-mentioned communication bottleneck.

Chapter 86

A Random Practice Schedule Provides Better Retention and Transfer Than Blocked When Learning Computer Mazes: Preliminary Results

Nancy Getchell, Alex Schilder, Emily Wusch, Amy Trask
University of Delaware, Newark, DE, United States

AIMS

Research in motor learning has shown that the ways in which practices are organized can affect the amount of learning that occurs. Blocked practices, in which the same skill is repeated many times before moving to the next skill, leads to better performance during acquisition than random practices, in which skills are interwoven. However, those in the random practice groups perform better than blocked on retention and transfer tests, indicating better motor learning. This effect of random practice has been termed "contextual interference." Our aim is to examine the neural activation differences in the prefrontal cortex associated with blocked and random practice schedules when learning and performing three different computer mazes.

METHOD

Prior to data collection, three acquisition and two transfer mazes were created using a virtual Maze Simulator (MazeSuite). Adult participants were randomly assigned to either a blocked (BLK; $n=2$) or random (RND; $n=2$) practice schedule group. They attended three acquisition sessions in which BLK completed 45 trials of one maze per session and RND completed 15 trials of each maze presented in a random order for 45 trials. In a fourth session, all participants performed a block of retention trials for each maze and then two novel mazes that were more difficult than the original mazes. Behavioral measures of path length and time in maze were obtained from MazeSuite. fNIRS data fNIR100A were taken throughout each condition, and relative changes in concentration of oxygenation (Δoxy-Hb) and deoxygenation (Δdeoxy-Hb) were measured.

RESULTS

During the acquisition trials, behavioral data indicated that both groups improved on all three mazes, decreasing both path length and time in the maze across the acquisition trials. By the final trial block, there were no significant differences between groups on these measures. However, on the retention trials, BLK behavioral performance significantly worsened on both measures, whereas RND performed similarly to their final acquisition block. On the two transfer mazes, RND performed better than BLK. Average activation differed significantly between the groups in both retention and transfer trials, with BLK showing greater overall activation and different focal activation areas than RND (see Fig. 86.1).

CONCLUSIONS

Although the results are preliminary, the data suggest that a random practice schedule provides a more efficient and effective learning environment than blocked. The comparatively lower activation levels seen in the RND group during retention and transfer suggest that these participants could traverse the mazes with more automaticity. With additional support, these results show the importance of practice structure and environment when learning a computer task.

Neuroergonomics. http://dx.doi.org/10.1016/B978-0-12-811926-6.00086-5
Copyright © 2019 Elsevier Inc. All rights reserved.

(A)　　　　　　　　　　　　　　　　　　　**(B)**

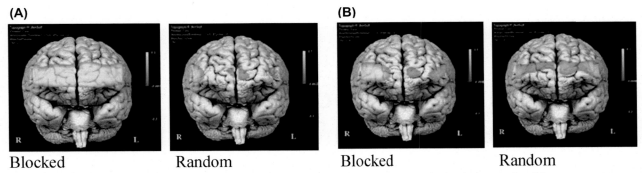

　　Blocked　　　　　　　Random　　　　　　Blocked　　　　　　Random

FIGURE 86.1　　(A) Average activation during retention trials. (B) Average activation during transfer trials.

Chapter 87

Detection of Mind-Wandering in Driving: Contributions of Cardiac Measurement and Eye Movements

G. Pepin[1], S. Malin[1], J. Navarro[2], Alexandra Fort[1], Christophe Jallais[1], C. Gabaude[1]

[1]Université de Lyon, IFSTTAR, TS2, LESCOT, Lyon, France; [2]Université de Lyon, Bron, France

In 2014 in France, there are still over 3300 killed on our roads. Distraction and inattention are considered contributing factors for these accidents. These two different states seem to have different negative impacts on drivers' behaviors but each would be responsible for an equivalent part of accidents.[1] Distraction can be defined as a diversion of attention away from critical driving activities due to external elements,[2] whereas inattention can be defined as an internal state resulting from an endogenous shift of attention.[3] The emergence of inattention state can be promoted by cognitive underload or Mind-Wandering (MW). To improve road safety, it is essential to identify solutions to avoid the emergence of these two states.

MW is defined as a shift in the contents of thought away from an ongoing task to self-generated thoughts and feelings **leading to a perceptual decoupling corresponding to the disengagement of attention from perception.**[4] It is a recurring phenomenon in daily living activities, because half of our thoughts are dedicated to MW[5] and it happens while driving.[6] **MW is associated with a visual attention focused narrowly on the road ahead** leading to a potential excess risk.[7] MW could also decrease speed microregulations, lead to less microregulation of vehicle lateral position[8] **and could change different cardiac markers.**[9] Studying this phenomenon to limit its impacts on driving is, therefore, essential and seems promising using oculometric[10] and cardiac measurements.[11]

In this study, the objective was twofold: (1) To identify physiological changes associated with MW; (2) to explore the influence of MW on drivers' driving performances and visual strategies. For that, electrocardiography (ECG) and eye tracking were recorded during simulated driving sessions. Participants were asked to (a) think prospectively about innovation in specific areas and (b) self-report their MW episodes. A previous indicator showing that it is possible to detect cognitive effort using evoked cardiac response[12] will be tested to detect MW. Some first results have shown that **gaze-fixity rate was higher 1 s before the declaration of MW episodes.** It is expected that **complementary results could foster new MW objective indicators.**

ACKNOWLEDGMENTS

This work was supported by IFSTTAR and the Valeo Innovation Challenge funds.

REFERENCES

1. Galéra C, Orriols L, M'Bailara K, Laborey M, Contrand B, Ribéreau-Gayon R, Masson F, Bakiri S, Gabaude C, Fort A, Maury B, Lemercier C, Cours M, Bouvard M-P, Lagarde E. Mind wandering and driving: responsibility case-control study. *British Medical Journal BMJ* 2012;**345**:e8105.
2. Regan MA, Hallett C, Gordon CP. Driver distraction and driver inattention: definition, relationship and taxonomy. *Accident Analysis and Prevention* 2011;**43**(5):1771–81.
3. Lemercier C, Cellier J-M. Les défauts de l'attention en conduite automobile : inattention, distraction et interférence. *Le Travail Humain* 2008;**71**(3):271–96.
4. Smallwood J, Schooler JW. The science of mind wandering: empirically navigating the stream of consciousness. *Annual Review of Psychology* 2015;**66**:487–518.
5. Killingsworth MA, Gilbert DT. A wandering mind is an unhappy mind. *Science* 2010;**330**(6006):932.

Neuroergonomics. http://dx.doi.org/10.1016/B978-0-12-811926-6.00087-7
Copyright © 2019 Elsevier Inc. All rights reserved.

6. Berthié G, Lemercier C, Paubel P-V, Cour M, Fort A, Galéra C, Lagarde E, Gabaude C, Maury B. The restless mind while driving: drivers' thoughts behind the wheel. *Accident Analysis and Prevention* 2015;**76**:159–65.

7. He J, Becic E, Lee Y-C, McCarley JS. Mind wandering behind the wheel: performance and oculomotor correlates. *Human Factors* 2011;**53**(1):13–21.

8. Lemercier C, Pêcher C, Berthié G, Valéry B, Vidal V, Paubel P-V, Cour M, Fort A, Galéra C, Gabaude C, Lagarde E, Maury B. Inattention behind the wheel: how factual internal thoughts impact attentional control while driving. *Safety Science* 2014;**62**:279–85.

9. Ottaviani C, Shahabi L, Tarvainen M, Cook I, Abrams M, Shapiro D. Cognitive, behavioral, and autonomic correlates of mind wandering and perseverative cognition in major depression. *Autonomic Neuroscience* 2015;**8**:433.

10. Uzzaman S, Joordens S. The eyes know what you are thinking: eye movements as an objective measure of mind wandering. *Consciousness and Cognition* 2011;**20**(4):1882–6.

11. Gabaude C, Baracat B, Jallais C, Bonniaud M, Fort A. Cognitive load measurement while driving. In: *Human factors: a view from an integrative perspective*. Human Factors and Ergonomic Society; 2012. p. 67–80.

12. Pepin G, Jallais C, Fort A, Moreau F, Navarro J, Gabaude C. Towards real-time detection of cognitive effort in driving: contribution of cardiac measurement. *Le Travail Humain* 2017;**80**:21. http://dx.doi.org/10.3917/th.801.0051.

Chapter 88

Electroencephalography and Eye Tracking Signatures of Target Encoding During Guided Search

Anne-Marie Brouwer[1], Maarten A.J. Hogervorst[1], Bob Oudejans[1], Anthony J. Ries[2], Jonathan Touryan[2]

[1]TNO Netherlands Organisation for Applied Scientific Research, Soesterberg, The Netherlands; [2]Army Research Laboratory, Aberdeen, MD, United States

AIMS

Perceiving a target stimulus amid a sequence of non-target stimuli elicits specific event-related potentials (ERPs) (i.e., P3). This reliable finding has typically been demonstrated in research in which observers do not move their eyes; however, recent research shows similar results when observers scan the environment with their eyes, in which the ERPs are locked to eye saccades or fixations rather than to stimulus onset (e.g.,[1]). These ERPs have been shown to distinguish between target and nontarget stimuli with greater accuracy than measures of fixation duration alone.[2] In the current study, we compared saccade-related potentials (SRPs) between targets that were subsequently reported (hits) and targets that were not (misses). Previous work[3] has shown similar SRPs between missed targets and nontargets, but in this study it is likely that the targets were not identified after fixation. Here we focused on targets that were missed due to memory encoding failures. In addition, we examine eye-related features (pupil size and fixation duration) that have been associated with the distinction between targets and nontargets as well as between hits and misses.

METHODS

21 observers were asked to scan 15 squares that displayed "####". They were successively highlighted for 1s in pseudorandom order. When highlighted, the four hashtags were replaced by either "#OK#" or (in two–four squares) by "#FA#". After all squares were highlighted, participants indicated which squares had displayed "#FA#" by clicking the squares using a mouse. To induce memory-encoding failures, participants concurrently performed an aurally presented math task (high-load condition). In a low-load condition, participants ignored the math task. Sixteen blocks of 11 trials were performed, alternating between high- and low-load blocks. Thirty-two channel electroencephalography (EEG) and electrooculography (EOG) were recorded. A Smart Eye eye-tracking system was used to determine pupil size and fixation duration.

RESULTS

As expected, more "#FA#" targets were missed in the high-load condition compared to the low-load (overall miss/hit ratios were 27 and 4% respectively). Single-trial classification and averaged results indicated that for both load conditions, eye features distinguish better between hits and misses than between targets and nontargets (62% classification accuracy for hits and misses in the high-load condition, with larger pupil size and shorter fixations for missed than for hit targets). In contrast, SRP features distinguish better between targets and nontargets than between hits and misses (66% classification accuracy for targets and nontargets in the low-load condition, with average SRPs showing larger P3 waveforms for targets than for nontargets). Average SRPs indicate that misses look similar to targets rather than to nontargets at least for the first 500 ms after which we observed deviations from hits as well as from nontarget SRPs.

Neuroergonomics. http://dx.doi.org/10.1016/B978-0-12-811926-6.00088-9
Copyright © 2019 Elsevier Inc. All rights reserved.

CONCLUSIONS

We replicated previous SRP results concerning the distinction between observing targets and nontargets. SRPs indicated similar initial processing between hit and missed targets, consistent with missed targets being less well stored rather than being less well identified. Results from eye variables were consistent with targets being missed when fixation duration is short and when momentary workload is high (large pupil size). This work suggests complimentary contributions of eye and EEG measures in potential applications to support search and detect tasks. SRPs may be useful to monitor what objects are relevant to an observer, and eye variables may indicate whether the observer should be reminded to them later.

ACKNOWLEDGMENTS

Research was sponsored by the U.S. Army Research Laboratory and was accomplished under Contract Number W911NF-10-D-0002.

REFERENCES

1. Ries AJ, Touryan J, Ahrens B, Connolly P. The impact of task demands on fixation-related brain potentials during guided search. *PLoS One* 2016;**11**(6):e0157260.
2. Brouwer A-M, Reuderink B, Vincent J, van Gerven MAJ, van Erp JBF. Distinguishing between target and nontarget fixations in a visual search task using fixation-related potentials. *Journal of Vision* July 17, 2013;**13**(3):17.
3. Dias JC, Sajda P, Dmochowski JP, Parra LC. EEG precursors of detected and missed targets during free-viewing search. *Journal of Vision* November 12, 2013;**13**(13):13.

Chapter 89

Effects of an Acute Social Stressor on Trustworthiness Judgements, Physiological and Subjective Measures– Differences Between Civilians and Military Personnel

Martijn Bijlsma, Alexander Toet, Helma van den Berg, Anne-Marie Brouwer
TNO Netherlands Organization for Applied Scientific Research, Soesterberg, The Netherlands

AIMS

Based on the interest in the effects of stress on social interactions, we here examined the effect of stress on the perceived trustworthiness of faces. Military operations involve working under highly stressful circumstances. We also tested the hypothesis that, either through training or selection, military personnel are more resilient to stress than civilians as indicated by several subjective and physiological measures. We used a well-controlled paradigm to induce (acute, social) stress: the Sing-a-Song Stress Test (SSST[1]). This paradigm was slightly modified to further shorten and simplify the procedure.

METHODS

Forty-five male military participants and 45 male civilian participants were asked to rate their current level of stress on a scale from 0 to 10. Then they rated trustworthiness of five neutral faces (Princeton facial set[2]). Subsequently, they were presented with the (modified) SSST. This entailed viewing eight sentences that were presented for 5 s and interleaved with a counter counting down from 60 to 0 s. The first seven sentences were neutral; the eighth sentence read "When the counter reaches zero, start singing a song." Participants were sitting still throughout the whole SSST, up to the moment that they started singing. After singing, participants judged five (other) neutral faces again. Finally, they were asked to rate their stress level as experienced during the countdown interval following the Sing-a-Song sentence. We recorded pupil size, heart rate, and skin conductance throughout the experiment. Analysis included only datasets of participants who complied with the instructions and whose physiological data were recorded well. This applied to 33 military and 32 civilian datasets.

RESULTS

Subjective stress was generally higher for civilian than for military participants. This was mainly due to higher post-SSST subjective stress; the difference between groups in pre-SSST subjective stress did not reach significance. We compared heart rate, skin conductance, and pupil size as recorded during the countdown interval following the last neutral sentence (baseline) to the countdown interval following the Sing-a-Song sentence (stress). There was a strong increase for all three variables. Baselines did not differ between the military and the civilian groups. However, the increase in heart rate, skin conductance, and pupil size was significantly stronger for civilian than for military participants. Both groups judged neutral faces less trustworthy after, compared to before the stressor. No differences between groups were found. For civilians, the increase in heart rate was correlated with the decrease in perceived trustworthiness.

Neuroergonomics. http://dx.doi.org/10.1016/B978-0-12-811926-6.00089-0
Copyright © 2019 Elsevier Inc. All rights reserved.

CONCLUSIONS

As hypothesized, military personnel show higher resilience to stress; not only subjectively but also (or especially) physiologically, in response to the social stressor that was used. This is interesting, as this type of social stressor is not what military personnel are specifically trained for. Whether the higher stress resilience is (mainly) caused by training, selection, or other differences between the groups remains to be investigated. In addition, we found that perceived trustworthiness of faces drops after presentation of a stressor, with a modest association between stress response and decrease in perceived trustworthiness. A stronger association might be found when faces are judged during the stress interval rather than after the stressor had ended. Finally, earlier results of the SSST of Brouwer and Hogervorst (2014) were replicated, even though in the current study we used a shorter and easier version by decreasing the number of neutral sentences and having one rather than two people present in the room to serve as audience. This facilitates application of the test for other purposes.

REFERENCES

1. Brouwer A-M, Hogervorst MA. A new paradigm to induce mental stress: the sing-a-song stress test (SSST). *Frontiers in Neuroscience* 2014;**224**(8):1–8.
2. Oosterhof NN, Todorov A. The functional basis of face evaluation. *Proceedings of the National Academy of Sciences of the United States of America* 2008;**105**:11087–92.

Chapter 90

Meta-Cognitive Skills Modeling: Communication and Reasoning Among Agents

Laurent Chaudron[1], Hélio Kadogami[1], Nicolas Maille[1], Guillaume Roumy[2]
[1]ONERA Provence Research Center, Salon-de-Provence, France; [2]French Air Force, France

The aim of this paper is to describe an ongoing research summarized as follows: to investigate the conjecture *"Speaking implies Reasoning."*

Indeed, in both professional situations and everyday life, any conversation requires hypothetico-deductive capabilities that are more or less ignored or relegated to a so-called context space. The purpose of the current study is then to formally describe these cognitive functionalities (ie, to design a theoretical model) as far as the reasoning is a permanent invisible process involved in any interaction between human agents.

The methodology refers to qualitative analysis and grounded theories.

Three fields are investigated: (1) everyday life situations (eg, instructions on highways, audio dialogs), (2) professional interactions (the study is focused on cockpit dialogs, eg, between captain and first officer), (3) pathological cases (eg, interactions with autistic persons).

A main conceptual result appeared: the reasoning process in communication always involves metacognitive skills so as to perform efficient interactions.

Thus metacognition and reasoning are deeply linked and the descriptive model has to capture this structure. Two consecutive results are currently emerging:

1. A first model DAK (for Deductively Assigned Knowledge) captures the minimal mental metacognitive and reasoning capabilities required in basic interactions. DAK is expressed through logic-based formalism compatible with classical multiagent approaches and empowers the classical components of speech acts theory (locutionary and illocutionary items).

2. Thanks to a specific work, some hints related to training are proposed for flight crews, in particular the enrichment of metacognition within already existing training activities. A six-component training program is proposed.

The current study consists in: the development of the DAK, the introduction of links with the neuroergonomics works about consciousness and mental states of the ONERA lab, the inclusion of this metacognition activity within a frame of the cognitive continuum to promote both methodologies for training and recommendations for new devices of complex systems.

FURTHER READING

1. Chaudron LR, Doux E, Ribière G. Positive dissonance and reasoning in operations. In: *Proceedings 'Dissonance Issues and Practices ADRIPS'*, Aix en Provence, France. 2015.
2. Markovits H, Thompson V, Brisson J. Metacognition and abstract reasoning. In: Banbury, Trembley, editors. *A cognitive approach to situation awareness*. 2006.
3. Austin J. *How to do things with words*. Oxford University Press; 1975.
4. Garbis C, Artman H. Team situation awareness as communicative practices. In: *A cognitive approach to situation awareness: theory and application*. Aldershot: Ashgate; 2004.
5. Flavell JH. Metacognitive aspects of problem solving. In: Resnick LB, editor. *The nature of intelligence*. Hillsdale, NJ: Erlbaum; 1976. p. 231–6.

Neuroergonomics. https://doi.org/10.1016/B978-0-12-811926-6.00090-7
Copyright © 2019 Elsevier Inc. All rights reserved.

Chapter 91

Assessing Differences in Emotional Expressivity Between Expert and Non Expert Video Game Players Using Facial Electromyography

Marc-André Bouchard, Jérémy Bergeron-Boucher, Cindy Chamberland, Sébastien Tremblay, Philip L. Jackson
Université Laval, Québec, QC, Canada

AIM

Recent advances in non invasive biometrics have sparked the interest for the development of emotion-based adaptive gaming in the scientific community. However, identifying correspondence patterns between affective states and physiological signals in the context of gaming remains a challenge given the significant amount of interindividual variability reported in physiological patterns of expert and non expert video game players.[1]

Among many disparities, expert players show reduced emotional expressivity as measured by facial electromyography (fEMG) compared to non experts when looking at pleasant, neutral or unpleasant pictures.[2] In the present study, we examined whether expert and non expert players also display similar emotional expressivity patterns in the ecological context of a gaming session. More precisely, we evaluated if these patterns remain the same between positive and negative emotional states, measured respectively through fEMG of the zygomaticus major and the corrugator supercilii. These data would provide more precise identification of player's affective states for the development of emotion-based adaptive gaming.

METHOD

Nine expert and nine non expert players aged between 18 and 35 years old were recruited to play two missions from Assassin's Creed Syndicate. All expert players met the following two criteria: (1) they played video games on a regular basis to get better; and (2) they considered themselves experts in terms of video game skills (not their knowledge of video games). In addition, they met one of these additional conditions: (a) they had already participated in an official video games tournament that led to an actual prize for the winners; (b) they considered themselves having already achieved a high ranking in a multiplayer video game; or (c) they had completed several video games at the highest difficulty level. All these criteria come from an accumulation of recommendations in several studies for the evaluation of expertise in video games. The muscular activity of the zygomaticus major and the corrugator supercilii were measured at 1000 Hz with a Biopac MP150 system during the game session to assess player emotional expressivity. After the game session, participants were administered the short version of the NASA-TLX subjective workload questionnaire on a 20-point Likert type scale.

RESULTS

Mean amplitude, maximum, frequency, area under the curve, and length of muscle activity were extracted from the preprocessed fEMG signals of both muscles. All measures were controlled for baseline activity. Preliminary results showed no group difference regarding the muscular activity of the zygomaticus major or the corrugator supercilii. Scores at the NASA-TLX were computed for both groups and revealed that expert players perceived the game session as less mentally and physically demanding than non expert players. Non experts also reported that they had to work harder than experts to accomplish their level of performance. Taken together, results suggest that experts subjectively found the game session less challenging than non experts despite the similarity of emotional expressivity between groups.

Neuroergonomics. http://dx.doi.org/10.1016/B978-0-12-811926-6.00091-9
Copyright © 2019 Elsevier Inc. All rights reserved.

CONCLUSION

Preliminary results from the present study suggest that expert and non expert players show similar positive and negative emotional expressivity during a video game session. As suggested by Weinreich and colleagues,[2] this absence of difference between groups could be explained by the fact that both expert and non expert players might reduce their emotional expressivity during a video game session to improve their performance. Furthermore, differences between experts and non experts reported in previous work were observed using pictures that were specially chosen to elicit positive and negative effects.[2] Thus, video games may elicit weaker emotional responses than these pictures chosen for this purpose. Finally, these results will provide recommendations for future research programs aiming to develop emotion-based adaptive games by describing differences in emotional expressivity of expert and non expert players during a gaming session.

REFERENCES

1. Mandryk RL, Inkpen KM, Calvert TW. Using psychophysiological techniques to measure user experience with entertainment technologies. *Behavior & Information Technology* 2006;**25**(2):141–58.
2. Weinreich A, Strobach T, Schubert T. Expertise in video game playing is associated with reduced valence-concordant emotional expressivity. *Psychophysiology* 2015;**52**(1):59–66.

Chapter 92

Mental Fatigue Assessment in Prolonged BCI Use Through EEG and fNIRS

Amanda Sargent[1], Terry Heiman-Patterson[1], Sara Feldman[1], Patricia A. Shewokis[1], Hasan Ayaz[1,2,3]
[1]Drexel University, Philadelphia, PA, United States; [2]University of Pennsylvania, Philadelphia, PA, United States; [3]Children's Hospital of Philadelphia, Philadelphia, PA, United States

Amyotrophic lateral sclerosis (ALS) is a neurodegenerative disease that causes loss of motor neurons and progressive weakness including loss of speech.[1] As a result, affected individuals become "locked in" and unable to communicate. The use of brain–computer interface (BCI) can enable communication and increase quality of life, but to date the effects of mental fatigue have been ignored. Mental fatigue due to prolonged BCI use can reduce the accuracy and increase the response time decreasing the utility of the BCI and reducing user satisfaction.

In this project, our aim is to identify neuroimaging-based biomarkers of mental fatigue associated with extended BCI use. Awareness about the factors underlying BCI performance and use could allow for development of flexible systems that are adaptable to different clinical profiles.[2] We have utilized both electroencephalography (EEG; for electrophysiological activity) and functional Near-Infrared Spectroscopy (fNIRS: cortical hemodynamic activity) to detect mental fatigue during use of P300-based BCI paradigm. The central hypothesis of this project is that mental fatigue during prolonged BCI use induces changes in brain activation as measured by fNIRS and EEG.

Eight participants took part in this study (6M, 2F), two ALS patients (2M) (Age = Mean + SD; 58.5 ± 16.3 years) and six healthy controls (Mean Age = 25.3 ± 2.7 years). Prior to the study, all participants consented and written informed consent was obtained based on the approved protocol by the Institutional Review Board of Drexel University. All subjects were asked to use P300-based BCI to complete a spelling task. For the first half of the BCI protocol, the subject's copy spelled three words. The three words they spelled for accuracy after calibration were DREXEL, UNIVERSITY, and BRAIN. In the second half of the BCI protocol, the subjects spelled three more words. The words were COMPUTER, INTERFACE, and 8675309. Sixteen channel EEG and prefrontal fNIRS were collected simultaneously. Before BCI calibration and between the BCI2000 protocol blocks, subjects completed a continuous performance task (CPT) that was designed to assess participants' vigilance, and behavioral performance.[3] CPT task was completed four times (repetitions) throughout the entire protocol. The task lasts for approximately 5 min followed by 5 min to complete the modified Chalder fatigue questionnaire to subjectively assess their mental fatigue level as a comparison to EEG and fNIR signals.[4] Linear Mixed Models with Repeated Measures were used for statistical comparison.

Preliminary results of fNIRS data during execution of the CPT (mean oxygenated-hemoglobin changes in left dorsolateral prefrontal cortex, optode 4) indicated a decline over sessions ($F_{2,3.4} = 23.15$, $P = .01$) but no group differences ($F_{1,15.5} = 1.25$, $P > .05$); see Fig. 92.1. Preliminary results of EEG data during the BCI sessions [average Parietal zero scalp electrode (Pz) P300 amplitude] indicated a significant difference between the two sessions ($F_{1,11.9} = 4.85$, $P < .05$) and significant difference for group ($F_{1,11.9} = 18.20$, $P < .01$). These preliminary results indicate that fNIRS and EEG can be used to assess brain activation changes during prolonged BCI use. Further analysis will be performed to identify biomarkers of mental fatigue. Future studies are needed to validate and eventually incorporate these into online use to improve BCI classifier performance and BCI usability.

Neuroergonomics. https://doi.org/10.1016/B978-0-12-811926-6.00092-0
Copyright © 2019 Elsevier Inc. All rights reserved.

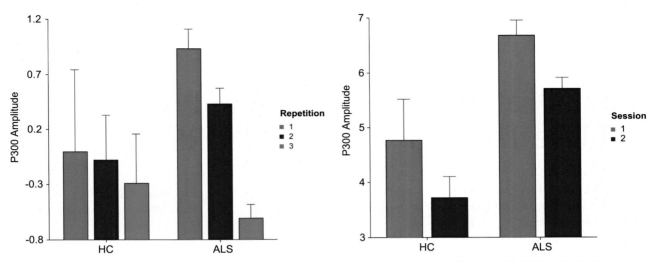

FIGURE 92.1 Average oxyhemoglobin changes during CPT task comparison across three repetitions (left). Average Pz P300 amplitude during the two BCI sessions (right).

REFERENCES

1. Wijesekera LC, Leigh PN. Amyotrophic lateral sclerosis. *Orphanet Journal of Rare Diseases* 2009;**4**:3.
2. Riccio A, et al. Attention and P300-based BCI performance in people with amyotrophic lateral sclerosis. *Frontiers in Human Neuroscience* 2013;**7**:732.
3. Dinges DF, Powell JW. Microcomputer analyses of performance on a portable, simple visual RT task during sustained operations. *Behavior Research Methods, Instruments, & Computers* 1985;**17**(6):652–655.
4. Chalder T, Berelowitz G, Pawlikowska T, Watts L, Wessely S, Wright D, Wallace EP. Development of a fatigue scale. *Journal of Psychosomatic Research* 1993;**37**:147–53.

Chapter 93

Neuroergonomic Multimodal Neuroimaging During a Simulated Aviation Pursuit Task

Robert J. Gougelet[1,2], Cengiz Terzibas[3], Daniel E. Callan[4,5]

[1]Department of Cognitive Science, University of California, San Diego, San Diego, CA, United States; [2]Swartz Center for Computational Neuroscience, University of California, San Diego, La Jolla, CA, United States; [3]Multisensory Cognition and Computation Laboratory, Universal Communication Research Institute, National Institute of Information and Communications Technology, Kyoto, Japan; [4]Center for Information and Neural Networks (CiNet), National Institute of Information and Communications Technology (NICT), Osaka University, Osaka, Japan; [5]ISAE-SUPAERO, Université de Toulouse, Toulouse, France

AIMS

We designed an aviation pursuit task wherein subjects followed a flying target plane at different speeds and distances, and were tasked to train crosshairs on the target plane by maneuvering their own plane as well as detect an auditory cue at perceptual threshold. The aim of this project was to collect multimodal neuroimaging data in the form of magnetoencephalography (MEG), functional magnetic resonance imaging (fMRI), and mobile electroenecephalography (mEEG) during such a task, including on a six-degree-of-freedom motion platform, to understand perception–action coupling in the brain, as mediated by neural oscillations and cross-frequency interactions. In so doing, we hope to inform the future design of brain–computer interfaces and adaptive flight systems that respond to the cognitive states of the user/pilot.

METHODS

The task was implemented using X-Plane 9 flight simulation software. Behavioral, MEG, and fMRI data were collected from 20 subjects. Three subjects performed the task while we measured EEG and MEG concurrently. One subject performed the task in a six degree-of-freedom motion platform flight simulator while wireless EEG data were recorded.

There were four behavioral conditions: flying with auditory cue, flying without, passive with auditory cue, and passive without. During the **flying condition**, the target plane's heading changed every 15 s as the subject followed behind. The distance of the subject's plane from the target plane was manipulated by controlling the throttle of the subject's plane as a function of the subject plane's distance and speed. The initial conditions of the task were reset if the subject's plane was too far, too high, or too low compared to the target plane, or if the subject was facing the wrong direction. Subject performance was evaluated based on their ability to keep the target plane in the center of the screen, under fixed crosshairs. During the **passive condition**, the subjects passively viewed prerecorded flying runs conducted by the experimenter while the subjects rotated their joystick to control for movement artifacts. During the **auditory-cued condition**, the auditory-response cue was played approximately every second. Each auditory cue was a chirp function lasting approximately 60 ms. Subjects only had 800 ms to respond to the cue by pulling on their joystick trigger. The volume of the auditory cues increased or decreased depending on whether they heard or missed the previous cue, keeping the cue at 50% perceptual threshold.

RESULTS AND CONCLUSIONS

Neuroimaging results are forthcoming. As for behavior, subjects were better able and faster to detect the auditory cue during the passive condition, as compared to the flying condition, even though subjects were performing motor manipulation of the joystick during the passive condition. We tested this using a Mann–Whitney U-test on heard auditory-cue trial volumes for flying (Med = 27.9) versus passive (Med = 27.1) conditions ($Z = 13.95$; $n_1 = 7163$, $n_2 = 3583$; $P \ll .01$), and reaction

Neuroergonomics. https://doi.org/10.1016/B978-0-12-811926-6.00093-2
Copyright © 2019 Elsevier Inc. All rights reserved.

time for flying (Med = 528) versus passive (Med = 517) to heard auditory cues ($Z = 2.36$; $n_1 = 7163$, $n_2 = 3583$; $P < .05$). These preliminary results suggest that pilots are at risk of missing auditory cues, such as warning alarms, due to inattentional deafness caused by increased workload[1,2]. The next step is to identify neuroimaging correlates of performance and inattentional deafness that can be measured using mEEG, setting the stage for enhancing flight cockpits with neuroergonomic brain–computer interfaces.

REFERENCES

1. Durantin G, Dehais F, Gonthier N, Terzibas C, Callan DE. Neural signature of inattentional deafness. *Human brain mapping* 2017;**38**(11):5440–5455.
2. Callan DE, Gateau T, Durantin G, Gonthier N, Dehais F. Disruption in neural phase synchrony is related to identification of inattentional deafness in real–world setting. *Human brain mapping* 2018;**39**(6):2596–2608.

Chapter 94

The Use of Neurometric and Biometric Research Methods in Understanding the User Experience of First-Time Buyers in E-Commerce

Tuna E. Çakar[1], Kerem Rızvanoğlu[2], Özgürol Öztürk[2], Deniz Zengin Çelik[2]

[1]MEF University, İstanbul, Turkey; [2]Galatasaray University, İstanbul, Turkey

AIM

User experience (UX) research has attracted increasing attention especially in the last decade as the demand for online shopping has increased by 30.7% from 2014 to 2015 in Turkey.[1] The traditional methods including surveys/questionnaires, think-aloud procedures, and in-depth interviews have contributed greatly for understanding the problems during the use of shopping internet sites. On the other hand, the use of neuroscientific methods, such as biometrics and neurometrics, has also grabbed attention with the exciting idea of providing an objective means of understanding cognitive and affective processes during the user experience during online shopping.[2,3] Despite significant/strong limitations, many researchers are interested in exploring actively its potential use in the field.

METHOD

This empirical study focuses on understanding the online shopping experience through the use of electroencephalography (EEG)/ event-related potentials (ERPs), eye-tracker, galvanic skin response (GSR), pulse rate (PR), facial coding (FC), as well as in-depth interviews and surveys. The participants were all males, 24–35 years of age, right-handed, and actively working as professionals at least for 3 years. The participants were also required not to have yet done any online shopping from a specific website. This requirement was significant to assess their first-time visit to a website. The participants were asked to sign up to this website, they were then to choose a specific item category to buy, provide three concrete options for this category, and then purchase one of these options. These participants were motivated by a gift card of 250 TL (approximately 84 USD) as a contribution to their shopping task. In the posttest session, the participants were given a survey about the experiment and also taken into an in-depth interview through a retrospective think-aloud protocol to be able to grasp their online shopping experience in more detail.

RESULTS

A 32-channel wireless dry-electrode EEG/ERP system has been used to calculate Frontal Alpha Asymmetry (FAA) that has been claimed to be an indicator of approach/withdrawal tendencies of the participants.[2,3] Thus, the valence dimension of the emotional experience has been estimated by these empirical outputs. The main scope has been to capture the instants with negative peaks including the issues related to personal information entry of the email addresses, log in to the email accounts, identity number and the issues related to the usability of the website including delay in the page loadings and unselecting items in the shopping cart. Secondly, GSR method, as a measure of physiological arousal, has been used for exploring the changes in arousal dimension. The changes in arousal have already been issued with experience of difficulties while using an interface thus have been associated with frustration and stress. The outputs obtained from the eye-tracker system have

Neuroergonomics. http://dx.doi.org/10.1016/B978-0-12-811926-6.00094-4
Copyright © 2019 Elsevier Inc. All rights reserved.

generally been used as a complementary method especially while attempting to understand the details (where and when information) of the user's experience. Meanwhile, the facial-coding results indicated that the participants have a dominantly neutral facial expression during online shopping that might have been due to experimental conditions. Interestingly enough, the second-dominant facial expression has been attributed to sadness.

CONCLUSION

Overall, the obtained empirical findings indicate that the users do experience a series of problems related to the design and implementation of the target website. Addressing these problems accurately is quite significant, especially for enhancing the usability of the website. There have been several findings from the neuroscientific/psychophysiological methods, and the traditional methods, that seem to support each other to a certain extent. Despite the fact the potential use of neurometric and biometric methods are still controversial and need more direct research in this specific area, it is also arguable that such multimodal use will provide fruitful contributions with the integrated use of traditional and neuroscientific methods.

REFERENCES

1. TUBISAD. *E-commerce market size report.* 2015.
2. Ohme R, Reykowska D, Wiener D, Chormanska A. Application of frontal EEG asymmetry to advertising research. *Journal of Economic Psychology* 2010;**31**:785–93.
3. Chai J, Ge Y, Liu Y, Zhou L, Yao L, Sun X. Application of frontal EEG asymmetry to user experience research. In: Harris D, editor. *Engineering psychology and cognitive ergonomics* 2014. p. 234–43.

Chapter 95

An Applied Driving Evaluation of Electrodermal Potential as a Measurement of Attentional State

Bruce Mehler, Ben D. Sawyer, Tom McWilliams, Bryan Reimer
Massachusetts Institute of Technology, AgeLab, Cambridge, MA, United States

Traditional electroencephalography (EEG) systems are the gold standard for measuring levels of alertness, have shown mixed but arguably promising results in measuring cognitive workload,[1] and can be used even in applied roadway tasks to detect neural correlates of complex decision-making.[2] However, high-quality, multichannel EEG arrays tend to be expensive and time-consuming to set up. Wrist-wearable electrodermal potential (EDP) devices promise classification of EEG states without such limitations. The lone description of such a device in the present literature describes a generalized measure of attentional state constructed from EDP that achieves accuracy of 84.2% in distinguishing, in a laboratory setting, between relaxed participants with eyes closed and patients actively engaging in visual search.[3] Such a measure, provided it remained reasonably reliable under actual field conditions, would have broad neuroergonomic potential. At the same time, such a system would have to overcome and address a number of technical challenges to practically provide the functionality suggested by the developers. Thus, a careful, realistic, and independent evaluation of this technology under actual driving conditions appeared warranted. As such, the present work sought to evaluate a Freer Logic BodyWave EDP wearable, and, in particular, the attentional index provided by this device, to determine if it could distinguish between levels of demand imposed on a driver. A gender-balanced and age-diverse sample of experienced drivers drove on-road while connected to the EDP wearable as well as a heart monitor. Participants experienced epochs of driving alone, and epochs of driving and being parked while completing a working-memory "n-back" task,[4] in which they were presented with a stream of single-digit numbers and were instructed to either verbalize the number (0-back), the immediately preceding number (1-back), or the number two-places back in the sequence (2-back). It was hypothesized that both heart rate and EDP would be able to distinguish between these levels of load. Heart-rate data indeed showed significant differences between baseline driving and each level, in a pattern suggesting a successful manipulation: the n-back task elevated demand in an orderly fashion, and demand was much higher in the "driving plus n-back" conditions than in the "parked plus n-back" conditions. The EDP attention metrics showed trends that appear to match these patterns, but none reached the level of significance ($\alpha = 0.05$). A closer look at the alpha, beta, delta, and theta indices, which underlie the BodyWave EDP wearable-reported attention metric,[3] show some promise, and help explain this orderly but nonsignificant pattern. Nonetheless, it appears that EDP wearables need more development before they are robust to noisy, applied contexts like on-road driving.

REFERENCES

1. Parasuraman R, Wilson GF. Putting the brain to work: neuroergonomics past, present, and future. *Human Factors* 2008;**50**(3):468–74.
2. Sawyer BD, Karwowski W, Xanthopoulos P, Hancock PA. Detection of error-related negativity in complex visual stimuli: a new neuroergonomic arrow in the practitioner's quiver. *Ergonomics* 2016:1–7.
3. Kim I, Kim M, Hwang T, Lee CW. Outside the head thinking: a novel approach for detecting human brain cognition. In: *International conference on human-computer interaction*. Springer International Publishing; 2016. p. 203–8.
4. Mehler B, Reimer B, Coughlin JF. Sensitivity of physiological measures for detecting systematic variations in cognitive demand from a working memory task: an on-road study across three age groups. *Human Factors* 2012;**54**(3):396–412.

Neuroergonomics. http://dx.doi.org/10.1016/B978-0-12-811926-6.00095-6
Copyright © 2019 Elsevier Inc. All rights reserved.

Chapter 96

Applied Potential: Neuroergonomic Error Detection in Single Electrode Electroencephalography

Ben D. Sawyer[1], Waldemar Karwowski[2], Petros Xanthopoulos[3], Peter A. Hancock[2]

[1]*Massechusetts Institute of Technology, AgeLab, Cambridge, MA, United States;* [2]*Department of Psychology, University of Central Florida, Orlando, FL, United States;* [3]*Decision and Information Sciences, Stetson University, United States*

The present state of the art in fixed, laboratory EEGs are multichannel devices, which provide superior spatial coverage of functional regions of the brain, as well as rich variance–covariance data to support postprocessing approaches such as independent component analysis (ICA). In contrast, functional design of wearable electroencephalographic (EEG) equipment represents a balancing act in which each additional channel adds weight, artifact-inducing moving parts, computational requirements, and associated power consumption. Brain processes per se are often already well studied, and optimal electrode placement known and described.[1,2] As such, applied efforts should leverage this knowledge to employ the lowest number of electrodes allowing consistent detection of the pattern of interest. The ideal here would be one. Nevertheless, can single-electrode EEG, devoid of ICA postprocessing, perform well enough in the electrically noisy real world? Here, we build on previous work to address this question. Sawyer et al.[2] described, for the first time, the detection of the error-related negativity (ERN) evoked-response potential (ERP) in a visual search for complex stimuli. In this work, participants completed tasks during eight-channel EEG recording, which was then analyzed using ICA postprocessing.[3] These same data, restricted to the central scalp electrode (Cz), the electrode closest to the focus of the ERN signal, and without ICA postprocessing, was here reanalyzed. Visual inspection (see Fig. 96.1) and subsequent statistical analyses of these resultant time-locked ERP data clearly demonstrate that the ERN was detectable under both analytic approaches. Further, a large effect size was seen for both analyses, clearly showing that ERN ERP may be robustly detected in aggregate single-electrode encephalography data.

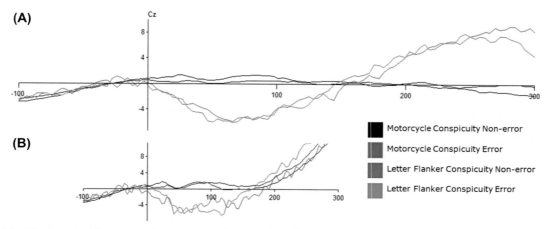

FIGURE 96.1 Waveform data for errors and nonerrors are shown across a simple letter flanker task and a complex motorcycle conspicuity task. Inspection of both (A) the ICA post-processed data and (B) raw single-electrode data reveals a pronounced negative deflection for error trials, as is typical for the ERN ERP. These data are plotted negative-down relative to a 50 ms baseline, whereas 100 ms of preresponse activity is shown for evaluative purposes.

Neuroergonomics. http://dx.doi.org/10.1016/B978-0-12-811926-6.00096-8
Copyright © 2019 Elsevier Inc. All rights reserved.

Successful elicitation and detection of this ERN in a visual search of complex images opens the door to applied neuroergonomics "in the field" (as in Fedota and Parasuraman, 2010),[1] enabling the investigation of the brain's error detection system in everyday life and work. Moreover, our present analytic approach can, and should, be applied to other brain processes. EEG features amenable to neuroergonomic analysis "in the wild" represent significant opportunities.

REFERENCES

1. Fedota JR, Parasuraman R. Neuroergonomics and human error. *Theoretical Issues in Ergonomics Science* 2010;**11**(5):402–21.
2. Sawyer BD, Karwowski W, Xanthopoulos P, Hancock PA. Detection of error-related negativity in complex visual stimuli: a new neuroergonomic arrow in the practitioner's quiver. *Ergonomics* 2016:1–7.
3. Bell AJ, Sejnowski TJ. An information-maximization approach to blind separation and blind deconvolution. *Neural Computation* 1995;**7**(6):1129–59.

Chapter 97

Negative Mood States in Neuroergonomics

Ahmad Fadzil M. Hani, Ying Xing Feng, Tong Boon Tang
Universiti Teknologi Petronas, Perak, Malaysia

Cognitive brain functions such as language, planning, and problem solving are highly reliant on working memory (WM),[1-3] a temporary buffer that stores limited amount of information for immediate manipulation.[2,4,5] Neuroimaging studies based on functional magnetic resonance imaging (fMRI)[6-8] and functional near-infrared spectroscopy (fNIRS)[9-11] have shown that negative-mood states impair the WM functions, which are related to activations in the prefrontal cortex (PFC) region. Although fMRI has become the gold standard for neuroimaging, fNIRS oxygenated hemoglobin (OxyHb) signal is found highly correlated (r = .94) with the fMRI blood-oxygen-level-dependent (BOLD) signal during a WM task.[12] In this study, we examine the relationship between induced-mood states and PFC activation using the International Affective Picture System (IAPS)[13] merged within the N-back task paradigm. Behavioral task performances by subjects who were under neutral- or negative-mood state bias were monitored for 0-back, 1-back, and 2-back tasks using e-Prime. Optical Topography,[14] a neuroimaging modality based on fNIRS, is used to measure the simultaneous PFC hemodynamic response. This technique allows the experimentation to take place in sitting position, thus imposing relatively lessened physical constraints on subjects compared to fMRI. Negative mood-induced subjects showed a significantly ($P < .05$) poor accuracy during the 2-back task, as compared to the control (neutral mood) group. In addition, a smaller region of the PFC is found activated in all n-back tasks and a significantly lower OxyHb level is obtained around the Broca language region ($P < 0.05$) during the 2-back task. These findings suggest that the influence of negative moods on an individual is significant when attentive resources are engaged with higher WM load. Therefore, we propose that similar work can be applied for neuroergonomics research, which demands natural contexts at work rather than laboratory settings.[15] For example, natural contexts allow investigation of the brain's responses toward mental stress in a field-specified workplace (e.g., oil and gas platform, corporate offices). Besides, the surrounding factors that may boost the cognitive performance and productivity of an employee (e.g., lighting, noise level, spaciousness, etc.) can be identified through virtual reality technology. These efforts will eventually contribute to promote mental health and economic growth through preventing losses related to human factors.

Neuroergonomics. http://dx.doi.org/10.1016/B978-0-12-811926-6.00097-X
Copyright © 2019 Elsevier Inc. All rights reserved.

SUPPLEMENTARY (RESEARCH FRAMEWORK)

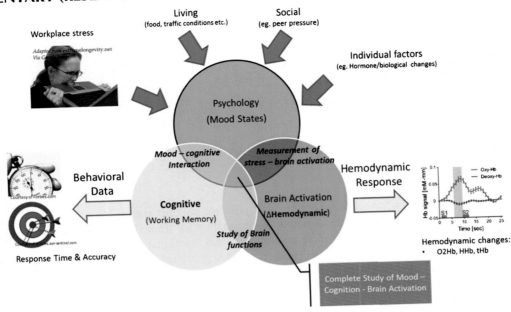

REFERENCES

1. Conway AR, Kane MJ, Bunting MF, Hambrick DZ, Wilhelm O, Engle RW. Working memory span tasks: a methodological review and user's guide. *Psychonomic Bulletin & Review* 2005;**12**(5):769–86.
2. D'Esposito M, Postle BR, Rypma B. Prefrontal cortical contributions to working memory: evidence from event-related fMRI studies. *Experimental Brain Research* 2000;**133**(1):3–11.
3. Cohen JD, Perlstein WM, Braver TS, et al. Temporal dynamics of brain activation during a working memory task. *Nature* 1997;**386**(6625):604–8.
4. Fletcher P, Henson RNA. Frontal lobes and human memory insights from functional neuroimaging. *Brain: A Journal of Neurology* 2001;**124**(5):849–81.
5. Baddeley A. The episodic buffer: a new component of working memory? *Trends in Cognitive Sciences* 2000;**4**(11):417–23.
6. Qin S, Hermans EJ, van Marle HJ, Luo J, Fernández G. Acute psychological stress reduces working memory-related activity in the dorsolateral prefrontal cortex. *Biological Psychiatry* 2009;**66**(1):25–32.
7. Fales C, Barch D, Burgess G, et al. Anxiety and cognitive efficiency: differential modulation of transient and sustained neural activity during a working memory task. *Cognitive, Affective, & Behavioral Neuroscience* 2008;**8**(3):239–53.
8. Gray JR, Braver TS, Raichle ME. Integration of emotion and cognition in the lateral prefrontal cortex. *Proceedings of the National Academy of Sciences* 2002;**99**(6):4115–20.
9. Aoki R, Sato H, Katura T, et al. Relationship of negative mood with prefrontal cortex activity during working memory tasks: an optical topography study. *Neuroscience Research* 2011;**70**(2):189–96.
10. Sato H, Aoki R, Katura T, Matsuda R, Koizumi H. Correlation of within-individual fluctuation of depressed mood with prefrontal cortex activity during verbal working memory task: optical topography study. *Journal of Biomedical Optics* 2011;**16**(12):126007.
11. Sato H, Dresler T, Haeussinger FB, Fallgatter AJ, Ehlis AC. Replication of the correlation between natural mood states and working memory-related prefrontal activity measured by near-infrared spectroscopy in a German sample. *Frontiers in Human Neuroscience* February 1, 2014;**8**.
12. Sato H, Yahata N, Funane T, et al. A NIRS–fMRI investigation of prefrontal cortex activity during a working memory task. *Neuroimage* 2013;**83**:158–73.
13. Lang PJ, Bradley MM, Cuthbert BN. International affective picture system (IAPS): technical manual and affective ratings. *NIMH Center for the Study of Emotion and Attention* 1997:39–58.
14. Maki A, Yamashita Y, Ito Y, Watanabe E, Mayanagi Y, Koizumi H. Spatial and temporal analysis of human motor activity using noninvasive NIR topography. *Medical Physics* 1995;**22**(12):1997–2005.
15. Parasuraman R, Hancock P. Neuroergonomics: harnessing the power of brain science for human factors and ergonomics. *Human Factors and Ergonomics Society Bulletin* 2004;**47**(12):1.

Chapter 98

EEG and FNIRS Connectivity Features for Mental Workload Assessment: A Preliminary Study

Raphaëlle N. Roy, Alexandre Moly, Frédéric Dehais, Sébastien Scannella
ISAE-SUPAERO, Université de Toulouse, Toulouse, France

INTRODUCTION

According to the literature, mental workload (MW) is defined as the interaction of mental demands imposed on operators by tasks they attend to and the effort they put into achieving these tasks. An interesting prospect to assess MW is to consider the use of electroencephalography (EEG) and functional near-infrared spectroscopy (fNIRS) which allow a nonintrusive probing of a user's state. Several features can be extracted from the signal collected by their respective sensors, such as the power spectral density in different frequency bands for the EEG, and the level of oxygenated hemoglobin (HbO) for the fNIRS. These features are classically found to vary depending on the MW imposed on the participants. Additional features can be derived from the EEG and fNIRS signals such as connectivity features. Recently, Charbonnier and collaborators[1] showed that connectivity features computed between EEG electrodes and the Frobenius distance between the matrices of connectivity features could be used to estimate quite accurately their mental state.

AIMS

The aims of this preliminary study were to (1) evaluate whether basic connectivity features such as Pearson correlation and coherence are modulated by MW for both EEG and fNIRS, (2) assess the impact of MW on the Frobenius distance between connectivity matrices of each MW condition, and (3) determine whether the connectivity features and the distance matrices computed between recording modalities also fluctuate with MW.

METHODS

Six volunteers (three men; mean age: 24 years old) equipped with EEG and fNIRS performed a Sternberg task in which they had to memorize and recall either three or seven consonants, or perform a control condition. Thirty trials per condition were pseudorandomly presented by blocks of five.

RESULTS

There was a significant decrease of performance (accuracy) with the increase in difficulty that validates the modulation of MW by the paradigm. Regarding cerebral features, the main results are:

1. A significant modulation of alpha and theta EEG power, as well as the classical theta at Fz over alpha at Pz ratio with mental MW;
2. Significant positive correlations of HbO values between frontal and temporal optodes, and negative correlations between temporal and occipital ones, but no modulation of these networks depending on MW. Also, different correlation patterns arise between EEG and fNIRS depending on the experimental condition;
3. Significant modulations of the Frobenius distance in EEG for both frequency bands and both the correlation and the coherence matrices that, respectively, help distinguish load levels or on-task versus off-task conditions. Additionally, EEG and fNIRS coupling also allowed for a discrimination of load levels (trend).

Neuroergonomics. http://dx.doi.org/10.1016/B978-0-12-811926-6.00098-1
Copyright © 2019 Elsevier Inc. All rights reserved.

DISCUSSION

This preliminary study demonstrated that connectivity features and distances between connectivity matrices could be used to differentiate MW conditions. These results have to be strengthened by adding more participants. In addition, this study should be pursued by applying classifiers or performing trend analysis to evaluate the relevance of such features for MW estimation.

REFERENCE

1. Charbonnier S, Roy RN, Bonnet S, Campagne A. EEG index for control operators' mental fatigue monitoring using interactions between brain regions. *Expert Systems With Applications* 2016;**52**:91–8.

Chapter 99

Using Cognitive Models to Understand In-Car Distraction

Christian P. Janssen

Utrecht University, Utrecht, The Netherlands

The aim of this presentation is to demonstrate the value of cognitive models to understand and predict in-car distraction. People distract themselves with various tasks in many situations,[1] including cars.[2,3] In-car distraction increases the risk of an accident and can take various forms. There is, therefore, a need to understand general principles of driver distraction that are independent of specific technology. Cognitive models have the potential to provide this general understanding by specifying theories of distraction within a single framework. This framework allows one to predict behavior in novel situations, such as interaction with novel technology. I will present research in which models were used to predict how the use of alternative strategies for interleaving two tasks impacts dual-task performance.[4–7] The models explain human behavior as making performance trade-offs: of the available strategies, people typically use strategies that are relatively efficient compared to other strategies. Sometimes, this leads to unsafe behavior, when strategies that achieve a trade-off do not align with the objectively safest strategies. Model results also suggest that task structure can impact task-interleaving strategy: if tasks contain "natural breakpoints," then interleaving here offers valuable speed–accuracy trade-offs.[7] This general understanding of human behavior (eg, as affected by performance trade-offs) can inform the design of novel interfaces.

ACKNOWLEDGMENTS

Christian Janssen is supported by a Marie Sklodowska-Curie fellowship of the European Commission (H2020-MSCA-IF-2015, grant agreement no. 705010, "Detect and React").

REFERENCES

1. Janssen CP, Gould SJ, Li SYW, Brumby DP, Cox AL. Integrating knowledge of multitasking and interruptions across different perspectives and research methods. *International Journal of Human-Computer Studies* 2015;**79**:1–5.
2. Klauer SG, Guo F, Simons-Morton BG, Ouimet MC, Lee SE, Dingus TA. Distracted driving and risk of road crashes among novice and experienced drivers. *New England Journal of Medicine* 2014;**370**(1):54–9.
3. Dingus TA, Guo F, Lee S, et al. Driver crash risk factors and prevalence evaluation using naturalistic driving data. *Proceedings of the National Academy of Sciences* 2016:201513271.
4. Janssen CP, Brumby DP. Strategic adaptation to performance objectives in a dual-task setting. *Cognitive Science* 2010;**34**(8):1548–60.
5. Janssen CP, Brumby DP. Strategic adaptation to task characteristics, incentives, and individual differences in dual-tasking. *PLoS One* 2015;**10**(7):e0130009.
6. Janssen CP, Brumby DP, Dowell J, Chater N, Howes A. Identifying optimum performance trade-offs using a cognitively bounded rational analysis model of discretionary task interleaving. *Topics in Cognitive Science* 2011;**3**(1):123–39.
7. Janssen CP, Brumby DP, Garnett R. Natural break points the influence of priorities and cognitive and motor cues on dual-task interleaving. *Journal of Cognitive Engineering and Decision Making* 2012;**6**(1):5–29.

Neuroergonomics. http://dx.doi.org/10.1016/B978-0-12-811926-6.00099-3
Copyright © 2019 Elsevier Inc. All rights reserved.

Chapter 100

Concurrent fNIRS and TMS for Neurocognitive Enhancement on a Speed-of-Processing Task

Adrian Curtin[1,3], Shanbao Tong[1], Yingying Tang[2], Junfeng Sun[1], Jijun Wang[2], Hasan Ayaz[3,4,5]

[1]Shanghai Jiao Tong University, Shanghai, China; [2]Shanghai Jiao Tong University School of Medicine, Shanghai, China; [3]Drexel University, Philadelphia, PA, United States; [4]University of Pennsylvania, Philadelphia, PA, United States; [5]Children's Hospital of Philadelphia, Philadelphia, PA, United States

Learning through practice is expected to result in the development of strategies that are more efficient at cognitive resource management.[1] Speed of processing represents a fundamental limiting step in performance of routine tasks. This measure is typically quantified using tests that require an elementary amount of cognitive effort, delineating them from purely sensorimotor activities. Because processing speed is so sensitive to cognitive changes associated with age, brain injury, or disease, it has frequently been studied as a benchmark for diagnosis, cognitive remediation, and enhancement.[2] Among tasks attempting to measure cognitive processing speed, the Symbol Digit Substitution Test (SDST), derived from the Wechsler intelligence scales, has been widely employed due to its ease of use and sensitivity. Despite the apparent simplicity and brevity of the task, the SDST can be taken as a general index of an individual's cognition due to the intersection of perception, encoding, working memory, and response selection.[3]

In this study, we sought to illuminate the dynamics of neural efficiency during adaptation to a common speed of processing task (i.e., SDST), and explore how metrics of neural efficiency might change with different types of transcranial magnetic stimulation (TMS) methods. Sixteen healthy volunteers were enrolled in this study and were compensated for their participation. Subjects participated in four sessions across 2 days separated by a minimum of 1 h. In each session, subjects performed four repetitions of a 90-s digitized SDST task interleaved with a 30 s baseline period. In between trials 2 and 3, the subject received 10 pulse trains of TMS at F3 (according to the International 10-20 System) separated by 40 s, according to a randomized paradigm (Sham, Single Pulse–110% Resting Motor Threshold (RMT), High Frequency 2s 15 Hz–110% RMT, and Theta Burst 2s–90% RMT) with a different stimulation type performed each session. Cortical activity in the dorsolateral prefrontal cortex (dlPFC) was measured continuously using functional Near-Infrared Spectroscopy (fNIRS).

Our preliminary results suggest that neurocognitive metrics of efficiency are sensitive to learning produced by repeated task performance over several sessions. We observed strong prefrontal involvement in SDST task performance that increased in efficiency with training. Additionally, we noted that intersession TMS paradigms influenced efficiency metrics despite subtherapeutic dosing (Fig. 100.1).

Neuroergonomics. https://doi.org/10.1016/B978-0-12-811926-6.00100-7
Copyright © 2019 Elsevier Inc. All rights reserved.

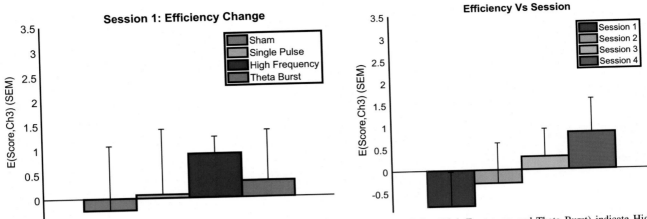

FIGURE 100.1 Efficiency change for Sham and three different TMS stimulation (Single Pulse, High Frequency, and Theta Burst) indicate High Frequency has highest impact for enhancement comparing before and after stimulation (left). Efficiency (performance normalized by neuroimaging-based mental effort) changes over four sessions (right).

REFERENCES

1. Neubauer AC, Fink A. Intelligence and neural efficiency. *Neuroscience and Biobehavioral Reviews* 2009;**33**(7):1004–23.
2. Rypma B, Berger JS, Prabhakaran V, Martin Bly B, Kimberg DY, Biswal BB, D'Esposito M. Neural correlates of cognitive efficiency. *Neuroimage* 2006;**33**(3):969–79.
3. Dickinson D, Ramsey ME, Gold JM. Overlooking the obvious: a meta-analytic comparison of digit symbol coding tasks and other cognitive measures in schizrenia. *Archives of General Psychiatry* May 2007;**64**(5):532–42.

Index

Printed in the United States
By Bookmasters